D1116534

Robust Design: A Repertoire of Biological, Ecological, and Engineering Case Studies

Santa Fe Institute
Studies in the Sciences of Complexity

Lecture Notes Volume

Author	*Title*
Eric Bonabeau, Marco Dorigo, and Guy Theraulaz	Swarm Intelligence: From Natural to Artificial Systems
Mark E. J. Newman and Richard Palmer	Modeling Extinction

Proceedings Volumes

Editor	*Title*
James H. Brown and Geoffrey B. West	Scaling in Biology
Timothy A. Kohler and George J. Gumerman	Dynamics in Human and Primate Societies
Lee A. Segel and Irun Cohen	Design Principles for the Immune System and Other Distributed Autonomous Systems
H. Randy Gimblett	Integrating Geographic Information Systems and Agent-Based Modeling Techniques
James P. Crutchfield and Peter Schuster	Evolutionary Dynamics: Exploring the Interplay of Selection, Accident, Neutrality, and Function
David Griffeath and Cristopher Moore	New Constructions in Cellular Automata
Murray Gell-Mann and Constantino Tsallis	Nonextensive Entropy—Interdisciplinary Applications
Lashon Booker, Stephanie Forrest, Melanie Mitchell, and Rick Riolo	Perspectives on Adaptation in Natural and Artificial Systems
Erica Jen	Robust Design: A Repertoire of Biological, Ecological, and Engineering Case Studies

Robust Design: A Repertoire of Biological, Ecological, and Engineering Case Studies

Editor

Erica Jen
Santa Fe Institute

Santa Fe Institute
Studies in the Sciences of Complexity

2005

OXFORD
UNIVERSITY PRESS

Oxford University Press, Inc., publishes works that further
Oxford University's objective of excellence
in research, scholarship, and education.

Oxford New York
Auckland Cape Town Dar es Salaam Hong Kong Karachi
Kuala Lumpur Madrid Melbourne Mexico City Nairobi
New Delhi Shanghai Taipei Toronto

With offices in
Argentina Austria Brazil Chile Czech Republic France Greece
Guatemala Hungary Italy Japan Poland Portugal Singapore
South Korea Switzerland Thailand Turkey Ukraine Vietnam

Published by Oxford University Press, Inc.
198 Madison Avenue, New York, New York 10016

www.oup.com

Library of Congress Cataloging-in-Publication Data
Robust design : repertoire of biological, ecological, and engineering case studies /
editor, Erica Jen.
p. cm. — (Santa Fe Institute studies in the science of complexity)
Includes bibliographical references and index.
ISBN-13 978-0-19-516532-6; 978-0-19-516533-3 (pbk.)
ISBN 0-19-516532-2; 0-19-516533-0 (pbk.)
1. System theory. 2. Robust control. 3. Adaptation (Biology) I. Jen, Erica.
II. Proceedings volume in the Santa Fe Institute studies in the science of complexity.
Q295.R62 2004
003—dc22 2004046502

9 8 7 6 5 4 3 2 1

Printed in the United States of America
on acid-free paper

About the Santa Fe Institute

The *Santa Fe Institute* (SFI) is a private, independent, multidisciplinary research and education center, founded in 1984. Since its founding, SFI has devoted itself to creating a new kind of scientific research community, pursuing emerging science. Operating as a small, visiting institution, SFI seeks to catalyze new collaborative, multidisciplinary projects that break down the barriers between the traditional disciplines, to spread its ideas and methodologies to other individuals, and to encourage the practical applications of its results.

All titles from the *Santa Fe Institute Studies in the Sciences of Complexity* series carry this imprint which is based on a Mimbres pottery design (circa A.D. 950–1150), drawn by Betsy Jones. The design was selected because the radiating feathers are evocative of the outreach of the Santa Fe Institute Program to many disciplines and institutions.

Contributors List

David Ackley, *University of New Mexico, Computer Science Department, Albuquerque, NM 87131; e-mail: ackley@cs.unm.edu*

John M. Anderies, *Arizona State University, Department of Biology, Tempe, AZ 85287; e-mail: marty.anderies@asu.edu*

Frances Arnold, *California Institute of Technology, Division of Chemistry and Chemical Engineering, 210-41, Pasadena, CA 91125; e-mail: frances@cheme.caltech.edu*

Justin Balthrop, *University of New Mexico, Computer Science Department, Albuquerque, NM 87131; e-mail: judd@cs.unm.edu*

Dirk Bucher, *Volen Center MS 013, Brandeis University, Waltham, MA 02454; e-mail: bucher@brandeis.edu*

Jean Carlson, *University of CA/Santa Barbara, Department of Physics, 6239 Broida Hall, Santa Barbara, CA 93106; e-mail: carlson@physics.ucsb.edu*

Steve Carpenter, *Center for Limnology, 680 North Park Street, University of Wisconsin, Madison, WI 53706; e-mail: srcarpen@wisc.edu*

John C. Doyle, *California Institute of Technology, Control & Dynamical Systems & BioEngineering, 107-81, Pasadena, CA 91125; e-mail: doyle@cds.caltech.edu*

Douglas H. Erwin, *Department of Paleobiology MRC-121, National Museum of Natural History, Washington, DC 20560; e-mail: erwin.doug@nmnh.si.edu*

Stephanie Forrest, *University of New Mexico, Computer Science Department, Albuquerque, NM 87131; e-mail: forrest@cs.unm.edu*

Matthew Glickman, *University of New Mexico, Computer Science Department, Albuquerque, NM 87131; e-mail: glickman@cs.unm.edu*

Peter J. E. Goss, *Harvard University, MCZ Labs 3rd Floor, 26 Oxford Street, Cambridge, MA 02138; e-mail: goss@mcz.harvard.edu*

Joachim Hermisson, *Department of Biology II, Ludwig-Maximilians University Munich, 80333 Munich, Germany; e-mail: joachim.hermisson@lmu.de*

Erica Jen, *Santa Fe Institute, 1399 Hyde Park Road, Santa Fe, NM 87501; e-mail: erica@santafe.edu*

Ann Kinzig, *Arizona State University, Department of Biology, Tempe, AZ 85287; e-mail: kinzig@asu.edu*

David C. Krakauer, *Santa Fe Institute, 1399 Hyde Park Road, Santa Fe, NM 87501; e-mail: krakauer@santafe.edu*

Simon Levin, *Department of Ecology and Evolutionary Biology, Princeton University, Princeton, NJ 08544; e-mail: slevin@princeton.edu*

Richard C. Lewontin, *Harvard University, Museum of Comparative Zoology, Department of Population Biology, 26 Oxford Street, Cambridge, MA 02138; e-mail: lewontin@oeb.harvard.edu*

Stephen H. Low, *California Institute of Technology, Control & Dynamical Systems & BioEngineering, 107-81, Pasadena, CA 91125; e-mail: lowcds.caltech.edu*

Eve Marder, *Volen Center MS 013, Brandeis University, Waltham, MA 02454; e-mail: marder@brandeis.edu*

Steven L. Mayo, *Howard Hughes Medical Institute and Division of Biology, California Institute of Technology, Mail Code 147-75, Pasadena, CA 91125; e-mail: steve@mayo.caltech.edu*

Fernando Paganini, *University of California—Los Angeles, Electrical Engineering Department, Los Angeles, CA 90095-1594 e-mail: paganini@ee.ucla.edu*

Pablo Parrilo, *Institut für Automatik, ETH Zentrum, ETL I 26, Physikstrasse 3, 8092 Zürich, Switzerland; e-mail: parrilo@control.ee.ethz.ch*

Garry Peterson, *University of Wisconsin, Center for Limnology, 680 N. Park Street, Madison, WI 53706; e-mail: gdpeterson@facstaff.wisc.edu*

Joshua B. Plotkin, *Princeton University, Institute for Advanced Study, Olden Lane, Princeton, NJ 08540; e-mail: plotkin@ias.edu*

Glenn Vinnicombe, *University of Cambridge, Department of Engineering, Trumpington Street, Cambridge CB2 1PZ, United Kingdom; e-mail: gv@eng.cam.ac.uk*

Christopher A. Voigt, *Biochemistry and Molecular Biophysics, California Institute of Technology, Mail Code 210-41, Pasadena, CA 911256; e-mail: cavoigt@picasso.ucsf.edu*

Günter P. Wagner, *Department of Ecology and Evolutionary Biology, Yale University, New Haven, CT 06520; e-mail: gunter.wagner@yale.edu*

Walter Willinger, *AT&T Labs—Research, 180 Park Avenue, Room B207, Florham Park, NJ 07932-0971; e-mail: walter@research.att.com*

Zhen-Gang Wang, *Division of Chemistry and Chemical Engineering, California Institute of Technology, Mail Code 210-41, Pasadena, CA 91125; e-mail: zgw@cheme.caltech.edu*

Brian Walker, *Sustainable Ecosystems, CSIRO, P. O. Box 284, Canberra, ACT 2601, Australia, and The Resilience Alliance; e-mail: brian.walker@cse.csiro.au*

Colleen T. Webb, *Princeton University, Department of Ecology and Evolutionary Biology, Guyot Hall, Princeton, NJ 08544; e-mail: ctwebb@princeton.edu; current address: Department of Biology, Colorado State University, Fort Collins, CO 80523-1878; e-mail: ctwebb@lamar.colostate.edu*

Contents

Introduction

Erica Jen

In a world of uncertainty, adversity, and rapid change, it might be thought that *failure* of systems, whether natural or designed, should prove the rule rather than the exception. And yet we are surrounded by examples of phenomena on all scales that we instinctively label as *robust* to failure, whether because of the agility with which they have responded to changing circumstances, or because of their resilience in the face of internal or external attack, or merely because they have proved so long-lived. Robustness is a term that captures our intuitive sense of one of the key determinants of long-term success or failure, but what do we mean by robustness, and what specific features of a phenomenon contribute to its robustness or fragility?

In the past few years, the concept of robustness has been the subject of growing interest in the natural and engineering sciences. Building on traditional fields such as stability, reliability, and control theory, the study of robustness focuses on the ability of a system to maintain specified features when subject to assemblages of perturbations either internal or external (see, for example, Chapter 1 in this volume, and the web site ⟨http://discuss.santafe.edu/robustness⟩).

In most cases, the system of interest is *not* in equilibrium, and the perturbations are typically such that it is unrealistic to attempt estimations of their supports and distributions.

Some questions of interest then include:

- What is meant by "robustness" in the various contexts in which the term is used? In what ways does robustness differ from stability, persistence, resilience, and recovery?
- What are the origins of robustness? Do biological organisms evolve robustness? What is the "null hypothesis" regarding robustness; in other words, what does a functionally fit but non-robust system look like and how does it evolve?
- What are the organizational principles—possibilities include spatial structure, redundancy, modularity, diversification, and hierarchy among others—that characterize highly robust entities? What are the costs of these organizational principles?
- What are the consequences of robustness for evolvability, adaptability, and degree of fitness of an entity to its environment?

Questions such as those listed above stimulate a rich set of research initiatives, and are contributing to new ways of thinking on issues ranging from the architecture of regulatory control, through the relation between performance and flexibility, to the evolution of general-purpose information-processing algorithms in these contexts. Examples of specific phenomena for which robustness is proving a useful "handle" include: the effects of structural mutation on cellular processes of metabolism and growth; directed evolution of enzymes with specified catalytic properties; disturbance regimes and recovery of ecosystems; computer network security systems subject to purposeful attack; and design of Internet protocol systems. Insights that have been achieved in these contexts relate to:

- The relationship between robustness and flexibility; in particular the role of robustness in reducing the potential lethality of mutations and increasing an organism's capacity to accumulate nonlethal genotypic and phenotypic variation. One consequence is to reduce the number of mutations needed to produce phenotypically novel traits [7, 9].
- The robustness of food web networks that consist of a large number of weak interactions together with a small number of strong interactions. Endogenous and exogenous changes are then more likely to result in local elaboration of new morphologies rather than in a catastrophic failure of global organization [7].
- Implications for robustness of what software engineers call "on-line management," namely, the need to maintain current functionality while implementing change or repair. Examples include the separation in protein evolution algorithms of modules governing function and structure so as to permit exploration

of new functionality while maintaining necessary structure (see the chapter by Voigt et al.).

- The existence of "congruence principles" that translate robustness at one level into robustness at another level. Examples include mutational robustness arising as a correlated response to selection on environmental robustness in RNA evolution [1], and the use of the same molecular mechanisms to realize both learning capabilities and developmental stability in neuronal systems [8].
- The existence of conservation principles for robustness implying that systems that have evolved or been designed to withstand specified thresholds of shocks are commensurately susceptible to shocks above those thresholds and to other forms of shocks [3].
- An analysis of biological sensory-processing systems that use redundancy of representation as a general-purpose algorithm and thereby gain multiple functionalities, including improved abilities to deal with poor signal-to-noise ratios, and to separate out multiple signal sources [11].
- The identification of an extensive repertoire of biological mechanisms that employ diverse strategies that can be characterized as: (1) buffering the organism from the effects of an insult; (2) amplifying the insult to enable the organism to purge it; and (3) identifying and repairing the damage from the insult.
- The establishment in ecosystem management of the need for policies and management that are flexible, adaptive, and experimental at scales compatible with the scales of critical ecosystem function [2, 6, 10], and recognition of the pathways by which fixed rules and rigid management structures lead to systems exhibiting a *loss* of robustness—i.e., that suddenly break down under disturbances that previously could be absorbed [5].

The purpose of the present volume is to present some of the recent advances made in the understanding of general principles of robustness especially in the context of evolutionary and developmental biology, ecology, and computer network design. Even more importantly, however, the volume attempts to provide examples of *how researchers are trying to think* about robustness. Each of the chapters is intended to highlight and to illuminate key issues in exploring robustness; namely, the tension between staying the same and responding to change; opportunities for innovation, and vulnerabilities to collapse, on multiple scales; effects of interactions among slow variables such as evolutionary change and fast variables such as ecological shifts; and, interwoven throughout, the role of adaptation and learning.

It is important to note, however, that while the volume attempts to survey and to integrate where possible diverse perspectives on robustness, it in no way seeks to unify all such perspectives, or to put forward "universality principles" for robustness that would be inconsistent with the patent diversity and distinctiveness of the range of processes to which the concept applies.

In fact, as the examples above indicate and the chapters that follow will illustrate in detail, diverse and sometimes conflicting interpretations of robustness are found to be useful at this point in different subcommunities in science and engineering. For example,

- In engineering applications, robustness of systems is the object of intense effort, so much so that regulatory and control features designed for the explicit purpose of achieving robust performance often outweigh in both complexity and cost the actual functional features. Typically, robustness in these contexts is understood to mean reliability of function in the presence of failures with estimable probabilities and supports. Morever, robust design is typically implemented "after the fact"; in other words, functionality comes first, with regulatory controls imposed subsequently to avoid failure modes. The incorporation of robustness as an integral component of the functional design process is a relatively new idea in engineering, and has yet to be realized to any real extent.
- In the context of software engineering, robustness is distinguished from correctness. A program is expected to perform correctly on cases covered by its specifications (which might include faulty data, user error, etc.), but programmers also explicitly design for robustness *outside of* specifications—for example, the program might be expected to fail gracefully in the event of a disk crash. The problem of designing robust computer network systems (as discussed in the chapters by Forrest and by Doyle et al.) is an extreme example illustrating the challenge of protecting against what may be inherently unforeseeable.
- Within developmental biology (see chapters by Lewontin and Goss, Krakauer and Plotkin, and Marder and Bucher), robustness typically refers to the ability of developmental processes to stay "on track" in the presence of perturbations such as environmental insult or developmental noise or knockout mutations. In recent years, robustness in cell biology has been used to describe the ability of certain metabolic and regulatory processes to perform correctly within a large range of parameters.
- In the context of ecosystems, robustness is often interpreted as what Gunderson and Holling (see the chapters by Webb and Levin, and by Walker et al. for more details) call "ecological resilience;" namely, the capacity of a system to undergo disturbance and still maintain its functions and controls. Note that no concepts of metrics or of equilibrium are implied in this definition of resilience, and the concept differs in this way from traditional notions of stability or other forms of ecosystem resilience, where the appropriate measure is the rate at which the system returns to equilibrium following a perturbation.

Essential differences notwithstanding, the twelve chapters in this volume can be said to adopt two complementary perspectives on robustness in biological, ecological, and engineering systems. The first (exemplified by ecological studies such as those by Webb and Levin, and Walker et al., and the neurophysiology work by Marder and Bucher) views robustness as characterizing a stage in the developmental history of a process. The process is analyzed as a set of dynamic interactions with feedback across multiple scales and in multiple dimensions on multiple networks. The question then is the role of these different dynamics in providing flexibility or rigidity in the response of the process to uncertainty and change, and in leading to any of the future possibilities of innovation, persistence, degradation, or collapse.

The second perspective (exemplified by the chapters by Krakauer, by Doyle et al., as well as by Marder and Bucher) treats the robustness of a process—whether it be cell growth, or Internet message routing—as a measure of its sensitivity to perturbations at a fixed point in time. To a large extent, this perspective treats robustness as a characteristic of a process that uses feedback control to perform distributed information-processing. The goal then is to understand the features such as error-correction or buffering that enable the process to perform successfully even with model uncertainty, unforeseeable consequences, conflicting data, and other complexities that could preclude the process from functioning as desired.

A challenge for future research in robustness is to use the insights being gained from perspectives such as those described above to generate actual *design principles* for robustness. A very different but equally important challenge is to construct new perspectives that incorporate useful aspects of the "developmental" and the "feedback control" views described above, but that are relevant to *social* processes. Any such perspective would necessarily include the uniquely social features—including the critical role of cognition and learning, intentionality and identity, evolving cultural repertoires, and the extraordinary human capacities for effective behavior including deliberate collective action or the envisioning of alternative realities—that lend social processes their distinctive flavor of complexity, and that are clearly key to any study of social robustness.

REFERENCES

[1] Ancel, L. W., and W. Fontana. "Plasticity, Evolvability, and Modularity in RNA." *J. Exper. Zool.* **288** (2000): 242–283.

[2] Costanza, Robert. *Institution, Ecosystems, and Sustainability.* Boca Raton, FL: Lewis Publishers, 2001.

[3] Doyle, J., and J. Carlson. "Highly Optimized Tolerance: Robustness and Power Laws in Complex Systems." *Phys. Rev. E.* **60** (1999): 1412.

[4] Gumerman, G., and T. Kohler. "Creating Alternative Cultural Histories in the Prehistoric Southwest." In *Proc. Durango Conference on Southwestern Archaeology*, 1995.

[5] Gunderson, L. H., and C. S. Holling. *Panarchy: Understanding Transformations in Human and Natural Systems.* Washington, DC: Island Press, 2001.

[6] Gunderson, L. H., C. S. Holling, and S. S. Light. *Barriers and Bridges to the Renewal of Ecosystems and Institutions.* New York: Columbia University Press, 1995.
[7] Kirschner, M., and J. Gerhart. "Evolvability." *PNAS* **95** (1998): 8420–8427.
[8] Marder, E., and R. L. Calabrese. "Principles of Rhythmic Motor Pattern Generation." *Physiol. Rev.* **76** (1996): 687–717.
[9] van Nimwegen, E., J. P. Crutchfield, and M. Huynen. "Neutral Evolution of Mutational Robustness." *PNAS* **96** (1999): 9716–9720.
[10] Walters, C. J. *Adaptive Management of Renewable Resources.* New York: Macmillan, 1986.
[11] Zweig, G. Private communication.

Stable or Robust? What's the Difference?

Erica Jen

Exploring the difference between "stable" and "robust" touches on essentially every aspect of what we instinctively find interesting about robustness in natural, engineering, and social systems. It is argued here that robustness is a measure of feature persistence in systems that compels us to focus on perturbations, and often *assemblages* of perturbations, qualitatively different in nature from those addressed by stability theory. Moreover, to address feature persistence under these sorts of perturbations, we are naturally led to study issues including: the coupling of dynamics with organizational architecture, implicit assumptions of the environment, the role of a system's evolutionary history in determining its current state and thereby its future state, the sense in which robustness characterizes the fitness of the set of "strategic options" open to the system, the ability of the system to switch among multiple functionalities, and the incorporation of mechanisms for learning, problem solving, and creativity.

1 INTRODUCTION

"What's the difference between stable and robust?" It is the first question that comes to mind, especially for researchers who work with quantitative models or mathematical theories. Answering the question is not made any easier by the fact that "robustness" has multiple, sometimes conflicting, interpretations—only a few of which can be stated with any rigor. (For a list of working definitions of robustness, see the Santa Fe Institute robustness website at ⟨http://discuss.santafe.edu/robustness⟩.)

But, in fact, the question of the difference between "stable" and "robust" touches on essentially every aspect of what we instinctively find interesting about robustness. It is worth trying to answer the question even if the answers are wrong (or, as is more likely, even if the answers are too vague to be either right or wrong).

It may help to ease into the topic by asking, in order, "What is stability?" "What do stability and robustness have in common?" and "What is robustness beyond stability?"

The concept of stability is an old one that derives from celestial mechanics and, in particular, from the study of the stability of the solar system. A readable treatment is provided by Wiggins [35]. Definitions will be paraphrased here for the sake of establishing some basic language.

Loosely speaking, a solution (meaning an equilibrium state) of a dynamical system is said to be *stable* if small perturbations to the solution result in a new solution that stays "close" to the original solution for all time. Perturbations can be viewed as small differences effected in the actual state of the system: the crux of stability is that these differences remain small for all time.

A dynamical system is said to be *structurally stable* if small perturbations to the system itself result in a new dynamical system with qualitatively the same dynamics. Perturbations of this sort might take the form of changes in the external parameters of the system itself, for example. Structural stability requires that certain dynamical features of the system, such as orbit structure, are preserved, and that no qualitatively new features emerge. Diacu and Holmes [7] give the example of flow on the surface of a river to illustrate the notion of structural stability. Assuming that the flow depends on an external parameter, such as wind speed, and ignoring other factors, the flow is structurally stable if small changes in wind speed do not qualitatively change the dynamics of the flow; for example, do not produce a new structure such as an eddy.

The two commonalities between stability and robustness are:

- Most if not all communities would agree that both concepts are defined for the *specified features* of a given system, with *specified perturbations* being applied to the system. It makes no sense to speak of a system being either stable or robust without first specifying both the feature and the perturbations of interest.

- Both stability and robustness are concerned with the *persistence*, or lack thereof, of the specified features under the specified perturbations. Persistence, therefore, can be seen as evidence of either stability or robustness.

So what is the difference between stability and robustness? It will be argued here that robustness is broader than stability in two respects:

- First, robustness addresses behavior in a more varied class of:
  - systems,
  - perturbations applied to the system of interest, and
  - features whose persistence under perturbations is to be studied.
- Second, robustness leads naturally to questions that lie outside the purview of stability theory:
  - organizational architecture of the system of interest;
  - interplay between organization and dynamics;
  - relation to evolvability in the past and future;
  - costs and benefits of robustness;
  - ability of the system to switch among multiple functionalities;
  - anticipation of multiple perturbations in multiple dimensions; and
  - notions of function, creativity, intentionality, and identity.

Arguments supporting the differences between stability and robustness follow. All arguments are based primarily on plausibility, and are in urgent need of both empirical and theoretical elaboration.

2 CONTEXTS FOR STABILITY VERSUS ROBUSTNESS

It is easy to list examples of systems, features, and perturbations for which the language and framework of traditional stability theory—whether addressing the stability of states or the structural stability of the system—seem inadequate. Table 1 includes examples for which stability theory is entirely appropriate, and others that arguably call for some notion of robustness different from stability. It will be left as an exercise to the reader to say which is which.

What characterizes the contexts in which robustness captures some aspect of a system different from those described by stability theory?

The first observation is that robustness is a measure of feature persistence for systems, or for features of systems, that are difficult to quantify, or to parametrize (i.e., to describe the dependence on quantitative variables); and with which it is, therefore, difficult to associate a metric or norm. The differences among various auction designs and their robustness to collusion, say, is difficult to describe from a traditional stability or structural stability perspective. It is easy even in the context of traditional dynamical systems to define qualitative features

TABLE 1 Examples of which stability theory is appropriate.

System	Feature of Interest	Perturbation
earth's atmosphere	temperature	increase in fluorides
rangelands	biomass	change in grazing policies
laptop	software performance	incorrectly entered data
laptop	software performance	disk crash
bacterial chemotaxis	adaptation precision	change in protein concentrations
bacterial chemotaxis	adaptation precision	bacterial mutation
human immune system	antibody response	new virus
human immune system	antibody response	autoimmune disorder
U.S. political system	perceived legitimacy	demographic changes
U.S. political system	perceived legitimacy	economic depression
religions	popularity	modernity
footbinding[1]	longevity	change in status of women
automotive market [2]	identity of Volkswagen "Bug"	changes in design

[1]Suggested by Sam Bowles.
[2]Suggested by Josh Epstein.

(equivalence classes of attractors, details of phase transitions, for example) that would be a stretch for stability analysis [12].

Second, robustness is a measure of feature persistence in systems where the perturbations to be considered are not fluctuations in external inputs or internal system parameters, but instead represent changes in system composition, system topology, or in the fundamental assumptions regarding the environment in which the system operates. Morever, as pointed out by David Krakauer [18], it is typical in stability theory to postulate a single perturbation; from the robustness perspective it is often ineluctably necessary to consider instead multiple perturbations in multiple dimensions. A biological signaling pathway for example may be robust to an entire assemblage of perturbations including not only fluctuations in molecular concentrations but also the "knocking out" of an array of different genes all of which *prima facie* appear essential in different ways to the functioning of the pathway.

Robustness moreover is especially appropriate for systems whose behavior results from the interplay of dynamics with a definite organizational architecture. Examples of organizational architectures include those based on modularity, redundancy, degeneracy, or hierarchy, among other possibilities, together with the linkages among organizational units. The redundancy and degeneracy of the genetic code [17], the functional modularity of ecosystems [19], and the hierarchical nature of regulatory regimes [22]—these are examples of organizational features not easily represented in a stability framework. Even more importantly, these organizational features are in many systems spliced together into what social scientists term "heterarchies" [28]; namely, interconnected, overlapping, often hierarchical networks with individual components simultaneously belonging to and acting in multiple networks, and with the overall dynamics of the system both emerging and governing the interactions of these networks. Human societies in which individuals act simultaneously as members of numerous networks— familial, political, economic, professional, among others—are one example of heterarchies, and signaling pathways in biological organisms are another, but, in fact, the paradigm is a powerful one with relevance to many natural, engineering, and social contexts.

Note that robustness is meaningful for heterarchical and hierarchical systems only when accompanied by specification of the "level" of the system being so characterized. In an ecosystem, for example, the individual constituent species may be robust with regard to certain disturbances, and interact in such a way as to give rise to a similarly robust aggregate. Even with species that are themselves robust, however, the ecosystem as a whole may not be robust. Even better, species that are not themselves robust can undoubtedly interact so as to create a robust aggregate. In other words, the presence or absence of robustness at one level does not imply its presence or absence at another level, and perhaps the most interesting cases are those in which the interconnections among components that are not themselves robust give rise to robustness at the aggregate level [21, 34].

Implicit in the above is the idea that robustness typically applies to what, for lack of better terminology, are often called "complex adaptive systems." As John Holland points out, "Usually we don't care about the robustness of a rock." In many of these cases, robustness may be interpreted as an index of the relative strengths and weaknesses— what might also be called the "fitness"—of the set of "strategic options" that either have been designed top-down or have emerged bottom-up for the system. The options available to the system serve in other words as a "strategy" for how to respond to perturbations.

The concept of robustness as applying to systems with strategic options is useful in unifying two ostensibly different interpretations of the term. Robustness is often thought of as reflecting the ability of a system to withstand perturbations in *structure* without change in *function*—in biological contexts, this is sometimes called "mutational robustness," and as argued above may be seen as measuring the fitness of a strategy that has either emerged, or has been selected, for responding to insult or uncertainty. However the dual interpretation of robustness is equally valid; namely, robustness may be seen as measuring the effectiveness of a system's ability to switch among multiple strategic options. Robustness in this sense reflects the system's ability to perform multiple functionalities as needed *without* change in structure—this might be called "phenotypical plasticity." As an example of the second type of robustness, simple invertebrate neuronal networks can be seen as heterarchies with the ability to reconfigure themselves and to modify the intrinsic membrane properties of their constituent neurons [24]—the result of this neural "strategy" is a form of robustness that enables a single network to perform, and to switch among, multiple functional tasks.

Strategy is associated with systems that are acting to some purpose, and indeed, robustness is usually ascribed to systems perceived as having a well-defined function or performance measure. Often the function or performance measure is seen as *the* central feature of the system. The concept of function is however problematical in almost all cases. Identifying function with a physical process such as a fluid flow is certainly in the mind of the beholder, and it's not clear that it brings any insight whatsoever. In the case of ecosystems, the concept of function is usually replaced by that of nutrient cycling, productivity, or other "functionals" of the system that may or may not correspond to intuitive ideas of ecological resilience.

As for social systems such as stock markets or religious institutions, the assignment of function may arguably be more natural but equally subjective, in part because such systems acquire, in the course of their development, multiple functions that are both consistent and inconsistent. Moreover, as Levin [20] has pointed out for ecosystems, "robustness" for social institutions often becomes synonymous with rigidity. The *survival* of social institutions such as firms or bureaucracies or governments may sometimes emerge as their primary "function"; adaptation and evolution in these cases largely represent attempts to maintain legitimacy rather than improvements in the institutional function as it was originally defined [9]. The relevance of robustness to the long-term future of social systems will be discussed in Jen [13].

Even for engineering or computational systems, however, the notion of "function" has pitfalls. Intentions of the designers notwithstanding, the systems may possess multiple functionalities in part as a function of its heterarchical nature, and as in the example of the neuronal networks, it may be the ability to switch among tasks that is the true feature of interest. Certainly the possibility cannot be excluded that the systems may develop new functionalities unanticipated in their design. "Function" may, therefore, not be as important a distinction between robustness and stability as it first appears, but it nevertheless remains an issue that requires careful attention.

3 ROBUSTNESS BEYOND STABILITY

Many questions that sit uneasily within a stability framework arise naturally in a study of robustness.

It's argued above that robustness is a concept appropriate for measuring feature persistence in certain contexts; namely, systems in which the features of interest are difficult to parametrize, where the perturbations represent significant changes either in system architecture or in the assumptions built into the system through history or design, or where the system behavior is generated through adaptive dynamics coupled to strong organizational architecture. The study of robustness then naturally prompts questions relating to organization, the role of history, the implications for the future, and the anticipation of insults, along with other questions that are even more difficult to formulate relating to creativity, intentionality, and identity.

3.1 INTERPLAY BETWEEN DYNAMICS AND ORGANIZATION

What is the interplay between the dynamics and organizational architecture of robust systems, and how does the architecture both facilitate and constrain the dynamics by which robustness may be achieved? Hartman et al. [11] argue, for example, that it is this interplay that permits "living systems to maintain phenotypic stability in the face of a great variety of perturbations arising from environmental changes, stochastic events, and genetic variation." One simple example of a coupling between dynamics and organization in a computational context is the use of a majority voting rule to resolve conflicting inputs and to provide an error-correcting capability [3, 31]. In biological contexts, a growing literature [11, 16, 17] provides empirical evidence and theoretical conjectures on more sophisticated dynamics—including for example neutral evolution, and positive or negative feedback—that serves a similar purpose. Much of this work describes the diverse and intricate array of molecular and cellular mechanisms that permit the accumulation of genetic variation while buffering the organism against deleterious phenotypic consequences, or that faciliate higher-level mechanisms in identifying and in correcting such variations as needed.

Arnold and Voigt [30] provide another perspective on the interplay between dynamics and architecture. They argue that in the context of directed evolution, organizational structure can be exploited to accelerate the discovery of proteins with novel or prespecified catalytic properties. In particular, the approach is based on separating the units that modulate function from those that maintain structure. As they find, "by making specificity-determining units structurally tolerant, the space of possible functions can be explored quickly."

The role of organizational architecture in generating robustness is obscure in part because the origins of organization are themselves murky. It may be that the converse question is in fact the correct one for many systems, especially in engineering and computational context; namely, "What is the role of robustness in generating organizational architecture?" Carlson and Doyle, for instance, argue that much of the complexity in sophisticated engineering systems stems not from the specifications for functionality, but from the exigencies of robustness [10]. They argue that in traditional engineering design, regulatory mechanisms for robustness are typically superimposed after the fact on the mechanisms for functionality, and that Rube Goldberg prevails as a consequence.

A different view of the interplay between organizational architecture and robustness emerges from the study of certain hierarchical systems. As pointed out in the previous section, the discussion of robustness for such systems has meaning only when the level of the system is clearly identified. Robustness may exist on the level of the individual components, or on an intermediate level, or on the level of the whole, or not at all. Robustness on one level need not imply robustness on any other level. Conversely, robustness at one level may—through processes seredipitous or otherwise—confer robustness at another level (see work by Ancel and Fontana [2] on mutational and environmental robustness in RNA evolution, for example, and by Marder [23] indicating the role of cellular mechanisms for plasticity in ensuring higher-level neuronal and circuit stability). Are there what Ancel and Fontana call "congruence principles" for translating robustness on one level to robustness at another level?

3.2 HISTORY AND FUTURE

The role of a system's history, and the implications for its future, represent a second set of questions that are stimulated by the study of robustness. In evolutionary and developmental terms, the specific nature of a system's robustness may both reflect the legacy of its history, and constrain the realizations possible in its future [27].

Intuition tells us for example that there are tradeoffs between robustness and evolvability. Robustness loosely speaking may be seen as insensitivity to external and internal perturbations. Evolvability on the other hand requires that entities alter their structure or function so as to adapt to changing circumstances.

Kirschner and Gerhart[16] argue, however, that the existence of redundancy and drift can introduce "useful" variability into a system. If the system can be protected from lethal mutations, the accumulation of variability may permit the system to move to a state within the same neutral network—sets of systems that are "genotypically" different although "phenotypically" equivalent—such that fewer subsequent mutations are needed to effect a major innovation. This form of robustness, thus, exploits the combination of redundancy and the dynamics of drift and neutral evolution in order to increase evolvability. The precise distribution of the neutral network within the space of all systems will determine not only the accessibility of the system from historical antecedents, but also the course and the robustness associated with future innovations [29].

What is the relation between robustness and historical persistence (mimicking those who ask the relation between conservation throughout evolutionary history and the adaptedness of genes)? The question is relevant to phenomena throughout the natural, engineering, and social spheres, but it is the rare case where even tentative conclusions can be drawn. One such case is provided by mammalian sensory processing systems, for which it can be inferred from empirical data that the systems that are oldest in evolutionary terms are in some sense the most robust [36]. The sensory system for taste, for example, can regenerate itself after destruction of the taste buds, and the olfactory system can similarly recover from loss of the chemoreceptors. However, the taste system—which predates the olfactory system in evolutionary terms—apparently can regenerate itself after loss of enervating nerves, whereas the olfactory system cannot.

Krakauer [18] argues in this context that persistence may result from any number of reasons—including constancy of environment, or constraints developmental and evolutionary—that do not necessarily imply robust design. Moreover feature persistence in a population may merely reflect the fact that the feature is deleterious in individuals and hence individuals with the feature do not survive. Robustness in that case can only be seen as an after-the-fact property of the population that is generated as a byproduct of the constraints on individuals. In general, what are the contexts in which persistence may be taken as evidence of robustness, and what are the mechanisms by which those cases of persistence are realized?

As alluded to above, perhaps the central question in the relation between robustness and history is to distinguish between robustness as a byproduct versus the direct object of selection in either natural or directed evolution. In engineering systems, robustness typically is seen as a direct object of selection (whether part of the initial design or an after-the-fact consideration). This view of robustness may however be oversimplified: Wagner et al. [33] for example describe a neural net algorithm that develops a high degree of functional modularity and robustness as a byproduct of its learning to solve a complex problem in robotic control.

In biological examples, the question of distinguishing between byproduct and direct object is even more murky. Have biological systems been expressly selected for robustness [32]? Or is robustness merely a characteristic of systems that contribute directly to high organismal fitness and that in addition survive successive rounds of selection due to their tolerance of high variability and their ability to generate phenotypic novelty? Has the chemotaxis pathway been selected for its robustness, or is robustness a byproduct of a highly fit functional design [1]? To what extent is the robustness of the pathway a feature of importance to the organism? What is the difference between robustness and regulation in this case?

3.3 ROBUSTNESS AS FITNESS OF STRATEGIC OPTIONS

If robustness can be viewed as characterizing the fitness of the "strategic options" open to a system (whether the options have been selected for or have emerged), then there are likely to be important system-wide consequences associated with that set of options. One such consequence is the balance of costs and benefits of a particular form of robustness. For example, in the auditory system there may be energetic costs to be traded off against the performance benefits of redundant processing of signals by arrays of hair cells. As a general principle, Carlson and Doyle [4] conjecture the existence of strict "conservation laws" for robust systems that require that high tolerance to certain insults be accompanied necessarily by low tolerance to other insults. An aircraft can be either highly manueverable or extremely resistant to anti-aircraft fire, but probably not both.

Other consequences might relate to the number and type of options opened up or closed off by the particular form of robustness. The political regime in Renaissance Florence established by Cosimo de'Medici (1389–1464), for example, is analyzed by Padgett [25] as deriving its robustness from a strategy of flexible opportunism that permits actions to be "interpreted coherently from multiple perspectives simultaneously," with the consequence of maintaining what Padgett calls "discretionary options across unforeseeable futures in the face of hostile attempts by others to narrow those options."

Moreover, as a different type of constraint, systems that are robust often are required to maintain their functions while exploring new functionality. Procedures to upgrade networking protocols on the Internet, for example, must be implemented without interrupting functionality. Software engineers refer to this principle as "online management," and Walter Fontana likens it to the "need to fix a boat out on the water rather than on dry dock." What are the implications for the evolvability of robust systems?

3.4 IDENTIFICATION OF AND RESPONSE TO INSULTS

The use of robustness as a design principle raises a deep set of questions as to the nature of, and the response to, assemblages of insults previously unencountered and in a real sense unforeseeable. What design principles should guide the construction of systems for which there exists an infinite number of possible insults to which the system may be subjected? The possibility of using joint probability distributions to estimate the likelihoods and consequences of failure is fairly dim here. What other tools can be developed to endow a system with "open-ended robustness," and what would such a system look like?

One comment in this regard is that robustness of this nature, if it exists, would share some characteristics with the higher-level cognitive processes of the brain (which is, of course, in the processes of development and learning, a quintessentially robust system). The conjecture is that "open-ended robustness" would depend on the performance of the system in dimensions such as induction and deduction, the emergence of innovation, and creative problem solving.

As a particular example of the challenge of modeling insults, what is the difference between designing robustness against "purposeless" versus "purposeful" perturbations or attacks? To first order, stability theory can be said to address the consequences of perturbations that lack intentionality. By contrast, as pointed out by Schneider [26] and Kearns [15], the robustness of computer network security systems is an example—as is the rule of Cosimo de'Medici mentioned above—in which it is necessary to posit instead the existence of attackers intimately familiar with the vulnerabilities of the specific system, and in possession of the expertise and resources needed to mount a coordinated attack explicitly designed to cripple or to destroy. Robustness to this form of attack clearly calls for design that includes the ability to learn, to anticipate, and to innovate.

3.5 ROBUSTNESS AND IDENTITY

Finally, in some important contexts the feature of a system that is robust to disturbances is the *identity* of the system itself. (See the Volkswagen Bug example in table 1, as well as examples from physical systems capable of self-assembly and self-repair [8].) In such cases, does there exist an instruction set or memory that permits the system to preserve its identity even under severe disruption? What are the mechanisms of repair and of self-maintenance? Why does an observer choose to perceive a system as robust despite its perhaps having undergone fundamental changes either structurally or functionally? The issue of identity—not one often highlighted in the natural sciences—underlies the study of robustness, and raises questions for which many disciplines lack even the language in which to pose them.

4 SUMMARY

In its weakest form, the argument for robustness as different from stability can be stated as follows:

> Robustness is an approach to feature persistence in systems for which we do not have the mathematical tools to use the approaches of stability theory. The problem could in some cases be reformulated as one of stability theory, but only in a formal sense that would bring little in the way of new insight or control methodologies.

In stronger form, the argument can be stated as:

> Robustness is an approach to feature persistence in systems that compels us to focus on perturbations, and assemblages of perturbations, to the system that are different from those considered in the design of the system, or from those encountered in its prior history. To address feature persistence under these sorts of perturbations, we are naturally led to study the coupling of dynamics with organizational architecture; implicit rather than explicit assumptions about the environment; the role of a system's evolutionary history in determining its current state and thereby its future state; the sense in which robustness characterizes the fitness of the set of "strategic options" open to the system; the intentionality P of insults directed at, and the responses generated by, the system; and the incorporation of mechanisms for learning, innovation, and creative problem solving.

The above "Strong Form" of the thesis might at first glance appear to rule out applicability to essentially any real system in any but a metaphorical sense.

"Strategic options" for biological systems? And yet the interpretation of biological systems acting "on their own behalf" has proved useful in several contexts [6, 14]. As for physical systems—which ostensibly lack organizational architecture along the lines typical of biological systems—it is not impossible that the insights from studying the robustness of hierarchical systems, say, will assist in the understanding of physical processes across multiple scales. (Note that the converse is also true, see for example [4].) Certainly with respect to engineering and computational systems, the evolutionary dynamics (both past and future) of these systems represents a topic with enormous potential for illuminating principles of robust design.

But the proof is in the pudding, or as is said in Chinese, what is needed is to "tou bi cong rong" ("throw down the pen to join the army"). Important distinctions between "robust" and "stable" notwithstanding, the study of robustness as a design principle of natural, engineering, and social systems will become meaningful only if its use in some specific context results in an interesting insight that couldn't have been gotten otherwise. Stay tuned.

ACKNOWLEDGMENTS

This chapter is the outgrowth of conversations with many participants in the SFI program on robustness. Discussions with Homayoun Bagheri, Sam Bowles, Leo Buss, Josh Epstein, Walter Fontana, Juris Hartmanis, Leo Kadanoff, Michael Kearns, Simon Levin, John Padgett, John Pepper, Eric Smith, Erik van Nimwegen, Gunter Wagner, and Brian Walker have been especially helpful. David Krakauer, John Holland, Steve Carpenter, and George Zweig have influenced the thinking of this essay. I thank Eve Marder for the opportunity to work in her neurophysiology laboratory at Brandeis on the robustness of crustacean stomatogastric systems. The work has been supported by core grants from SFI together with funds from the David and Lucile Packard Foundation.

REFERENCES

[1] Alon, U., M. G. Surette, N. Barkai, and S. Leibler. "Robustness in Bacterial Chemotaxis." *Nature* **397** (1999): 168–171.

[2] Ancel, L. W., and W. Fontana. "Plasticity, Evolvability, and Modularity in RNA." *J. Esp. Zool.* **288** (2000): 242–283.

[3] Blahut, R. E. *Theory and Practice of Error-Control Codes.* Reading, MA: Addison-Wesley, 1983.

[4] Carlson, J. M., and J. Doyle. "Highly Optimized Tolerance: A Mechanism for Power Laws in Designed Systems." *Phys. Rev. E* **60** (1999): 1412.

[5] Carpenter, S. "Robustness in Ecosystems." In Robustness in Natural, Engineering, and Social Systems, Santa Fe Institute, October 2001. Robustness Site ⟨http://discuss.santafe.edu/robustness/stories/StoryReader$20⟩.

[6] Dawkins, R. *The Selfish Gene.* New York: Oxford University Press, 1976.

[7] Diacu, F., and P. Holmes. *Celestial Encounters.* Princeton, NJ: Princeton University Press, 1996.

[8] Dierker, U., A. Hubler, and M. Dueweke. "Self-Assembling Electrical Connections Based on the Principle of Minimum Resistance." *Phys. Rev. E* **54** (1996): 496.

[9] DiMaggio, Paul J., and Walter W. Powell. "The Iron Cage Revisited: Institutional Isomorphism and Collective Rationality in Organizational Fields." *Am. Soc. Rev.* **48** (1983): 147–160.

[10] Doyle, J., and J. Carlson. "Robustness and Complexity." This volume.

[11] Hartman, J. L., B. Garvik, and L. Hartwell. "Principles for the Buffering of Genetic Variation." *Science* **291** (2001): 1001–1004.

[12] Hubler, A., and C. Tracy. "Robust Unstable Systems." Private communication.

[13] Jen, E. "Robustness in Social Processes." In preparation.

[14] Kauffman, S. *Investigations.* New York: Oxford University Press, 2000.

[15] Kearns, M. "Network Security." In preparation.

[16] Kirschner, M., and J. Gerhart. "Evolvability." *PNAS* **95** (1998): 8420–8427.

[17] Krakauer, D. C. "Genetic Redundancy." In *Encyclopedia of Evolution,* edited by Mark Pagel. New York: Oxford University Press, 2002.

[18] Krakauer, D. C. "Principles of Robust Design." This volume.

[19] Levin, S. *Fragile Dominion: Complexity and the Commons.* Reading, MA: Addison-Wesley, 1999.

[20] Levin, S. "Resilience in Natural and Socioeconomic Systems." *Envir. & Dev. Econ.* **3** (1998): 225–236.

[21] Levin, S. "Robustness and Structure in Ecosystems." RS-2001-011 In Robustness in Natural, Engineering,a nd Social Systems, Santa Fe Institute, October 2001. Robustness Site ⟨http://discuss.santafe.edu/robustness/stories/StoryReader$11⟩.

[22] Low, B. S., E. Ostrom, C. P. Simon, and J. Wilson. "Redundancy and Diversity in Governing and Managing Common-Pool Resources." Presented at "Constituting the Commons: Crafting Sustainable Commons in the New Millenium," Eighth Conference of the International Association for the Study of Common Property, Bloomington, IN. May-June 2000. ⟨http://dlc.dlib.indiana.edu/documents/dir0/00/02/95⟩

[23] Marder, E. "Robustness in Neuronal Systems: The Balance between Homeostasis, Plasticity, and Modulation." This volume.

[24] Marder, E., and R. L. Calabrese. "Principles of Rhythmic Motor Pattern Generation." *Physiol. Rev.* **76** (1996): 687–717.

[25] Padgett, J., and C. Ansell. "Robust Action and the Rise of the Medici, 1400–1434." *Am. J. Sociol.* **98(6)** (1993): 1259–1319.

[26] Schneider, F. "Robustness in Computing Systems." RS-2001-036. In Robustness in Natural, Engineering, and Social Systems, Santa Fe Institute, November 2001. Robustness Site ⟨http://discuss.santafe.edu/robustness/stories/StoryReader$36⟩.

[27] Stadler, B. M. R., P. F Stadler, G. Wagner, and W. Fontana. "The Topology of the Possible: Formal Spaces Underlying Patterns of Evolutionary Change." *J. Theoret. Biol.* **213(2)** (2001): 241–274.

[28] Stark, D. "Heterarchy: Distributing Authority and Organizing Diversity." In *The Biology of Business: Decoding the Natural Laws of Enterprise*, edited by J. H. Clippinger III, 153–179. San Francisco, CA: Jossey-Bass Publishers, 1999.

[29] van Nimwegen, E., J. P. Crutchfield, and M. Huynen. "Neutral Evolution of Mutational Robustness." *Proc. Natl. Acad. Sci. USA* **96** (1999): 9716–9720.

[30] Voigt, C., and F. Arnold. "Utilizing Robustness to Accelerate Protein Engineering." This volume.

[31] von Neumann, J. "Probabilistic Logics and the Synthesis of Reliable Organisms from Unreliable Components." In *Automata Studies*, edited by C. Shannon and J. McCarthy. Princeton, NJ: Princeton University Press, 1956.

[32] Wagner, A. "Mutational Robustness in Genetic Networks of Yeast." *Nature Genetics* **24** (2000): 355–361.

[33] Wagner, G., J. Mezey, and R. Calabretta. "Natural Selection and the Origin of Modules." In *Understanding the Development and Evolution of Complex Natural Systems*, edited by W. Callabaut and D. Rasskin-Gutman. Cambridge, MA: MIT Press, 2001.

[34] Webb, C., and S. Levin. "Cross-System Perspectives on the Ecology and Evolution Ecosystems." This volume.

[35] Wiggins, S. *Introduction to Applied Nonlinear Dynamical Systems and Chaos.* Berlin: Springer-Verlag, 1990.

[36] Zweig, G. Private communication, 2001.

Developmental Canalization, Stochasticity, and Robustness

R. C. Lewontin
Peter J. E. Goss

Evolution by natural selection involves three essential conditions. First, individuals must vary one from another in their phenotypic characters. If there is no variation there can be no natural selection. Second, some of the variant types must leave more offspring than other variants as a consequence of that phenotypic variation. That is, the phenotypic variation must be a cause of different probabilities of survival and fertility for the different types. Third, the offspring of the different types must themselves have phenotypes that resemble those of their parents more closely than they resemble the phenotypes of other types in the population. That is, there must be some mechanism for the passage of phenotypic differences from parent to offspring, a mechanism of the inheritance of differences. Although there are instances of the direct passage of antibody proteins from mother to offspring and there may be cellular passage of infective agents, these are exceptions to the general pathway of heritability by means of the organism's genome. We take it as a generality of biology that differences in phenotype must be a consequence of differences in genotype, if there is to be evolution by natural selection.

The requirement that differences in phenotype between individuals reflect difference in their genotypes would be unproblematical if it were true, as modern molecular and developmental genetics so often claims, that the information necessary to specify an organism is completely contained in the organism's DNA. We could then map phenotypic variation directly into genotypic variation and apply the theoretical apparatus of population genetics to give a total explanation of the observed variation within and between species. Unfortunately, the gross simplification of genotype-phenotype relations implied by the program of molecular and developmental genetics, is incorrect. The true general picture of this relation is illustrated by the classic experiment of Clausen, Keck, and Hiesey [9]. Individual plants of the species *Achillea millefolium* were collected from nature and cloned by cutting them into three pieces. A piece of each plant was then grown at low, intermediate, and high elevations, with the result shown in figure 1. Plants pictured horizontally are seven genetically different plants grown at a given elevation. Plants pictured vertically are the three clones of the same original plant, grown at the three elevations. It is immediately obvious that clones of the same genotype, when grown in different environments differ as much or more than different genotypes grown in the same environment. Moreover, the rank order of growth among the seven lines is not preserved across environments. The outcome of this experiment, far from being exceptional, is usual for phenotypic variation. The phenotype of an organism is not determined by its genotype, but is a consequence of a unique interaction between the genotype and the sequence of environments in which the organism develops and functions. There are, of course, cases in which a genetic difference is manifest irrespective of the environment, as in the classical mutations in *Drosophila*, used in research precisely because they are insensitive to environmental variations. But even these represent a small minority of the known mutations in that organism. The relationship between genotype, environment, and phenotype is typically expressed in a mapping, the *norm of reaction*, exemplified in figure 2 for abdominal bristle number in *Drosophila*, showing the phenotypes of different genotypes as a function of environment.

The dependence of phenotype on the genotype and environment, expressed in norms of reaction, does not exhaust the causal factors involved in phenotypic determination. The fingerprints of humans differ between the left and right hands and it is generally the case that bilaterally "symmetrical" organisms are, in fact, bilaterally asymmetrical to different degrees. The genotype of left and right hands is the same and the developmental environments of the left and right hands in the liquid environment within the amniotic sac are not different in any sense in which we ordinarily think of environment, yet the outcome of development on the two sides is clearly different. This asymmetry is a manifestation of random events in the movement and folding of sheets of tissue, which are the consequence of small differences in growth and timing of division of cells. These variations in turn, are the result of cell division and metabolism being dependent on a small, limited population of molecules of any particular kind which must come into

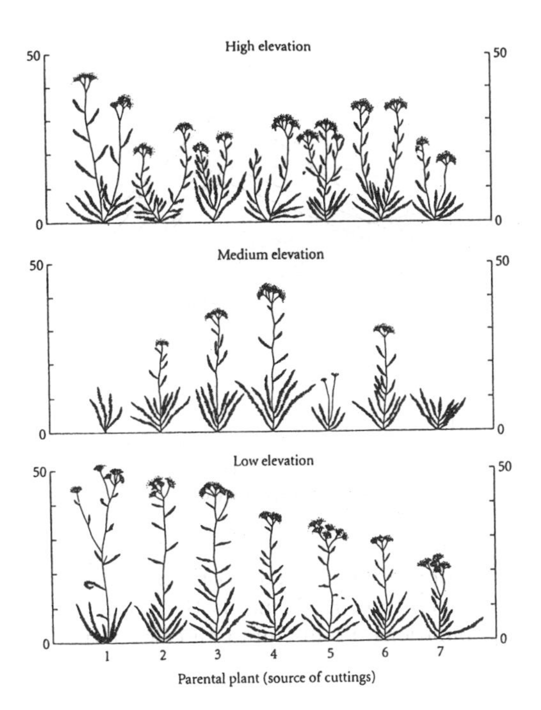

FIGURE 1 Growth of seven cloned plants of *Achillea millefolium* grown at three elevations. The seven plants shown horizontally within an elevation are genetically different. The three plants shown vertically are clones of a single plant. From Clausen, Keck and Hiesey [9]. Reprinted by permission.

contact with each other in the right three-dimensional juxtaposition. For many cellular processes the laws of mass action are irrelevant.

The indeterminate relation between phenotype and genotype means that we must reformulate the relation between the fitness of phenotypes and the pressure of natural selection on genotypes. It is incorrect to say that natural selection favors genotypes that correspond to a particular optimal phenotype. Rather, natural selection favors those genotypes whose bearers have the *highest probability* of having the optimal phenotype. Moreover, it is not only that environment enters into the determination of phenotype but the phenotype that is optimal may be different depending upon which environment the organism experiences at a particular moment in its life history. So, finally, the genotypes that are favored by natural selection are those that have the highest probability of bearing the phenotype that is optimal in each environment experienced by the organism.s

In mammals, a genotype corresponding to a fixed body temperature, despite fluctuations in ambient temperature, would be favored by natural selection

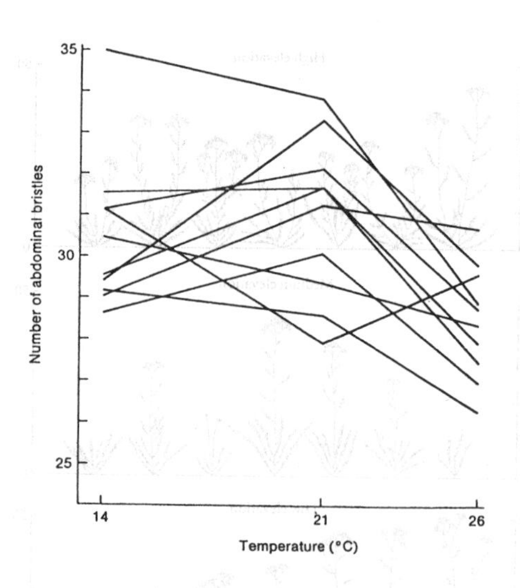

FIGURE 2 Norms of reaction of abdominal bristle number in *Drosophila pseudoobscura* as a function of temperature. Each line is the norm of reaction of a different genotype from natural populations. From Gupta and Lewontin [27]. Reprinted by permission.

(provided that fixed temperature corresponded to the temperature optimum of the organism's enzymes), but a genotype corresponding to a fixed heart rate irrespective of exertion would be at a selective disadvantage.

The probability formulation of selective advantage means that genotypes may be selected not merely for their effect on the mean phenotype over a range of environments but also for the variance of that phenotype and for its covariance with the environment. It is entirely possible that a genotype whose phenotypic mean is close to an optimum value, but whose phenotypic variance is very large would have a lower probability of having phenotypes at the optimum than a genotype whose mean was farther off but whose environmentally induced variance was smaller. Selection would then favor that latter genotype. An example of this phenomenon can be seen in the artificial selection of maize hybrids by plant breeders. Figure 3 shows the yields per acre of an older variety of hybrid maize and a newer selected variety, as a function of environmental quality.

On the average over all conditions the newer variety is superior, but not in the best environments. The newer variety, however, is less sensitive to environmental variation. This reflects the policy of commercial plant breeders to choose varieties in their selection programs that are more uniform over different farms and different years.

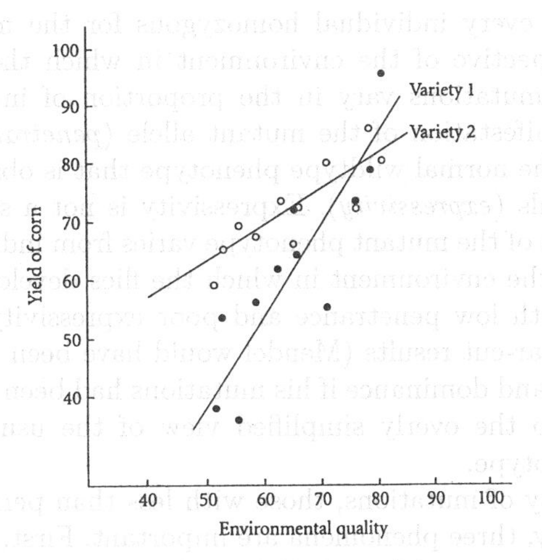

FIGURE 3 Yields per hectare of an older variety (1) and a newer selected variety (2) of maize as a function of environmental quality. (Data of Russell [49].)

1 PENETRANCE, EXPRESSIVITY, AND DOMINANCE

The textbook picture of the relationship between phenotype and genotype is that there are two classes of genetic variation, qualitative and quantitative. In the qualitative case a mutation of a single gene results in a discrete phenotypic change that is in one-to-one correspondence to the genotype. The famous variations in the garden pea studied by Mendel are the model of this variation: tall or short plants, red or white flowers. Each difference is the consequence of a single gene difference and the only ambiguity in the genotype-phenotype relation is that red is dominant to white and tall to short so that a red flowered plant may be homozygous for the "red" allele or heterozygous for the "red" and "white" alleles. The mutations of a model organism like *Drosophila* are also represented as belonging to this qualitative kind: white eyes or red eyes, normal or vestigial wings. The qualitative model is universal when phenotype is described at the level of the RNA or protein product of gene transcription and translation. In contrast, the quantitative case is concerned with continuous or quasi-continuous variation in phenotype which is supposed to be a consequence of the cumulative effects of genetic variation at multiple loci, each of small effect, acting together to produce the phenotype. In fact, this textbook dichotomy falsifies the actual situation. In the standard published catalogue of mutants of *Drosophila* (Lindsley and Zimm [34]), mutations are assigned a rank from 1 to 4 depending on how clear-cut and reliable the phenotypic manifestation of the mutation may be. Of 2300 listed mutations, only 20% are assigned rank 1, typified by the white eye mutation in which every individual homozygous for the mutation has totally colorless eyes irrespective of the environment in which the flies develop. The remaining 75% of mutations vary in the proportion of individuals displaying any detectable manifestation of the mutant allele (*penetrance*) and the degree of deviation from the normal wildtype phenotype that is observed in detectably abnormal individuals (*expressivity*). Expressivity is not a single value, but the degree of expression of the mutant phenotype varies from individual to individual and is sensitive to the environment in which the flies develop. Experimentalists avoid mutations with low penetrance and poor expressivity when they can in order to achieve clear-cut results (Mendel would have been unable to derive his laws of segregation and dominance if his mutations had been of poor penetrance) and this has led to the overly simplified view of the usual relation between genotype and phenotype.

For the majority of mutations, those with less than perfect penetrance and variable expressivity, three phenomena are important. First, the penetrance and expressivity can be increased by a judicious choice of environmental conditions such as temperature and crowding during development. Second, as laboratory stocks of these mutations are kept over many generations, the penetrance and expressivity decrease, with more and more of the individuals appearing as wild-type or very close to normal. Third, the penetrance and expressivity of these stocks can be increased by crossing them to wildtype strains that have not carried the mutation and then reextracting the mutant alleles from the segregating generation and also by selective breeding of the mutant strain from those individuals with the most extreme expression. In the process of selection, both the penetrance and the expressivity increase.

These results lead to the following model. There is variation in the development of phenotype as a consequence both of variation in environmental conditions and of genetic variation at other loci, "modifier" loci, besides the mutated gene. This variation results in a distribution among individuals of phenotypic expression that includes a threshold of detectability of the mutant effect. Thus, penetrance is simply a reflection of the proportion of individuals below this detectability threshold. Mutant phenotypes have lower reproductive fitness than wildtypes, and the more extreme the mutant manifestation, the lower the fitness. In successive generations of laboratory culture there is an increase in the frequency of those alleles at modifier loci that bias development toward the normal wildtype. This accumulation of wildtype biasing alleles can then be temporarily reversed either by outcrossing to a line that has never undergone selection in the presence of the mutation or by artificial selection against such alleles. At the level of phenotype, natural selection moves the distribution of developmental outcomes toward those that result in a wildtype phenotype despite the disturbing effect on development of the mutation. This then suggests a generalized model of natural selection on developmental pathways that buffers them against large perturbations from mutations.

Two questions are raised by this model of natural selection on modifier loci. First, what domain of major gene effects and developmental pathways is affected by a particular set of modifier loci? Is buffering against mutations specific to genes or developmental processes? On the one hand, the same mutation may have pleiotropic effects on different aspects of development that share some common metabolic or signalling molecule. On the other hand, development of a particular organ or tissue can be affected by mutations at very different loci. For example, the size of the *Drosophila* eye, determined by the number of facet cell clusters, can be reduced by the *Bar* mutation on X chromosome, by recessive ey^2 mutations on the second chromosome, and by a dominant ^{eyed} mutation on the fourth chromosome. Can the same set of modifier alleles be selected to buffer against all of them? If so, how large is the domain of developmental processes that can be buffered by the same set of modifier loci? Is it possible to select for alleles at modifier loci that will affect the expression of both mutations that affect, say, eye size and wing development? Our present understanding of the development of, say, wings and eyes in insects makes it unlikely that mutations affecting these different organs could be influenced by the same set of modifier genes. Moreover, what observations exist (see below) point to effects of modifiers that are quite specific not only to mutations at particular loci, but even to particular pleiotropic effects of a given mutation. Nevertheless, it seems impossible that every pleiotropic effect of every mutation at every locus could be modifiable by a different set of modifier genes. This is an area of experimental research that is not part of the present paradigm (research program) of developmental genetics but could easily be implemented.

The second neglected although experimentally accessible question is whether the same modifier genes that increase or decrease mutant expression in a given environment have a parallel effect on the sensitivity of phenotype to environmental variation in the absence of a mutation. The number of adult eye facets decreases with increasing developmental temperature in wildtype *Drosophila* and in a mutant, Ultrabar, that has very small eyes. Do modifier alleles that increase eye facet number in the mutant also affect the sensitivity of facet number to temperature in the wildtype?

In addition to the penetrance and expressivity, yet another phenotypic characteristic of mutations that varies from one case to another is their degree of dominance and recessivity. In Mendel's experiments a single "red" allele was sufficient in a heterozygote with a "white" allele to produce a phenotype indistinguishable from homozygous "red." While Mendel enunciated a Law of Dominance, it is not the case that one allele is always dominant over another, although there are certain patterns that appear. Wildtype alleles are most often dominant to abnormal mutants and loss of function mutations are almost always completely recessive, while mutations leading to a gain of function usually have a detectable phenotypic effect in heterozygotes. That is, most mutations away from the wildtype lead to a failure of some metabolic or developmental activity and so most are recessive. The asymmetry in dominance relations between wildtype alleles and their abnormal mutations led to the first serious attempts to understand the genetical evolution of developmental buffering. In his paper on the evolution of dominance Fisher [19] argued that dominance and recessiveness of alternate alleles was a consequence of evolution by natural selection for each allele at each locus. His claim was based on these observations:

1. The wildtype was usually dominant to an abnormal mutant allele.
2. Dominance of an allele may be different for different pleiotropic effects of the same allele. In particular, those phenotypic effects that were manifest at a level that was likely to be susceptible to natural selection would show dominance of the wildtype, whereas a hidden phenotypic effect would show intermediacy in heterozygotes. For example, red eye pigment is an adaptive trait in *Drosophila*, so heterozygotes for a white eye mutation still have fully red eyes, but the amount of pigment deposited in the internal Malphigian tubules, where it is irrelevant to their excretory function, is intermediate in heterozygotes.
3. Alleles that are recessive in a wild species may be of intermediate dominance in a closely related domesticated species. Fisher provides examples contrasting dominance in the jungle fowl with dominance of the same gene differences in the domesticated chicken. His argument is that domesticated forms are no longer under natural selection, a least for the same characters of morphology as their wild relatives.

Based on these observations, Fisher proposed that modifier genes are selected in nature to suppress the phenotypic effects of abnormal mutations in heterozygous condition, since individuals carrying a rare new abnormal mutation in heterozygous condition would be at a selective disadvantage if the mutation were manifest in the heterozygote. Thus, selection would favor alleles at modifier loci that suppressed the mutant effect in heterozygotes by causing the wildtype allele to be dominant.

Fisher's proposal was attacked by Wright [63], who pointed out that the selection for such modifiers would be ineffective since it would occur only in individuals that were heterozygous for the abnormal mutation and these would be in extremely low frequency in the population. Wright's own theory of dominance [64] was a purely biochemical, not an evolutionary one. He pointed out that mutations that reduced or abolished function were recessive and this was precisely what is to be expected from the known nonlinear relationship between enzyme concentration and the rate of formation of product in an enzymatic reaction. One unit of enzyme would result in about as much product as two for a given amount of substrate, so a heterozygote would have a normal phenotype. The exception would be in cases where the mutation caused a gain of function. As an example, albinism is almost always recessive since it is a loss of enzyme activity in the chain of pigment formation, but there does exist a dominant albinism in rodents which is the consequence of the appearance of an inhibitor of pigment formation.

While Fisher's evolutionary theory did seem to be contradicted by Wright's dynamical argument, Wright's biochemical theory did not account for the observations on pleiotropic effects or on observations in related wild and domesticated species. The resolution of this problem and the origin of a general theory of the evolution of developmental buffering was contained in an alternate theory of dominance provided by Haldane [28, 29]. This "safety factor" hypothesis argued that selection for the wildtype phenotype would result in an extreme wildtype allele at each locus that buffered the development of a trait against *all* perturbations, whether of genetic or environmental origin. It accepted Wright's argument about the nonlinear relationship between enzyme concentration and product formation, but assumed that different wildtype alleles could result in the production of different amounts of enzyme or of molecules of different activity. Alleles would then be selected that resulted in such high total product formation that even a single allelic copy would result in a wildtype morphology. However, the selective force to produce such an extreme allele is not the buffering against rare mutations at the locus, as supposed by Fisher, but as a safety factor against environmental perturbations to development. If we add the importance of random developmental noise perturbations at the cellular level, then the incidence of this selection would be extremely high, since environmental and noise perturbations to development are universal. Nor is it necessary to assume that the selection operates exclusively to substitute one wildtype allele for another, since modifiers at other loci can affect the rate of transcription and translation of an allele and the kinetics of the metabolic pathways in which any gene product is involved. On the other hand, the possibility that alternative wildtype alleles at the locus of interest may be a basis for selection in addition to modifiers at other loci helps to resolve the difficulty that arises if every locus can independently evolve dominance.

2 CANALIZATION

Biologists have long been concerned with the apparent teleological nature of individual development. Each species, beginning with a zygote, develops a morphology and physiology that is characteristic of the species and does so despite a variety of natural and experimental perturbations. The development of various organs of an adult from a larva in holometabolous insects occurs from clusters of larval embryonic cells, imaginal discs, that are developmentally independent of each other. Wings develop from wing discs, eyes from eye discs, and the experimental removal of a wing disc will have no effect on the eventual shape of an eye. Nevertheless, if in moths, for example, a wing disc is wounded, the entire development of the adult form is temporarily halted while the damaged disc is repaired and then proceeds normally after the repair. A major problem for modern developmental biology has been to replace the nineteenth century teleological notion of an "entelechy," an inner drive to a final end that resists perturbations, with a genetical and mechanical explanation of the apparent goal-directedness of the developmental process. Such an explanation must take into account not only the apparent invariant nature of the development of species characteristics within a species, but the switch from one invariant pattern to another at the time of formation of new species. The Darwinian explanation of evolution is built on the existence of variation among individuals, yet the characteristics that differentiate closely related species often show no individual variation within species. In the genus *Drosophila*, all individuals have four large bristles on the scutellum, but other Dipteran species are characterized by two or no scutellar bristles. How are we to explain the origin of variation between species in scutellar bristle number, if there is no variation within species that would be the raw material for evolution?

The concept that is in current use is that of *canalization*, which was introduced and developed by C. H. Waddington [56, 59]. Waddington's metaphor for development is that of a ball rolling down a potential surface. The surface is characterized by a number of grooves or canals that branch out from each other and at each branch point the ball rolls down one canal or another toward a final state that characterizes one of a number of alternative possible outcomes of development. The grooves are deep enough so that most perturbations of the ball's movement are insufficient to divert it into a new pathway, but that major perturbations may nevertheless push it over into a new pathway.

The shape of the potential surface is presumed to be determined by the species genotype. Large changes in the configuration of the surface are not required to move a species from one canalized pathway to another. Small changes in species genotype may result in small changes in the developmental surface in the vicinity of canal branch points so that a new species may now follow a different canalized pathway. This still leaves open the question of how the genotypic change of configuration of the surface comes to be selected in forming the new species if there is no genotypic variation of the surface in the first place. It is to this problem that experiments on canalization in the 1950s and 1960s by Waddington, Rendel, Fraser, Milkman, and others were devoted.

The first major analytical experiment on canalization was Waddington's [58] demonstration that an environmentally induced developmental abnormality could be selected to be genotypically caused. It has long been known that drastic environmental shocks during development can produce phenotypes that mimic the developmental effect of drastic mutations (phenocopies). By treating *Drosophila* at a critical stage in pupal development with a high temperature shock, Waddington produced adults, 34% of which had interruptions in one of the wing veins, mimicking the effect of the mutation *crossveinless*. He selected as parents of the next generation those flies that had an interrupted vein and continued this selection for 23 generations. In each generation he repeated the heat treatment but also observed untreated sibs of the treated flies. As the selection program proceeded, an increasing fraction of treated flies lacked crossveins until, in generation 16, when 85% of treated flies were crossveinless, 1% of crossveinless flies appeared among the *untreated* sibs. By selectively breeding these untreated abnormal flies he finally produced lines of flies that produced between 68% and 95% of crossveinless flies in the absence of any heat shock. In the words of the title of his paper, Waddington had achieved the "genetic assimilation of an acquired character." Waddington's explanation of this result was that in the original population of flies there was genetic variation for vein-making ability, but that this variation was not manifest at the phenotypic level because of the canalization of wing development. Minor genetic variations in vein-making ability were insufficient to deflect the developmental "ball" from the deeply canalized groove of vein formation. The heat treatment destroyed the canalization system and allowed the minor genetic variation to be manifest phenotypically and, therefore, provided a basis for selection to increase the alleles favoring crossveinless development under conditions of heat shock. Continued selection of these genotypes also created a new canalization of development leading to a stable absence of cross veins in the absence of any shock. What differentiates this experiment from a long history of successful selection for changes in environmental sensitivity in agricultural species is the appearance of the crossveinless flies in the absence of heat shock.

The critical elements of Waddington's explanation are that abnormal environmental conditions destroy canalization and that the selection which is now possible not only changes the mean expression of vein formation but also canalizes it around a new mode. While the explanation is consistent with the observations, the claim that the selection for crossveinless response to heat increased the sensitivity to the shock and also created a new mode of canalization despite the shock seems to demand a great deal of the selected genotypes. This mode of explanation is challenged by a series of experiments of much more analytic sophistication by Rendel [47, 46] and Rendel and Sheldon [48]. In place of an environmental shock to deflect normal development, Rendel's experiments used a drastic mutation, *scute*, which causes a variable loss of all or nearly all of the normally invariant four scutellar bristles in *Drosophila*. The genetic scheme was so arranged that in each generation both mutant *scute* and nonmutant sibs were produced from each mating. This is the equivalent of Waddington's heat shocked flies and their notreated sibs. Flies bearing the *scute* mutation that had the most scutellar bristles were selected for breeding in each generation.

The result was an increase in the number of bristles among the mutant flies so that by generation 17 the mean bristle number of males was 2.23 as contrasted with a mean in the unselected mutant stock of about 0.9. At this generation there were also a substantial number of four-bristle flies, whereas there were none in the original unselected line. In this generation there also appeared 5% of flies with five scutellar bristles among the nonmutant sibs. As selection then continued, more and more five-bristle flies and some six-bristle flies appeared among the nonmutants while the distribution of bristle numbers among mutants continued to move toward the four bristle class.

Rendel's explanation for these results, which parallel Waddington's observations on wing veins, was radically different from Waddington's. Rendel's hypothesis was that wildtype individuals differed genetically in the amount of some underlying bristle-making potential (called, for lack of a more concrete model, "Make"), but that developmental buffering prevented these variations in Make from being manifest phenotypically. The introduction of the mutant *scute* did not disrupt the system of canalization, but simply destroyed some fraction of the Make produced by each fly. The amount of Make was moved downward by the mutation so drastically that it was no longer within the buffered range, so that now the genotypic differences in Make production potential were revealed. The selection program simply increased the Make production in both *scute* flies and their nonmutant sibs. But the nonmutant sibs were now producing so much Make that it surpassed the upper limit of the canalization buffer, and now five- and six-bristle flies began to appear. The essential feature of this theory is that the creation of a buffering system is the consequence of a different set of genes than those that govern the basic quantitative basis of bristle-forming potential and it is the variation in these latter genes that is masked by the buffering. The buffering, like a pH buffer, has a limited range and if the inputs to the buffer are below the lower threshold or above the upper one, variation appears, just as in a pH buffer, the addition or subtraction of too many hydrogen ions surpasses the ability of the buffer to hold pH constant. In particular, it is assumed that the transformation is such that the width of the four-bristle class is so wide that normal amounts of Make all fall within it. Only by moving outside this zone of canalization where bristle classes are narrower on the Make scale, can phenotypic variation be observed. What was original in this proposal is the introduction of the concept of a genetically determined transformation or mapping function that converts, in a nonlinear way, the distribution of an underlying variable that is under a different genetic influence. That is, it distinguishes the genetics of the genotype-phenotype map from the genetics of underlying quantitative variables.

To support this hypothesis, Rendel analyzed the observed phenotypic classes in his experiment by means of an inverse probability function. He assumed that the genetically determined Make distribution is normal and that the scute mutation simply translates this normal distribution downwards. In wildtype flies the normal distribution of Make cannot be observed, since the entire distribution falls in the four-bristle class, but by introducing the scute mutation, the distribution is translated downward into a region where the narrower one-, two-, and three-bristle classes are included within the distribution of Make. Using an inverse normal function (probit distribution), it is now possible to estimate the width of the one-, two-, and-three bristle classes in standard deviations. If the model of rigid translation is correct then, as selection toward higher bristle numbers succeeds, even though the mean bristle number will increase, the estimated width of these classes in probit widths should remain constant. Moreover,

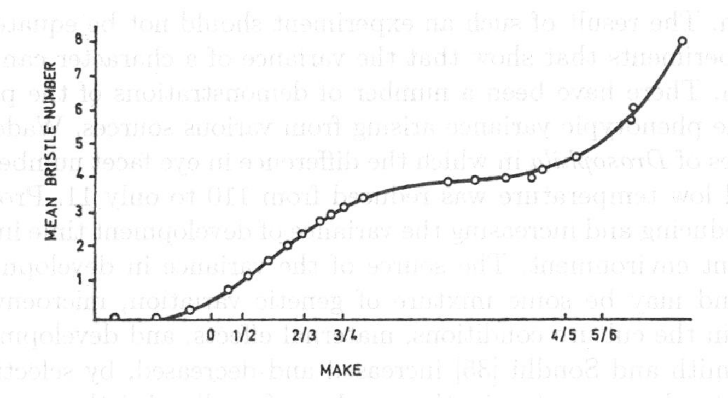

FIGURE 4 Transformation of underlying Make into bristle number in *Drosophila melanogaster*. Boundaries of bristle classes are shown on the abscissa which is on a probit scale of standard deviations. After Rendel [46].

it should be possible to estimate the width of the four-, five-, six-, etc. bristle classes as these bristle classes begin to appear in the nonmutant sibs of the selected scute flies. The estimated probit widths of the various bristle classes did indeed remain constant in the course of the experiments and it was possible to reconstruct the curve of transformation from Make into bristle number over the entire range from zero to six bristles. This transformation is shown in figure 4.

The model of an underlying distribution of a basic morphogen, which is then transformed nonlinearly to mold the actual variation of phenotype, has been applied to other cases. Milkman [40] successfully used this sort of analysis on the original problem of *Drosophila* cross veins, equating Rendel's Make with some (unidentified) protein that was inactivated by the heat shock. The data of Dun and Fraser [12, 13] on the effect of the Tabby mutation on the whiskerlike vibrissae of the head and paws of mice also show the same pattern of a lateral displacement of an underlying, normally distributed "Make" and a nonlinear transformation of that variable into phenotype.

This dual system of genes underlying a quantitative morphogen independently of genes underlying the mapping function implies that it ought to be possible to change the latter system as well as the former by an appropriate selection experiment. In terms of the probit analysis, it should be possible to change the relative probit widths of classes without otherwise changing the mean phenotype. Again, using the scute gene to reduce scutellar bristle number, Rendel and Sheldon [48] selected among families those that had the lowest variance in bristle number, while holding the mean constant at two bristles. The result was a doubling of the probit width of the two-bristle class. In this case, then, a new zone of partial canalization was created around the average phenotype and the experiment presumably serves as a model of the way in which evolution creates canalization. The result of such an experiment should not be equated necessarily with experiments that show that the variance of a character can be reduced by selection. There have been a number of demonstrations of the possibility of reducing the phenotypic variance arising from various sources. Waddington [57] selected lines of *Drosophila* in which the difference in eye facet number developed in high and low temperature was reduced from 110 to only 11. Prout [44] succeeded in reducing and increasing the variance of development time in *Drosophila* in a constant environment. The source of the variance in development time is unknown and may be some mixture of genetic variation, microenvironmental differences in the culture conditions, maternal effects, and developmental noise. Maynard Smith and Sondhi [35] increased and decreased, by selection, the degree of bilateral asymmetry in the number of ocellar bristles on the head of *Drosophila*. These various experiments in themselves do not enable us to distinguish general reductions in genetic variation underlying the character or general reductions in environmental sensitivity of development from specific canalization of development around a particular phenotypic endpoint for a given character. It is this last process that must underlie the character shifts in species evolution. Presumably, such shifts occur in three stages. First a change in environment or a genetic change somewhere in the genome has the effect of moving the distribution of an underlying morphogenetic product outside its current zone of canalization. This then reveals underlying standing genetic variation influencing the product that was previously not manifest at the phenotypic level. At the second stage there is selection on the newly revealed variation to some new adaptive mean, but there remains phenotypic variation around that mean. Finally, selection to reduce this variance while holding the mean results in a new canalization around the new state.

3 MOLECULAR NOISE AND CANALIZATION

As discussed in the first part of this chapter, developmental canalization is observed and has been experimentally manipulated at the level of phenotype, with various theoretical explanations for how canalization is mediated and how it evolves under the natural selection. With the advances in molecular biology and our ever-increasing understanding of the developmental interactions between genes and proteins, we can now start to address the question of how developmental canalization is mediated at the molecular level by modeling the detailed interactions of genes and proteins in what are becoming known as "gene regulatory networks."

Once we start considering genetic interactions at the level of single cells, we need to consider random effects caused by the small numbers of many molecules *in vivo*. This *biochemical noise* is an important source of variation in addition to genetic and environmental variation, and potentially accounts for much of the developmental noise observed at the phenotypic level, such as variation in bristle numbers in certain species of flies. In turn, biochemical noise has an evolutionary impact: it can be a trait to be canalized itself, or, by generating phenotypic differences around a mean, it can reveal väriation at other loci and thereby facilitate natural selection for canalizing selection to act on other (modifier) genes.

The recognition of a central role for randomness at the level of gene regulation leads to a very different view of molecular and cellular biology than the deterministic viewpoint of much of developmental biology, where genes produce proteins which lead to a fixed phenotype. Before we can explore the link between biochemical noise and canalization, we need to substantiate two basic claims: (1) Biochemical noise is a necessary concept at the molecular level, and (2) Biochemical noise at the molecular level affects organisms at the phenotypic level.

We define biochemical noise as molecular level variation arising between genetically identical cells sharing identical environments. The cause of this variation is the stochastic nature of events within the cell caused by the small numbers of many molecules *in vivo*. Biochemical noise can arise from two potential sources of variation that lie outside the usual definitions of genetic or environmental variation: (1) Internal state biochemical noise caused by differences in the number or state of biologically active molecules within the cell (for example, regulatory proteins segregate randomly at cell division), and (2) Brownian motion biochemical noise caused by the random timing of molecular interactions. The timing of individual reactions is affected by the motion of water molecules, and all of the other small molecular and atomic forces characterized as Brownian motion.

The reason for introducing the notion of noise arising from Brownian motion is to prevent a philosophical retreat from biochemical noise via a redefinition of environment to include the precise internal state of the cell (see, for example, Kitcher [31]). Because biochemical noise can arise from Brownian motion, it is irreducible. Nothing less than a complete description of the environment of the cell and its external environment, including the position and momentum of all molecules, would be required to exclude stochastic variation arising from Brownian motion—a biochemical variation of LaPlace's demon.

Next we must show that this concept, whilst clearly plausible in principle, is necessary in practice to understand the details of gene regulatory networks. In order to do so, it is worthwhile briefly exploring historical concepts of stochastic noise in molecular biology and genetics, where the standard metaphor is of deterministic machinery. An extreme illustration of the deterministic paradigm is the nematode, *C.elegans*, where a standard textbook gives the complete cell lineages of all 945 cells present in the adult along with the 131 cells which undergo apoptosis [61]. Even in this apparently most highly determined system, however, there is randomness in the choice of cells that differentiate to form the vulva [26].

There is, in fact, a long history both of theoretical considerations and experimental evidence for the importance of random variation at the molecular and cellular levels, and the necessity for cells and organisms to be robust to such noise. Erwin Schrödinger considered the issue in his classic book, *What is Life*? [50]. Based on fundamental physics and chemistry, he argued that the need to be robust against stochastic events constrains the minimum size of the cell. In chemistry itself, the importance of the stochastic timing of individual molecular reactions has been recognized for just as long [11]. Stochastic representations of chemical reactions at low concentrations have stronger theoretical justification than deterministic representations such as differential equations [33]. There is a large body of literature describing how to model stochastic molecular interactions (see Gillespie [23]; van Kampen [55]; Érdi and Tóth [17]; and Gardiner [20]). Since these "first principles" considerations from physics and chemistry, an abundance of biological evidence for biochemical noise at the molecular and cellular levels has been gathered; empirically in numerous systems, and theoretically using detailed quantitative models.

Observations of gene expression in individual cells illustrate the stochastic nature of transcription [1, 4, 32, 60, 62]. The presence of enhancer DNA sequences increases the probability but not the rate of DNA transcription [62]. As early as 1964, investigation of stem cell haematopoiesis suggested that differentiation is a stochastic process [53]. Differentiation in human liver cells requires two copies of a hormone molecule that is present at a concentration equal to one molecule per six cells, so that, on average, only one in 36 cells will differentiate [39]. Swiss T3 cells contain a stochastic mechanism for regulating cell cycle progression [8]. Extensive studies of plasmid replication show that the timing of replication is random for at least some plasmids, including ColE1 [41].

Finally, in the last few years, the experimental manipulation of artificial gene networks has allowed direct visualization of variation arising from biochemical noise (see Becskei and Serrano [5]; Elowitz and Leibler [14]; and Gardner et al. [21]). Ozbudak et al. [42] directly explored how changes to the kinetic parameters of gene expression (transcription and translation rates) influence the level of variation observed. Elowitz et al. [15] empirically explored the potential sources of stochasticity inherent in gene expression. They generated *E. coli* strains with genes encoding the cyan and yellow alleles of the green fluorescent protein (GFP), with both genes under the control of identical promoters. In the absence of biochemical noise, the expression of the two genes would be perfectly correlated, and cells would appear yellow. When one gene is expressed more than the other the cells would appear red or green. The results of the experiment clearly showed individual cells varying in color, with red, green, and yellow cells all present. Despite the fact that both copies of the GFP gene were exposed to the same internal cellular environment within any given cell, individual transcription and translation events were not synchronized, and this biochemical noise leads to gene expression becoming uncorrelated.

4 PHENOTYPIC EFFECTS OF BIOCHEMICAL NOISE

At the organismic level there is also an abundance of evidence for noise arising from neither genetic nor environmental variation; despite the best efforts of laboratory biologists to remove sources of variation in experiments, by working with clonal bacteria, inbred fruit flies and mice, and by standardizing the environment, differences between individuals still occur. The coefficient of variation in the length of the bacterial cell cycle is around 0.22, even for clonal cells that are initially synchronized and raised in a homogeneous environment [6, 43, 54]. In eukaryotes, evidence for the effects of developmental noise at the phenotypic level comes from a 20-year effort to standardize laboratory mice and rats [22]. Gärtner concluded that 70%–80% of the total phenotypic variation in an inbred line must be ascribed to noise.

The advent of stochastic models and computer simulations based on detailed and realistic molecular interactions has shown the link from biochemical noise at the molecular level to developmental noise manifested as phenotypic variation. A number of gene regulatory networks have been modeled stochastically (see de Jong [10]; Endy and Brent [16]; Hasty et al. [30]; Smolen et al. [52]; McAdams and Arkin [37]; and Goss and Peccoud [25]), and biochemical noise has also been shown to be important in processes such as bacterial chemotaxis (for example, Barkai and Leibler [3]). A variety of related modeling and simulation approaches have been proposed, with broadly similar underlying assumptions and mathematics. No standard modeling approach covers all problems, and Endy and Brent [16] outline the major issues facing quantitative models if they are to become core tools for modeling cellular behavior.

When regulatory proteins are involved, and particularly when there is a choice of developmental pathways—a "genetic switch"—biochemical noise can lead to qualitative phenotypic differences. The lysis/lysogeny decision of Lambda phage is the best characterized example of a genetic switch (see Ptashne [45]), and is often regarded as a model of how differentiation in development can be mediated by regulatory genes leading to bifurcations in cell state. Both the state of the host cells and the number of lambda phages that infect a particular bacterium affect the frequency of lysogeny. The molecular interactions involved in this switch have been simulated using a stochastic model [2]. The model showed that the random timing of reactions can lead to different outcomes, either lysis or lysogeny, even when the initial state of the host and the multiplicity of infection are fixed.

In some cases there may well be a selective advantage to biochemical noise affecting the outcome of a genetic switch, for example to generate variation (McAdams and Arkin [38]—TIG). In many cases, however, particularly where the switch is part of a cascade of developmental decisions, randomness would be disadvantageous to the organism. We can, therefore, postulate that the gene regulatory networks of many organisms will have mechanisms specifically designed to minimize the effect of biochemical noise—molecular canalization.

The use of computational models, particularly when linked with experimental evidence, allows us to explore how organisms respond to biochemical noise. Several studies have addressed the question of how regulatory stability is achieved in the face of randomness [3, 38]. A number of common themes have emerged, including the use of feedback, redundancy, checkpoints, and the choice of biochemical parameters such as transcription rate. These themes can be grouped into two major approaches for minimizing the impact of biochemical noise, or molecular canalization: (1) selection of biochemical parameters to promote robustness, and (2) design and selection of regulatory networks to promote robustness.

As a counterpoint to this approach, Siegal and Bergman [51] question the assumption (originally by Waddington) that canalization is a consequence of stabilizing selection. They present an abstract model of an evolving developmental-genetic system, and argue that the extent of canalization depends on the complexity of the network, and that canalization may be an inevitable consequence of complex developmental-genetic processes.

However, the possibility that canalization could be an epiphenomenon of the evolution of complex systems does not preclude a role for selection in the evolution of some, if not most, actual cases of canalization. One way to address this question is by computer simulations that explore in detail the molecular basis for canalization in a range of specific examples. In turn, these models require both a deep knowledge of the biochemical and genetic interactions in the network, and the facility to build stochastic models of these networks. We, therefore, look in detail below at results obtained using an approach called Stochastic Petri Nets, which allows biologists to focus on the specifics of the molecular interactions they want to include in the model rather than the underlying simulation methodology [25]. The example involves modeling the gene regulatory network controlling the replication of plasmid ColE1. One of the key findings of this work is that the observed rate of plasmid loss, a key phenotypic characteristic of plasmids for many experimentalists, can best be explained by developmental noise. The model also allows us to examine the relative importance of these two approaches to molecular canalization—selection of biochemical parameters or network structure—as well as showing evidence for canalization using a modifier gene.

5 PLASMID REPLICATION AS AN EXAMPLE OF MOLECULAR CANALIZATION

Plasmid replication is well-understood experimentally and has been well-studied theoretically. A commonplace observation in the laboratory is that, without the use of antibiotic resistance markers, plasmids are lost from bacterial cell cultures, even where the plasmids maintain relatively high copy numbers on average. Once plasmids are lost from even one bacterium, they are rapidly lost from the entire population, since plasmid-free bacteria replicate more rapidly than bacteria with plasmids. Standard laboratory practice involves the insertion of antibiotic resistance markers, so that plasmid-free bacteria are selected against.

Making the reasonable assumption that plasmid copy number is under selective control to prevent plasmid loss, and in light of the above discussion, two obvious questions arise:

- How is the average plasmid copy number regulated?
- How is the regulatory mechanism designed to cope with biochemical noise?

The first question has been extensively studied using a variety of approaches. Brendel and Perelson [7] built a deterministic model for ColE1 plasmid replication, taking into account the known regulatory effects of two main elements, RNA I and Rom protein. The reactions included in the model are shown in figure 5. Plasmid replication is controlled by two main feedback mechanisms. The primary feedback loop involves an antisense RNA call RNA I, which in turn is stabilized by a protein called Rom (RNA-one modulator). For a wildtype plasmid strain in bacteria doubling every 80 minutes, the average plasmid copy number immediately before cell division predicted by the model is about 38, in good agreement with empirical observations of about 40 plasmids per bacterium. Under binomial segregation, the probability of a daughter cell being plasmid-free is 7.3×10^{-12}, much rarer than is experimentally observed. When segregation is uneven, and one daughter cell starts a new cell cycle with less than the average of 19 plasmids, the negative feedback in the network causes the rate of plasmid replication to increase rapidly in that cell, returning the system to equilibrium. Feedback in the gene regulatory network buffers the system against the internal state developmental noise generated by random segregation of plasmids. However, even taking unequal segregation into account, the deterministic model still predicts a much lower rate of plasmid loss than is experimentally observed.

What effect does Brownian motion developmental noise, arising from the timing of plasmid replication and the associated reactions, have on the system? Goss and Peccoud [24] describe a stochastic model with the same structure and biochemical parameters as the deterministic model described above. The predicted mean plasmid number per bacterium is very similar for the deterministic and stochastic models. However, because it takes into account the random timing of individual reaction events, the stochastic model predicts a distribution of plasmid number around the mean even given an identical internal state for each bacterium at the beginning of the cell cycle. The 95% confidence interval for plasmid copy number is 38 ± 11.6. When the distribution of plasmid copy number before bacterial replication is combined with binomial segregation, the predicted probability of plasmid loss is 2.3×10^{-8}, four orders of magnitude greater than in the deterministic case. In a bacterial culture of 10^9 cells this means that four new plasmid-free bacteria are expected every generation, which explains the experimental observation that the loss of plasmids from the population is common. Biochemical noise, both internal state and Brownian motion,

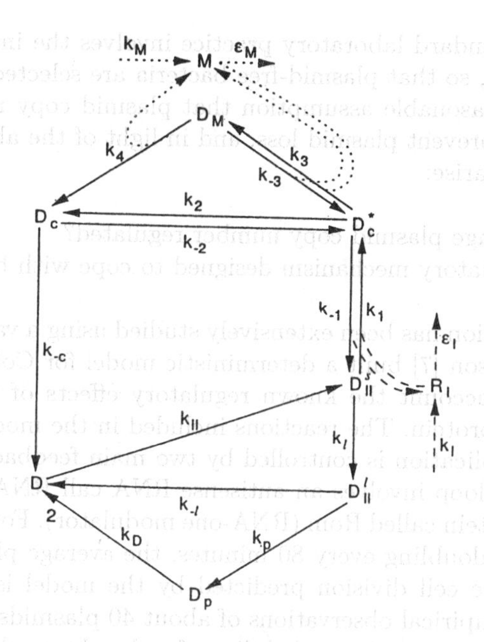

FIGURE 5 Diagram of control pathways in the replication of ColE1 plasmid replication. Plasmid DNA occurs in free form (D), complexed with short RNAII molecules(D^{s}_{II}), with long RNA molecules(D^{l}_{II}), an RNAII-DNA complex primed for replication(D_p) an unstable complex (D^{*}_{c}), in a stable complex with RNAI and RNAII(D_c), and with RNAI, RNAII and Rom protein(D_M). Free RNAI(RI) and Rom protein(M) are also produced. From Goss and Peccoud (1998) based on Brendel and Perelson (1993). Reprinted by permission.

is affecting an important phenotype of plasmid ColE1, the probability of loss at cell division.

What features of the regulatory network controlling plasmid replication enable it to cope with the variance in copy number caused by biochemical noise? The specific values of the biochemical parameters affect the mean plasmid copy number, and negative feedback in the network means that deviations from the average will automatically be countered. However, a deeper examination of the system reveals important limits to the negative feedback produced by RNA I. In fact, a primary role of the secondary regulatory mechanism, Rom, is to limit the variation caused by biochemical noise. In other words, the apparently redundant feedback loop involving Rom in the regulatory system is there for molecular canalization.

To show this, we need to consider how the plasmid copy number would behave in the absence of Rom. Rom-minus plasmid strains are observed in nature, but are rare. Experimental observations of mutant plasmid strains in which Rom protein is nonfunctional show roughly a twofold increase in the number of plasmids per bacterium. Why then do Rom-minus plasmids not replace wildtype plasmids? The answer lies in a subtle combination of selective factors both within bacterial lineages and between lineages, where plasmid copy number is under selection to be both low and canalized (i.e., low variance of copy number across individual bacterial lineages). Simulation of competition between a wildtype strain and a Rom-minus strain within a single bacterial lineage suggests that the Rom-minus strain is competitively neutral against the wildtype. However, a higher plasmid number imposes a metabolic cost on the host bacteria. Selection between bacterial lineages would favor those lineages which only have wildtype plasmids, which, on average, will have lower plasmid copy numbers, and thus quicker replication, than lineages with Rom-minus plasmids.

To compensate for this potential metabolic cost, simulations were run for Rom-minus plasmid strains where every individual kinetic parameter was modified in turn so that the mean plasmid numbers in the mutant strain and in the wildtype strain were equal. No matter which kinetic parameter was modified, the variance in plasmid number was greater in the mutant strain than in the wildtype strain, leading to a twofold to sixfold increase in the rate of plasmid loss. Artificially constraining the model so that plasmid segregation is equal rather than binomial shows that the majority of this variance is due to the specific timing of individual reactions, Brownian motion developmental noise. Competition simulations between wildtype plasmid and modified Rom-minus plasmids show that when mean plasmid numbers are equalized, Rom-minus plasmids are out-competed by wildtype plasmids. The results from the stochastic models suggest that Rom protein has a canalizing function, decreasing the rate of plasmid loss due to developmental noise [24]. If the heterogeneity introduced by developmental noise were not biologically significant, or if it were possible to produce the same result by optimization of the biochemical parameters, such molecular canalization would not be maintained.

Overall then, the regulatory mechanism controlling plasmid replication copes with biochemical noise in two ways: using first-order feedback (RNA I) to revert the plasmid copy numbers in bacteria that inherit more or fewer plasmids on average (internal state developmental noise) towards the mean, and using an additional layer of control in the system (Rom) specifically to constrain the variance around the mean generated by Brownian motion developmental noise. In this particular case, Brownian motion developmental noise accounts for a much greater portion of the variance in the phenotype (plasmid loss) than internal state developmental noise. More importantly, the structure of the network, in terms of the function of Rom, has a greater role in canalizing the system at a molecular level than does parameter optimization.

6 CONCLUSION

Although a single example of a very simple gene regulatory network is far from conclusive, there are more and more examples in which gene regulatory networks are well enough characterized to be modeled stochastically. As models become more sophisticated, we will be able to link phenotypic variation to a combination of genetic and environmental variation and biochemical noise, and to understand the molecular basis of developmental canalization. As we do, we may well find that many of the control elements in gene regulatory networks function primarily to reduce variation due to biochemical noise—molecular canalization.

REFERENCES

[1] Abkowitz, J. L., S. N. Catlin, and P. Guttorp. "Evidence that Hematopoiesis may be Stochastic In Vivo." *Nature Med.* **2** (1996): 190–197.

[2] Arkin, A., J. Ross, and H. H. McAdams. "Nongenetic Diversity: Random Variations in Gene Expression Reaction Rates Determine which Phage Lambda Infected Cells Become Lysogens." *J. Mol. Biol.* **149(4)** (1998): 1633–1648.

[3] Barkai, N., and S. Leibler. "Robustness in Simple Biochemical Networks." *Nature* **387** (1997): 913–917.

[4] Baroffio, A., and M. Blot. "Statistical Evidence for a Random Commitment of Pluripotent Cephalic Neural Crest Cells." *J. Cell Sci.* **103** (1992): 581–587.

[5] Becskei, A. and L. Serrano. "Engineering Stability in Gene Networks by Autoregulation." *Nature* **405** (2000): 590–593.

[6] Bremer, H. "A Stochastic Process Determines the Time at which Cell Division Begins in *Escherichia coli.*" *J. Theoret. Biol.* **118** (1986): 351–365.

[7] Brendel, V., and A. S. Perelson. "Quantitative Model of ColE1 Plasmid Copy Number Control." *J. Mol. Biol.* **229** (1993): 860–872.

[8] Brooks, R. F. "The Transition Probability Model: Successes, Limitations, and Deficiencies." In *Temporal Order*, edited by L. Rensing and L. I. Jaeger, 304–314. Berlin: Springer-Verlag, 1985.

[9] Clausen, J., D. D. Keck, and W. W. Hiesey. "Environmental Responses of Races of *Achillea.*" *Carnegie Inst. Wash. Pub.* **581** (1958): 1–129.

[10] de Jong, H. "Modeling and Simulation of Genetic Regulatory Systems: A Literature Review." *J. Comp. Biol.* **9** (2002): 67–103.

[11] Delbrück, M. "Statistical Fluctuations in Autocatalytic Reactions." *J. Chem. Phys.* **8** (1940): 120–124.

[12] Dun, R. B., and A. S. Fraser. "Selection for an Invariant Character—"Vibrissa Number"—in the House Mouse." *Nature* **181** (1958): 1018–1019.

[13] Dun, R. B., and A. S. Fraser. "Selection for an Invariant Character, Vibrissa Number, in the House Mouse." *Australian J. Biol. Sci.* **12** (1959): 506–523.

[14] Elowitz, M. B., and S. Leibler. "A Synthetic Oscillatory Network of Transcriptional Regulators." *Nature* **403** (2000): 335–338.

[15] Elowitz, M. B., A. J. Levine, E. D. Siggia, and P. S. Swain. "Stochastic Gene Expression in a Single Cell." *Science* **297** (2002): 1183–1186.

[16] Endy, D., and R. Brent. "Modelling Cellular Behaviour." *Nature* **401** (2001): 391–395.

[17] Érdi, P., and J. Tóth. *Mathematical Models of Chemical Reactions: Theory and Applications of Deterministic and Stochastic Models.* Princeton, NJ: Princeton University Press, 1989.

[18] Fisher, R. A. "The Possible Modification of the Response of Wildtype to Recurrent Mutations." *Amer. Natur.* **62** (1928): 115–126.

[19] Fisher, R. A. "The Evolution of Dominance." *Biol. Rev.* **6** (1931): 345–368.

[20] Gardiner, C. W. *Handbook of Stochastic Methods for Physics, Chemistry and the Natural Sciences*, 2d. New York: Springer-Verlag, 1990.

[21] Gardner, T. S., C. R. Cantor, and J. J. Collins. "Construction of a Genetic Toggle Switch in *Escherichia coli*." *Nature* **403** (2000): 339–342.

[22] Gärtner, K. "A Third Component Causing Random Variability Beside Environment and Genotype. A Reason for the Limited Success of a 30-Year-Long Effort to Standardize Laboratory Animals?" *Laboratory Animals* **24** (1990): 71–77.

[23] Gillespie, D. T. "Exact Stochastic Simulations of Coupled Chemical Reactions." *J. Phys. Chem.* **81** (1977): 2340–2361.

[24] Goss, P. J. E., and J. Peccoud. "Analysis of the Stabilizing Effect of Rom on the Genetic Network Controlling Plasmid ColE1 Replication." *Pacific Symp. Biocomp.* (1999) 65–76.

[25] Goss, P. J. E., and J. Peccoud. "Quantitative Modeling of Stochastic Systems in Molecular Biology using Stochastic Petri Nets." *Proc. Natl. Acad. Sci. USA* **95** (1998): 6750–6755.

[26] Greenwald, I., and G. M. Rubin. "Making a Difference: The Role of Cell-Cell Interactions in Establishing Separate Identities for Equivalent Cells." *Cell* **68** (1992): 271–281.

[27] Gupta, A., and R. C. Lewontin. "A Study of Reaction Norms in Natural Populations of *Drosophila pseudoobscura*." *Evolution* **36** (1982): 934–938.

[28] Haldane, J. B. S. "The Part Played by Recurrent Mutation in Evolution." *Amer. Natur.* **67** (1933): 5–19.

[29] Haldane, J. B. S. "The Theory of the Evolution of Dominance." *J. Genet.* **37** (1939): 365–374.

[30] Hasty, J., J. Pradines, M. Dolnik, and J. J. Collins. "Noise-based Switches and Amplifiers for Gene Expression." *Proc. Nat. Acad. Sci.* **97** (2000): 2075–2080.

[31] Kitcher, P. "Battling the Undead: How (and how not) to Resist Genetic Determinism." In *Thinking About Evolution*, edited by R. Singh, C. Krimbas, J. Beatty, and D. Paul. Cambridge, MA: Cambridge University Press, 1998.

[32] Ko, M. S. H., N. Nakauchi, and N. Takahashi. "The Dose Dependence of Glucocorticoid-Inducible Gene Expression Results from Changes in the Number of Transcriptionally Active Templates." *EMBO J.* **9** (1990): 2835–2842.

[33] Kurtz, T. G. "The Relationship between Stochastic and Deterministic Models for Chemical Reactions." *J. Chem. Phys.* **57** (1972): 2976–2978.

[34] Lindsley, D. L., and G. G. Zimm. *The Genome of Drosophila melanogaster.* San Diego, CA: Academic Press, 1992.

[35] Maynard Smith, J., and K. C. Sondhi. "The Genetics of a Pattern." *Genetics* **45** (1960): 1039–1050.

[36] McAdams, H. H., and A. Arkin. "Stochastic Mechanisms in Gene Expression. *Proc. Natl. Acad. Sci. USA* **94** (1997): 814–819.

[37] McAdams, H. H., and A. Arkin. "Simulation of Prokaryotic Genetic Circuits." *Ann. Rev. Biophys. Biomol. Struct.* **27** (1998): 199–224.

[38] McAdams, H. H., and A. Arkin. "It's a Noisy Business! Genetic Regulation at the Nanomolar Scale." *Trends in Genetics 15* (1999): 65–69.

[39] Michaelson, J. "Biology of Disease: Cellular Selection in the Genesis of Multicellular Organisms." *Laboratory Investigation* **69** (1993): 136–151.

[40] Milkman, R. D. "The Genetic Basis of Natural Variation—III. Developmental Lability and Evolutionary Potential." *Genetics* **46** (1961): 25–38.

[41] Nordström, K., and S. J. Austin. "Cell-Cycle Specific Initiation of Replication." *Mol. Microbiol.* **10** (1993): 457–463.

[42] Ozbudak, E. M, Thattai, I. Kurtser, A. D. Grossman, and A. van Oudenaarden. "Regulation of Noise in the Expression of a Single Gene." Published in *Nat. Genet.* 2002 May; 31(1): 69–73. Epub 2002 Apr 22. ⟨http://www.ncbi.nlm.nih.gov/entrez/query.fcgi?holding=npg&cmd=Retrieve&db=PubMed&list_uids=11967532&dopt-Abstract⟩

[43] Plank, L. D., and J. D. Harvey. "Generation Time Statistics of *Escherichia coli b* Measured by Synchronous Culture Techniques." *J. Gen. Microbiol.* **115** (1979): 69–77.

[44] Prout, T. "The Effects of Stabilizing Selection on the Time of Development in *Drosophila melanogaster.*" *Genet. Resh.* **3** (1962): 364–382.

[45] Ptashne, M. *A Genetic Switch: Phage and Higher Organisms*, 2d. Cambridge, MA: Cell Press and Blackwell Scientific Publications, 1992.

[46] Rendel, J. M. *Canalization and Gene Control.* London: Academic, Logos Press, 1967.

[47] Rendel, J. M. "Canalization of the Scute Phenotype of *Drosophila.*" *Evolution* **13** (1959): 425–439.

[48] Rendel, J. M., and B. L. Sheldon. "Selection for Canalization of Scute Phenotype in *Drosophila melanogaster.*" *Australian J. Biol. Sci.* **13** (1960): 36–47.

[49] Russell, W. A. "Competitive Performance of Maize Hybrids Representing Different Eras of Maize Breeding." *Proc. Annu. Corn Sorghum Resh. Conf.* **29** (1974): 81–101.

[50] Schrödinger, E. *What is Life?: The Physical Aspect of the Living Cell.* New York: Macmillan, 1945.

[51] Siegal, M. L., and A. Bergman. "Waddington's Canalization Revisited: Developmental Stability and Evolution." *Proc. Natl. Acad. Sci. USA* **99** (2002): 10528–10532.

[52] Smolen, P., D. A. Baxter, and J. H. Byrne. "Modeling Transcriptional Control in Gene Networks—Methods, Recent Results and Future Directions." *Bull. Math. Biol.* **62** (2000): 247–292.

[53] Till, J. E., E. A. McCulloch, and L. Siminovitch. "A Stochastic Model of Stem Cell Proliferation, Based on the Growth of Spleen Ccolony-Forming Cells." *Proc. Natl. Acad. Sci. USA* **51** (1964): 29–36.

[54] Tyson, J. J., and K. B. Hannagen. "The Distributions of Cell Size and Generation Time in a Model of the Cell Cycle Incorporating Size Control and Random Transitions." *J. Theoret. Biol.* **113** (1985): 29–62.

[55] Van Kampen, N. G. *Stochastic Processes in Physics and Chemistry.* Amsterdam: North-Holland, 1981.

[56] Waddington, C. H. "Canalization of Development and the Inheritance of Acquired Characters." *Nature* **150** (1942): 563–565.

[57] Waddington, C. H. "Experiments on Canalizing Selection." *Genet. Resh.* **1** (1960): 140–150.

[58] Waddington, C. H. "Genetic Assimilation of an Acquired Character." *Evolution* **7** (1953): 118–126.

[59] Waddington, C. H. *The Strategy of the Genes.* London: Allen and Unwin, 1957.

[60] Walters, M. C., S. Fiering, J. Eidemiller, W. Magis, M. Groudine, and D. I. K. Martin. "Enhancers Increase the Probability but not the Level of Gene Expression." *Proc. Natl. Acad. Sci. USA* **892** (1995): 7125–7129.

[61] Watson, J. D., N. H. Hopkins, J. W. Roberts, J. Argetsinger Steitz, and A. M. Weiner. *Molecular Biology of the Gene*, vol. II, 4th ed. Menlo Park, CA: Benjamin/Cummings, 1987.

[62] Weintraub, H. "Formation of Stable Transcription Complexes as Assayed by Analysis of Individual Templates." *Proc. Natl. Acad. Sci. USA* **85** (1988): 5819–5823.

[63] Wright, S. "Fisher's Theory of Dominance." *Amer. Natur.* **63** (1929): 274–279.

[64] Wright, S. "Physiological and Evolutionary Theories of Dominance." *Amer. Natur.* **68** (1934): 25–53.

Evolution of Phenotypic Robustness

Joachim Hermisson
Günter P. Wagner

1 INTRODUCTION

Evolutionary biology, in the neo-Darwinian tradition, is based on the study of genetic and phenotypic variation and its fate in populations. Thus, the observation that the genetic variability of a trait is itself influenced by the genotype has obvious theoretical implications [49, 55]. It is in this context that the robustness of phenotypic traits was first conceptualized as *canalization* and became the focus of a significant research effort (reviewed in Scharloo [39]). With increasing awareness of the intricate molecular mechanisms maintaining the life of cells, the ubiquity of buffering and compensatory mechanisms came into focus [18, 37, 63]. Recent years have seen a confluence of the classical concept of canalization and new research in molecular biology that resulted in a sharp increase in the interest in canalization and related phenomena. One can speak of an emerging field of *biological robustness* research that is able to draw on a sophisticated arsenal of technical and theoretical tools that were developed over the last ten years [8].

In the present chapter we summarize the principal findings of the classical and the newer literature on robustness of phenotypes and identify the issues that require attention in future research. We will start with an attempt to formalize the notion of phenotypic robustness to clarify the criteria that need to be met in order to experimentally demonstrate phenotypic robustness. Then we review the experimental evidence about environmental and genetic robustness and find that genetic robustness is particularly difficult to demonstrate. Finally an overview of theoretical models for the evolution of canalization is provided which shows that the existing literature is strongly biased toward a few scenarios. We conclude that there are major unsolved questions both in the experimental demonstration of canalization as well as in understanding the evolutionary dynamics of canalization.

2 DEFINING PHENOTYPIC ROBUSTNESS

Phenotypic robustness is about the sensitivity of the phenotype (e.g., some quantitative trait) with respect to changes in the underlying variables (genotype and environment) which determine its expression in an individual. Intuitively, this seems to be a clear notion. However, although the concept originated more than forty years ago, there is a conspicuous lack of a formal definition. Such a definition should include a clear criterion for the detection and characterization of robustness in empirical *and* theoretical study. For empirical work, robustness must be formalized as an operational concept. In order to be of conceptual use for an understanding of evolutionary processes, robustness should be classified into types that provide information about its evolutionary role and the circumstances of its origination. In this section, we will point out some key elements for such a formal definition and discuss demarcations to related concepts.

Our starting point is the formalization of robustness as a state of reduced impact from a given source of variation (such as mutations or environmental change) on the trait, in other words, as reduced variability due to that source [59, 61]. Let us state this first in an informal way: *A character state that has evolved under natural selection is phenotypically robust if the variability of the character under a given source of variation is significantly reduced in this state as compared to a set of alternative states.* Classification of different types of robustness now goes along with the formal clarification of three key notions of this definition, namely the character state, the source of variation, and the reference set of alternative states.

Genotypic and Population States. Since phenotypic robustness describes how variation in genotypic or environmental values is translated into variation on the level of the phenotype, a formal notion of the mapping between these two levels is needed. Each individual phenotype in a population, through development, is determined by a number of heritable and non-heritable factors, which we will refer to as its genotype and the environment experienced. Among the environmental factors, two types need to be distinguished in the following, commonly referred to as macro- and micro-environmental factors [37]. Macro-environmental factors, such as temperature and light intensity, describe the identifiable environment external to the organism. They are, at least in principle, experimentally controllable. Micro-environmental (or developmental) factors, in contrast, describe the uncontrollable developmental noise that is, to a large extent, internal to the organism [16]. Imagine now the space G of all possible genotypes $\boldsymbol{g}$ in a population, and sets E and I of all experimentally controllable environmental and uncontrollable developmental conditions, $\boldsymbol{e}$ and $\boldsymbol{i}$. Each individual is then represented by the micro-state vector $\boldsymbol{y} := (\boldsymbol{g}, \boldsymbol{e}, \boldsymbol{i})$. Following Rice [36], we may view a character ϕ as evolving on a phenotype landscape, defining phenotype as a function of the underlying variables, $\phi = \phi(\boldsymbol{g}, \boldsymbol{e}, \boldsymbol{i})$. For a given trait ϕ we now introduce the *genotypic character state* (or genotypic state) $(\boldsymbol{g}, \boldsymbol{e}, \rho_I)_\phi$ which describes the trait values of a genetically homogeneous population under controlled environmental conditions $\boldsymbol{e}$. The uncontrollable developmental factors are represented by a distribution ρ_I over I. For a natural (outbred) population with genotype distribution ρ_G, we define the *population (character) state*, $(\rho_G, \boldsymbol{e}, \rho_I)$, which describes the phenotype statistics of a population on the landscape in an averaged micro-environment.

Sources of Variation and Measures of Variability. Phenotypes can be buffered against variation from very different sources and reduced variability with respect to one source may or may not correlate with robustness under a different mode of variation. In order to explain what is actually buffered in a robust state, we will therefore need to specify the source of variation. The distinction with the most important evolutionary consequences is certainly the one between genetic and environmental types of robustness [61], depending on whether variation in heritable or non-heritable variables is buffered. Both types can assume a variety of forms. While genetic sources include different types of mutation and recombination, environmental sources may affect various macro-environmental factors. An important (and unavoidable) environmental source is further given by the uncertainties of development, namely developmental noise. Formally, any source S can be described by a mapping which assigns every micro-state $\boldsymbol{y}$ a distribution of variant micro-states $\mu_{\boldsymbol{y}}^{(S)}$. Usually, we can assume that genetic and macro-environmental sources and developmental noise are all independent of each other, and hence $\mu_{\boldsymbol{y}}^{(S)} = \mu_{\boldsymbol{g}}^{(S)} \mu_{\boldsymbol{e}}^{(S)} \mu_{\boldsymbol{i}}^{(S)}$. This does not imply that the *phenotypic* variation effects are independent. Since developmental variations are not heritable, the new variation per generation due to developmental noise is equal to the distribution of developmental factors in the population, $\mu_{\boldsymbol{i}}^{(S)} \equiv \rho_I$. It is important to note that this is, in general, not the case for new genetic variations, $\mu_{\boldsymbol{g}}^{(S)} \neq \rho_G$.

In order to quantify variability due to a source S, consider a trait ϕ equipped with some scale for the measurement of variation effects. The choice of the appropriate scale is an important, but subtle issue (cf. e.g., Lynch and Walsh [28, Chap. 11]). As long as the mean effect of the variations on the trait can be neglected, the variance of the variation effects is a valid measure of variability,

$$v_{\phi}^{(S)}(g,e) = \int dg'de'di\Big(\phi(g',e',i) - \bar{\phi}\Big)^2 \mu_g^{(S)}(g')\mu_e^{(S)}(e')\rho_I(i)$$

for the variability of a genotype state and

$$v_{\phi}^{(S)}(\rho_G,e) = \int dg\, v_{\phi}^{(S)}(g,e)\rho_G(g)$$

on the population level. If S is purely genetic or purely environmental, this is just the mutational or environmental variance, respectively. The mutational variance is an insufficient measure of variability if mutational effects are strongly asymmetric or even unidirectional. In this case, the mean-square effect or the (absolute) mean are more appropriate.

Reference Genotypes. Defined as a state of reduced variability, robustness is a relative concept. Characterizing a character state as robust always presupposes that we compare it with a set of alternative states which are, on average, more variable. For a general notion of robustness, many choices of this reference are possible. If we are interested in robustness as a physiological property, we might compare the character states of various traits and ask for the elements of the genetic architecture that lead to observed differences in variability. From an *evolutionary* point of view, however, the alternative states should reveal the reduction of variability as a property that has evolved under natural selection. Consequently, the reference must be chosen from the states that have been (or still are) segregating in the population, that is from the naturally occurring variants or the ancestral genotypes. Within this set, the specific choice of the reference depends on a precise reformulation of the evolutionary question: In the last paragraph, we have distinguished different types of robustness according to the source of variation. In the following section, we will further refine our definition by asking whether, how, and why robustness originated in evolution. Comparison with the properly chosen set of reference states then provides relevant statistical information for these questions. Note that this choice of a reference does not depend on whether these states are actually available for variability measurements. For ancestral genotypes, this is usually not the case. One then must resort to indirect information, such as comparing homologues across different species as a standard method of reconstructing the state of a common ancestor.

2.1 ADAPTIVE OR INTRINSIC?

It is a trivial truth that any cellular or organismal property is evolved in the sense that it is the outcome of the evolutionary process. Evolutionary biology, however, is concerned with the level of organization where it first arose, asking for the causes and circumstances of its origination. In the case of phenotypic robustness, this is made explicit in the distinction of *adaptive* and *intrinsic* forms of robustness (see also Gibson and Wagner [20] and deVisser et al. [8]).

Adaptive Robustness. We call robustness *adaptive* if the buffering of the trait with respect to some source of variation has been a *target of natural selection*; in other words, the robust character states have been selected because of their reduced variability. In order to come to an operational criterion for adaptive robustness, we test if the target state shows less variability than its mutational neighbors in genotype space that lead to the same phenotype. More precisely, we restrict our reference to mutational neighbors with a mean-associated phenotype of equal or even higher fitness. This restriction ensures that the target state has not been selected because of a correlated direct effect on fitness. Note that the definition does not exclude the possibility that robustness with respect to a given source of variations evolves as correlated response to selection for reduced effects of variations from a different source. This may have important evolutionary consequences in particular if buffering for genetic and environmental variations is coupled [61]. Phenomena of this kind have been observed in computational models of RNA secondary structure and have been called plasto-genetic congruence [1]. In order to account for congruence effects, we may further refine our definition by restricting the reference set to states with equal variabilities from other sources.

In the following, we will use the term *canalization* as synonymous with adaptive robustness. This is in line with the classic literature on this subject: According to Schmalhausen [43] and Waddington [55] (see also Scharloo [39]), canalization is a property that evolves for its own sake. The natural force assumed to be responsible for its evolution is stabilizing selection acting directly on the character or on some highly correlated pleiotropic trait (in which case buffering of the correlated trait is the evolutionary target).

Within adaptive robustness, two main types may be distinguished, marking the endpoints of a scale. The first, *mechanistic* type corresponds to the classic view of canalization. Here, the trait and its buffering mechanism are genetically independent. Since the selective advantage of buffering depends on the primary trait, the evolution of the buffering mechanism is secondary to the character adaptation itself. Empirically, one might think of a high-level feedback mechanism or special "canalizing genes" (such as, perhaps, certain heat-shock proteins); in theoretical modeling this corresponds to approaches where variability is regulated by independent modifier loci. The other endpoint of the scale is the *cooperative* scenario. In this case, the primary trait and the variability are still uncorrelated in genotype space (structurally independent *sensu* [48]), but are influenced by the same genes and mutations that will, in general, have effects on both. A distinction between primary and buffering genes is no longer possible. Buffering may then evolve much more in parallel with character adaptation, although there may still be primarily adaptive and buffering phases. Theoretical approaches to cooperative canalization include the UMF model [61] and the phenotypic landscape models by Rice [36].

Intrinsic Robustness. Intrinsic robustness refers to cases where evolution leads to states of reduced variability, although buffering is not directly selected for. Robustness then arises as a by-product of selection for some other, correlated property of the state. The variability of intrinsically robust states is not necessarily lower then at its mutational neighbors with the same phenotype. It is only reduced in comparison with some ancestral states that exhibit different phenotypes (which then form the reference set). Therefore, in order to detect intrinsic robustness, the correlation of variability to some other trait must be demonstrated and an ancestral value of this trait must be known.

Intrinsic robustness is sometimes rather a system-level property than an attribute of a particular state. Robustness and buffering may be understood as emergent phenomena of regulatory networks. Here, a scenario favoring intrinsic robustness could be brought about if complex tasks for adaptation require fine-tuned regulation of gene expression which in turn require sufficiently complex gene networks—with robustness arising as a by-product of complexity or connectivity (see Wagner [56] who shows that higher connectivity leads to higher robustness in certain networks). In this case, we would assume buffering to evolve for traits which use genes with multiple pleiotropic functions and in landscapes where adaptation is restricted by tradeoffs and physical or biochemical constraints. Taking a wildtype pattern of gene expression as the evolutionary target, a mechanism of this kind has recently been described in a model for the segment polarity network in *Drosophila* [53]. Here, the first model network found to produce the desired expression pattern at all also led to significantly increased robustness with respect to variations in the network parameters and the initial conditions as a non-selected by-product. Note, however, that the phenotype itself is not contained in this model. Instead, selection is assumed to act directly on gene expression. Interpretation of the result as an example of intrinsic *phenotypic* robustness, therefore, still awaits further study.

Another potential example of intrinsic robustness is the diminishing returns function between enzyme activity of a given enzyme and metabolic flux. This relationship was first proposed as an explanation for the dominance of wildtype genotypes against loss of function mutations by Wright [64]. Dominance can be seen as an indication of phenotypic robustness. In this case the phenotypic character is the steady-state flux of a metabolic pathway. If selection favors increased flux, it will reach a genotype where most variants have small effect on the character as a result of saturation phenomena. Experiments in *E. coli* have shown that this is, in fact, the case [10]. The famous controversy between R. A. Fisher, who thought that dominance is an adaptive phenomenon, and Wright, who thought it is a passive consequence of enzyme biochemistry, is essentially a controversy over whether dominance is a consequence of canalization (adaptive

TABLE 1 Typology of phenotypic robustness. Classification with respect to ...

1. the source of variation
   - genetic (mutational, recombinational, ...)
   - environmental (macro-env., micro-env. = developmental, ...)
2. the evolutionary origin
   - adaptive (robustness as evolutionary target = canalization)
   - intrinsic (robustness as correlated by-product)

robustness) or intrinsic robustness of states with maximal flux. Wright's position later got strong support from metabolic control theory [26] and Fisher's adaptive hypothesis is commonly dismissed today. A close examination, however, shows that this conclusion may not be as straightforward [3].

Summarizing our formal considerations, we have introduced phenotypic robustness for genotypic and population states, describing genetically homogeneous and outbreed populations on a phenotypic landscape. Phenotypic robustness, however, can not be defined as a property of the phenotypic landscape alone (as it appears in Rice [36]), but depends on a specific source of variation and the appropriate set of reference genotypes. Types of robustness are collected in table 1.

2.2 DEMARCATIONS

Studies of phenotypic robustness or certain aspects of it have been pursued by many authors not only in evolutionary biology, but from various research traditions, with the focus on different problems. This entails the possibility for a fruitful exchange of ideas, but also led to a confusing variety of terminologies for similar and overlapping concepts. In order to sharpen the above notion of phenotypic robustness, we briefly discuss its relation to three similar concepts.

Robustness and Stability. In our formal definition of sources of variation, we have not required variations to be especially large or unusual. As a consequence, we do not distinguish robustness and stability (i.e., reduced variability under large and small perturbations) as is sometimes done in other fields, such as engineering and control theory [46]. We feel that there are good biological reasons for this choice: Clearly, the restriction to unusual change (such as meteorites or genome duplications) would severely narrow the possibility of adaptive robustness from the outset. But also for the understanding of intrinsic robustness, the interrelation of buffering on different levels of organization and with respect to very different sources should play a key role. Within this larger scope, it is, of course, an interesting question to ask which type of robustness could also lead to reduced variability under major changes as a correlated by-product. If this is the case, and if robustness is needed to guarantee function and survival, major change could further advance the spread of robustness by means of population or species level selection. Note, however, that the *primary* cause for the evolution of robustness must always be found by selection among individuals and under more "usual" circumstances.

Robustness of the Trait, Genic Buffering, and Cryptic Variation. Genic buffering, in other words, a reduced level of expressed variation for single genes, is a widely observed phenomenon [21]. In many cases double or multiple mutants produce much stronger phenotypic changes than each of the single mutants. Buffering of single gene functions in this sense and phenotypic robustness are closely related concepts, and the former has often been taken as evidence of the latter [63, cf.]. It is important, however, to distinguish phenotypic robustness from cases where the variability of the contribution of a single gene to the trait is reduced, but not of the trait as a whole. Genic buffering that is not connected with phenotypic robustness may naturally arise as gene networks grow and become increasingly more complex: (1) Share of control between an increasing number of genes may reduce the effects of mutations in every single gene without changing the total impact of natural mutation on the trait. (2) Epistasis and genotype × environment interactions make the effects of mutations in single genes dependent on the genetic background and the environmental conditions. If interactions are strongly variable without a net directional effect, as found, for instance, for transposon insertions in *E. coli* [13], the increase of mutational effects at some loci after the background change is compensated by a decrease at other loci. Evidence of single genes that are buffered under wild conditions, therefore, does not necessarily imply reduced variability of the trait as a whole. Under this scenario, large amounts of hidden ("cryptic") genetic variation can build up in mutation-selection equilibrium which can be expressed after a change in the environmental conditions or the genetic background. Clearly, this means that the release of cryptic variation can not be regarded as a sufficient criterion for robustness or canalization (see below).

Robustness, Neutrality, and the Mutation Rate. Following our definition of genetic robustness as a state of reduced variability with respect to mutations, the evolution of complete invariability of the trait under some of these mutations is clearly a special form of robustness. Increased robustness, therefore, can manifest itself in an increasing fraction of neutral vs. non-neutral mutations on the molecular level or in a decrease of the mutational target size (i.e., number of genes that affect a trait) or the trait-specific deleterious mutation rate U on the level of gene-loci. This must also be taken into account if evolving systems with variable genome size are considered: Since decreasing the genome size may reduce the mutational target for a given trait, this may lead to a reduced overall impact of this source of genetic variation on the trait and, therefore, to increased robustness. This effect may, of course, be overcompensated if mutations in the remaining genome have more severe effects than before, for example, due to a reduction of genetic redundancy.

Phenotypic robustness, however, is still about variation effects, not about rates of variation on the level of the underlying variables. Therefore, we do not consider decrease in the molecular mutation rate μ as a mode of genetic robustness or canalization. Genetic canalization and the evolution of mutation rates are "sister problems" dealing with indirect selection on the genotypic level that does not increase the maximum fitness in the population (for a recent review on mutation rate evolution see Sniegowski et al. [47]). In both cases, mutational variation has a double function: it produces the genetic variation needed for any kind of selection, but also controls the selective advantage of mutants with lower mutation rate or higher robustness. Nevertheless, for phenomenological as well as conceptual reasons, both problems should be clearly distinguished. From a phenomenological point of view, entirely different mechanisms are responsible for these two processes. Mechanisms that regulate the copying fidelity and mutational repair act directly on the DNA level and may affect the genome as a whole. They have little in common with buffering mechanisms like genetic redundancy or feedback regulation, which act on an intermediate level between genotype and phenotype and often are trait-specific. But also the evolutionary forces that favor the respective adaptations are different: Selection for modifications of the molecular mutation rate μ entirely depends on linkage disequilibria between the modifier locus and directly fitness related loci [47]. Buffering mechanisms, on the other hand, may also be selected in linkage equilibrium since they increase the fitness of suboptimal mutants and, therefore, have a higher marginal fitness [61].

3 EMPIRICAL EVIDENCE

Although environmental and genetic robustness of phenotypes seem to be palpable phenomena in nature, genetic robustness, in particular, is exceedingly difficult to measure and to prove in empirical work. In this section we will briefly discuss different ways robustness could be detected experimentally, also pointing out some of the practical problems. We will present a short review of the experimental literature on phenotypic robustness and discuss the findings in light of our formalization above. We find in particular that there is no convincing proof for adaptive genetic robustness.

3.1 DETECTING ROBUSTNESS

Obviously the most direct approach is to measure the variability in wildtypes and mutants. Variability measurements are relatively straightforward for environmental sources. Here, superimposed variation from other sources (such as mutations) can be controlled or even ruled out. For the measurement of developmental noise this is often done by considering fluctuating asymmetry. The visible phenotypic variation in a population then provides a direct measure of variability. Measurement of genetic variability (i.e., the mutational variance), on the other hand, is connected with major technical difficulties (cf. Gibson and Wagner [20] for a discussion of some of the problems). There is, as yet, no data available that allows for the comparison of mutational variances in different genetic backgrounds. Conclusions in empirical studies, therefore, rely on indirect evidence (see below). A second practical problem for the detection of robustness lies in the difficulty of experimentally accessing the set of reference genotypes. Ideally, the decrease of variability would be observed during the evolution of robustness under natural or artificial selection. There are several such attempts in the literature. Convincing results, however, only exist for environmental robustness. In most cases, the reference point for the variability also has to be determined indirectly.

An alternative route for the detection of robustness is to identify particular buffering mechanisms first. If these mechanisms have no advantageous phenotypic effect themselves, but rather are physiologically costly, this would provide good evidence for evolved robustness. Intrinsic robustness can be demonstrated if a function of the mechanism is found that is directly related to fitness and *necessarily* leads to buffering as a by-product. Of course, only mechanistic forms of robustness (as opposed to cooperative forms, see above) can be detected that way.

3.2 DISCUSSION OF EXPERIMENTAL RESULTS

Evolution of Environmental Robustness. The classic example of natural selection for robustness against developmental noise comes from measurements of fluctuating asymmetry (FA) in Australian blowflies [7]. Here, a mutation leading to insecticide resistance had deleterious pleiotropic effects on development when it first occurred, increasing FA. However, in subsequent generations, modifiers reducing the pleiotropic effect were selected for, leading to an increase of developmental robustness. This clearly shows that environmental robustness can (and does) evolve in the wild, either for its own sake or as a correlated response to selection.

Heat-Shock Proteins. The expression of heat-shock proteins (Hsps) is an important mechanism to guarantee protein function under environmental stress. By impairing the function of Hsp90 by a heterozygous mutation or an inhibitor in *Drosophila*, Rutherford and Lindquist [38] found that large amounts of previously silent polygenic variation with effects on many different traits were expressed. Hsp expression, therefore, seems to entail a multifunctional buffering mechanism with a clearly defined molecular basis. Conclusions about the evolutionary origin and maintenance of Hsp function remain nevertheless difficult. Many Hsps have multiple, often vital functions in the cell even under "ideal," unperturbed conditions which could be primary targets for selection. Certain levels and patterns of Hsp expression, however, seem to have evolved in response to the phenotypic effects of environmental variations, one example being Hsp70 expression in *Drosophila* [15].

Release of Cryptic Variation. Following the original work of Waddington [54], the increase of the phenotypic variation after a major mutation or exposure to an environmental challenge (e.g., heat shock) during development is the primary observation from which genetic canalization has been inferred. The classical examples are vibrissae number in mice [9] and scutellar bristles [35] and cross-vein formation [54] in *Drosophila*, all reviewed in Scharloo [39]. In all these experiments, a character with almost vanishing variance in the wildtype showed significant variation after mutational or developmental perturbation. Much of the released variation was shown to be *genetic* (since it responds to artificial selection) and based on unexpressed (cryptic) variation already present in the base population (since inbred lines showed no selection response after a similar treatment). The increase in variance is then interpreted as reduced variability of the wildtype, hence canalization.

The model that has commonly been used for the quantitative analysis of these experiments has been developed by Rendel [34]. It assumes that the observed phenotypic value z is the function of some (unobserved) underlying character (sometimes called *liability*), $z = \phi(y)$. The *developmental map* ϕ rescales the underlying liability into the phenotype and determines the degree of canalization: flat regions of the developmental map correspond to so-called *zones of canalization*. Assuming that the liability is one dimensional and normally distributed in the base population, the developmental map can be reconstructed. A measure for canalization is then given by the relative slope (wildtype/mutant) of the developmental map (cf. fig. 1).

One may ask how these interpretations relate to the formal criteria for phenotypic robustness stated above. As far as the reference point is concerned, the mutant genotypes considered in the experiments are neither ancestral states, nor mutational neighbors with the same phenotype (the mean phenotype is changed in all cases). Nevertheless, since many perturbations of the wild conditions (with changes of the wild phenotype in both directions) lead to coherent results, it does not seem too farfetched to regard the mutant variabilities as typical for large regions of the genotype space. Problems, however, arise from the variability measurement itself: Because of the difficulties involved in determining *mutational* variances directly, the *genetic* variance in the population under wild and perturbed conditions is taken as a measure of genetic variability in all these experiments. As it turns out, however, this indirect measure is biased. This bias comes about because stabilizing selection during many generations has formed the genotype distribution in the wild population, reducing the genetic variance. After the experimental perturbation, the population is not in mutation-selection equilibrium, but still carries the genetic variation shaped by selection in the old background. Since this shape is not adapted to the new background, it will,

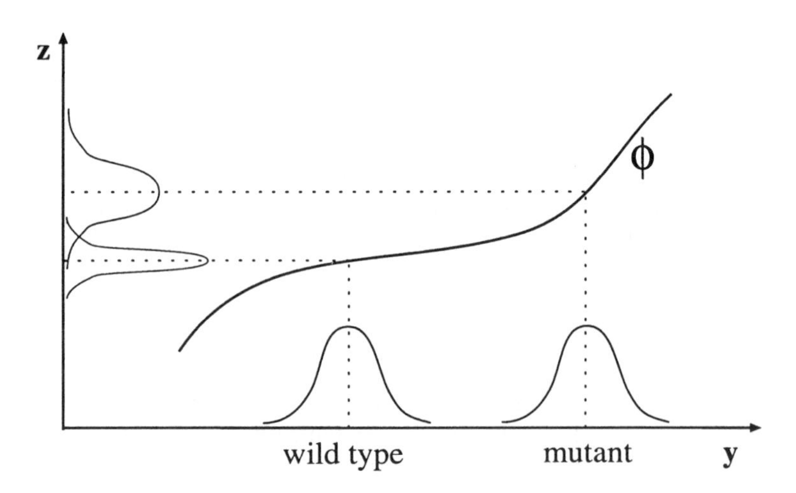

FIGURE 1 Rendel model. The distribution of the liability y in the wild population is shifted due to a background mutation (or environmental change). Since the developmental map ϕ is nonlinear, this results in a change of the genetic variance of the phenotype z. If the variance in the liability has a larger phenotypic effect in the mutant, cryptic variation is released and canalization of the wildtype (relative to the mutant) can be concluded.

in general, be stronger expressed here, even if the mutational variance (which measures variability due to *new* mutations) may not be increased. Theoretical estimates of this effect in population genetic models show that this bias can easily account for all observed effects if there is sufficiently strong variation in the epistatic interactions among loci [24]. A detailed reading of the original experimental literature shows that variable epistatic interactions are indeed present in all cases where canalization has been suspected.

This problem is not apparent in the Rendel model where any release of variation translates into an increase in slope of the developmental map, which again implies an increase in the mutational variance. This connection, however, is a result of the simple model assumption of a one-dimensional underlying character. This assumption implies that either all mutational effects are increased, or all are decreased by the background shift. The argument breaks down if there is variance in the epistatic interactions. In this case, the underlying parameter space must be higher dimensional, with the slope of the developmental map increasing in some dimensions, but decreasing in others (cf. fig. 2). An increase in the expressed variation then also arises if the mutational variance remains unchanged.

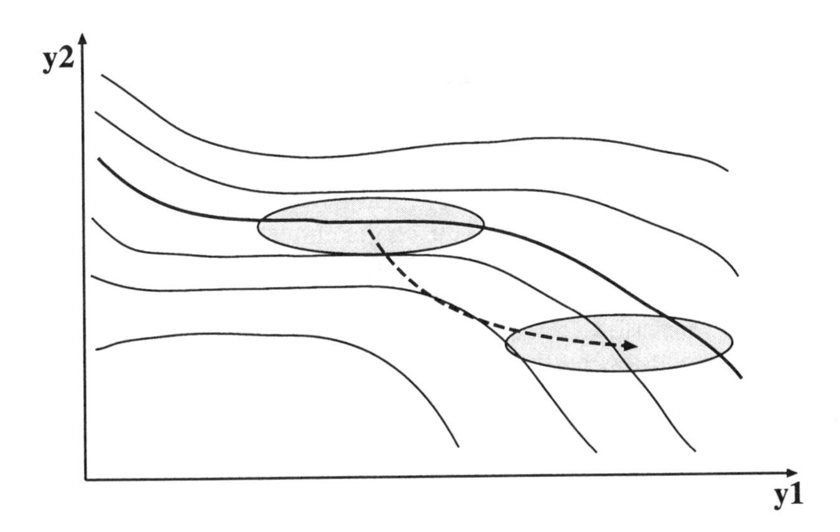

FIGURE 2 Release of cryptic variation and canalization for multidimensional liability. The figure shows a 2-dim liability map with iso-phenotype lines and wild and mutant liability distributions (shaded areas). The wild distribution is concentrated around the optimal phenotype (bold line). As in the one-dimensional case, the liability distribution is shifted by a background mutation. In the mutant background, the phenotypic variation is increased (distribution crosses more iso-phenotype lines). If the variance of new mutations in both liability directions is equal, however, the mutant background is genetically more robust (developmental map is less steep).

Artificial Evolution. Artificial evolution of genetic canalization has been reported in the above experimental settings by imposing mutant strains to stabilizing selection around a new phenotypic optimum [34, 40]. The criterion for canalization, however, again relies on the assumptions of the Rendel model. Alternative explanations are possible. For example, the experiments can not exclude the possibility that selection created a leptocurtic distribution of the underlying variables. In a recent study of artificial *E. coli* evolution, in which effects of new mutations in an ancestral and a derived background are directly measured, no evidence for the evolution of genetic robustness has been found [12].

Introgression Experiments. An interesting extension of the aboves experimental set-up is to study the effects of introgression of mutations into a panel of inbred lines. With this technique, dramatic increases of the variances among mutant relative to wildtype lines have been found for *Drosophila* bristle number [31] and photoreceptor differentiation [33]. Note, however, that interpretation of the results as robustness of the wildtype faces the same problems as discussed above, since all wildtype lines have experienced the same selection pressure toward a common optimum. A notable exception is the study of Gibson and van Helden [19] of *Drosophila* haltere characters in wildtype and *Ubx* mutant flies. Here, the variation (in shape and size) among the wildtype lines is rather due to drift or different optimal trait values in these lines, which were taken from different populations. The combined distribution is, therefore, not constrained by selection toward a *common* optimal trait value and is thus not biased as compared with the variation among mutant strains. Interestingly, no increase in the phenotypic variance, and hence no canalization, is found in this study.

Comparing Trait Variabilities. Stearns and co-workers have related the insertional variabilities to the fitness sensitivities of life-history traits in *Drosophila* by measuring the phenotypic effects of *P-element* insertions [50, 51]. They find smaller variabilities in traits that are more strongly coupled to fitness and explain this result as evidence for canalization: traits under stronger selection are able to evolve higher levels of robustness (note that this expectation is not supported by theoretical work, see below). In contrast to the above cases, variability measurements are direct and unbiased here. The conclusion, however, has been challenged by Houle [25], who suggests that the observed differences among traits are better explained by different numbers of genes affecting these traits. We have argued above that changes for a smaller mutational target size could well be a form of canalization. Nevertheless, the direct comparison of trait variabilities is biased if the mutational target sizes differ not only in the wildtype, but also on the set of reference genotypes. In particular, this will be the case if traits on a inherently different level of complexity are compared, such as, perhaps, adult size and fecundity. In general, the variability of one trait can not be used as a reference point for the variability of another trait without further argument.

RNA Secondary Structure. Using program packages for RNA secondary structure prediction, Wagner and Stadler [58] compared the stability of conserved and nonconserved elements in the secondary structure of RNA viruses with respect to point mutations. They find a trend toward a lower variability of conserved elements. Conserved elements probably have conserved functions and thus behave as "traits" under stabilizing selection. Since the nonconserved elements, which are used as the reference set in this approach, do not show obvious differences in size or thermodynamic stability, adaptive genetic robustness is inferred. Open questions remain, however, with regard to the conditions under which robustness might have evolved: Since mutational stability seems to be correlated to stability of the structure with respect to suboptimal folds [65], either genetic or environmental variations may have been the driving force for selection. As for the level of organization, selection could have acted among structures with equal functional properties (favoring more robust ones) or among sequences that fold into a given structure. This last point could be clarified in a study comparing the mutational robustness of evolved "character" states with sequences that fold into the same structure, as was done in Ancel and Fontana [1].

4 MODELS FOR THE EVOLUTION OF ROBUSTNESS

In recent years a large number of papers have been published about models for the evolution of robustness. The models differ with respect to the kind of robustness considered and assumptions about the population genetic mechanisms involved. The consensus of these papers is that, in principle, selection for robustness is possible by conventional natural selection, that is, it does not require exotic mechanisms like group selection. There is, however, no consensus with regard to which of the proposed mechanisms is likely to be effective in natural populations [8, 20]. This conflict is somewhat hidden in the literature and is not explicitly discussed, in part because the literature is fragmented with studies pursued in disconnected fields ranging from physics to artificial life research. The purpose of the present section is to sketch the emerging consensus about the population genetic mechanisms for the evolution of robustness and to identify the remaining unresolved questions. Here, we will concentrate on models that explain the evolution of robustness for its own sake (hence on canalization according to our definition). This excludes intrinsic robustness, which needs to be explained by modeling the physiology of the trait rather than by population genetics. We will also omit models for congruence, which basically need to explain a physiological coupling of genetic and environmental robustness.

The standard model for the evolution of adaptive robustness is a quantitative trait under stabilizing selection [55]. The reason is that, under stabilizing selection and if the mean phenotype coincides with the optimal phenotypic state, any form of variation, genetic or environmental, decreases fitness. Any mechanism that buffers the trait by decreasing the phenotypic effects of the variations should thus be favored by natural selection. However, the dynamical properties of population genetic models for the evolution of environmental and genetic canalization are nevertheless quite different and will, therefore, be discussed in separate sections.

4.1 MODELS FOR ENVIRONMENTAL CANALIZATION

Since environmental canalization is the stability of the phenotype against environmental or developmental perturbations, it is the complement to phenotypic plasticity, in other words, the ability to realize different phenotypes with the same genotype [42]. In the literature on phenotypic plasticity, differences between macro- and micro-environmental variations have been stressed [41]. The need to distinguish these two forms of non-genetic variation comes from the observation that they are influenced by different genetic factors. As it turns out, the potential population genetic mechanisms for the evolution of macro- and micro-environmental canalization are also different.

Waddington [55] proposed a model for the evolution of canalization against macro-environmental variation. He considered a scenario in which variation, across habitats or time, of an environmental factor changes the phenotype associated with a genotype but leaves the optimal phenotype unchanged. In this situation it would be beneficial to suppress phenotypic plasticity to this environmental factor and evolve a genotype that always realizes the optimal phenotype. This scenario is a special case of the evolution of phenotypic reaction norms. An extensive literature exists on the evolution of reaction norms [42] that will not be reviewed here. Instead we will briefly discuss models about the evolution of micro-environmental (or developmental) canalization that are usually not considered in the literature on reaction norms. In quantitative genetic theory, micro-environmental variation is usually thought of as the component of phenotypic variance that is not accounted for by genetic differences and interactions between the genotype and the observed macro-environment. Assuming stabilizing selection where the fitness is a negative quadratic function of the phenotype $m(z) = m_{\max} - kz^2$, the mean fitness in equilibrium is $\bar{m} = m_{max} - k(V_G + V_E)$. Here, the micro-environmental variance V_E may depend on the genotype, but not on the strength of selection, k. Any reduction in environmental variance increases mean fitness and will thus be favored by selection. With realistic estimates of V_E, the predicted selection coefficients in favor of environmental canalization are large enough for moderately strong selection, making this a plausible mechanism for the evolution of environmental canalization [14, 17, 61]. Indeed, the main question is why V_E is still as large as it is for most characters. There has to be an as yet unidentified mechanism that limits the level of environmental canalization reached in natural populations. Several scenarios are discussed in Gibson and Wagner [20].

4.2 MODELS FOR THE EVOLUTION OF GENETIC CANALIZATION

Models for the evolution of genetic canalization are more complicated and more ambiguous in their implications than those for environmental canalization. There are various selection scenarios that have been shown to favor genetic canalization, and there are different modeling approaches used with apparently different implications as to the mechanisms involved in the evolution of genetic canalization (e.g., compare Wagner et al. [61] and van Nimwegen et al. [52]). Most of the work has been done on models of mutation selection balance. Here, the vast majority of mutations that have a phenotypic effect on a trait are deleterious. Reduction of mutational effects should increase mean fitness and, therefore, be favored by selection independently of the relation of trait and fitness. But mutation selection balance is not the only scenario that may occur in nature, thus more work is needed to explore alternative situations. Several possible directions for further exploration are mentioned at the end of this section.

To our knowledge, the first study demonstrating the evolution of genetic robustness in a computational model is the paper by A. Wagner [56]. In this paper the phenotype is a gene activation state and evolution proceeds by changing the strength of regulatory interactions among the genes. Andreas Wagner demonstrated that networks acquire higher genetic robustness under stabilizing selection. He interpreted the results as an evolution of the population along neutral networks (fitness ridges) toward regions where the ridge is broader (see fig. 11(a) in Wagner [56]). This interpretation of the evolution of genetic canalization was further formalized in the work of VanNimwegen [52] and anticipates the results of Bornholdt and Sneppen [5]. In a quantitative genetic context this is paralleled by the concept of genetic canalization as evolution toward flat regions on an iso-phenotype contour [36]. Here, canalization may evolve whenever two or more underlying factors of a phenotype interact in a nonlinear manner.

A population genetic theory of the selection for genetic canalization only exists for populations in or close to mutation selection balance [61]. The analysis revealed a catch–22 for the selection for genetic canalization. The selection coefficient for a canalizing effect increases with the intensity of stabilizing selection, the degree of canalization caused by the allele and the amount of genetic variation affected by the canalizing effect, $s_c = kCV_G$. Since, however, the amount of genetic variation in mutation selection equilibrium is inversely related to the intensity of stabilizing selection, the resulting selection coefficient can actually remain the same with increasing selection intensity. The reason is that stabilizing selection is eliminating genetic variation and thus eliminates the effects for which canalizing alleles are selected. The magnitude of the selection coefficient depends to a great degree on the mutation rate in these models. With per locus mutation rates of less than 10^{-4}, the selection coefficients are very low. Only with high mutation rates can genetic canalization be effectively selected in mutation selection balance. The same conclusion is reached in similar models of genetic *redundancy* [32, 57]. A model with fluctuating stabilizing selection also predicts low selection coefficients [27].

The model analyzed in Wagner et al. [61] assumes that the phenotype landscape allows for simultaneous buffering of all genes. This, however, is only possible at isolated points in genotype space which usually do not correspond to the optimal phenotype [60]. A model that takes this into account was analyzed recently. Surprisingly, the equilibrium points in mutation-selection balance do not coincide with these maximally canalized points in many cases, but show rather high variability. Buffering, however, is found on the genic level for genes with high mutation rates [23].

Most of the models cited above use the standard population genetic approach in which one considers the frequencies of genes influencing the trait of interest. A somewhat different model is discussed by van Nimwegen et al. [52]. In their model the elements are genotypes on a fitness landscape which change by mutation in the absence of recombination. As already observed by A. Wagner [56], evolution of canalization can be described, in models with and without recombination, as the movement of the population along a neutral network toward regions of higher neutrality. Nimwegen et al. call their model "neutral evolution" of robustness since movement on the ridge does not change fitness. However, the results presented in the paper show that genetic robustness is rather reached as a result of indirect natural selection against the effects of genetic variation. After all, the equilibrium frequencies clearly depend on selection and are distinct from a model without fitness differences among genotypes. Consistent with the population genetic models [32, 57, 61], selection for robustness in the model is driven by mutation, with a selective advantage that is independent of the strength of selection and proportional to the mutation rate. Consequently, the robust state breaks down if small fitness differences among states on the ridge occur.

Given these similarities of models with and without recombination, it is interesting to highlight some of the conceptual differences. In population genetic models with recombination, the unit of selection clearly is the gene. Selection for canalizing can be conceptualized as genic selection and is driven by the higher marginal fitness of alleles with specific (epistatic) gene effects [61]. For clonal reproduction, the potential unit of selection is the genotype. If mutation can be neglected, selection among the genotypes only depends on differences in death and reproduction rates. Clearly, there is then no selective advantage among genotypes, canalized or not, on a neutral network. Including mutations in the model, the rate of growth of a genotype depends on the mutational loss in each generation, which again does not discriminate genotypes on the network, and on the gain due to *back mutations*. These depend in subtle ways on the fitness of the genotypes in a mutational neighborhood [6, 22]. If mutant genotypes are more fit, that is, if the wildtype is more robust, the contribution of back mutations to the growth rate of the optimal genotype is higher than if the mutants are strongly selected against. Perhaps this is an example of where we reach the limits of the unit of selection concept. In fact, what is selected are population distributions, not individual genotypes or quasispecies sensu [11]. Quasispecies can increase their fitness by moving to a new, more robust quasi-equilibrium even if the optimal fitness in the population decreases (broader peaks can outcompete higher, but sharper peaks, cf. Schuster and Swetina [44] and Wilke et al. [62]). The maximum gain in fitness, however, is small: as long as the trait-specific mutation rate U_x is small against fitness differences, the selective advantage is of the order U_x^2, and in general may not exceed U_x [6, 22].

Mutation selection theory thus predicts that, in mutation-selection balance, genetic canalization is unlikely to evolve by selection directly for genetic canalization if the mutation rate is low. This can, in fact, easily be understood in terms of the genetic load. Since canalization does not increase the optimal fitness in the population, the selective advantage that any canalizing genetic effect may have is limited by the load component L_x of the character x under consideration. If the load is entirely due to deleterious mutations in mutation-selection balance, it is very well approximated by twice the trait-specific mutation rate, $L_x \approx 2U_x$ (resp. $L \approx U_x$ for haploids), largely independent of the fitness landscape and mutation schemes [6, cf.]. If mutation rates are not higher than thought, this can mean three things: (1) genetic canalization is rare or does not exist at all. In this case, genetic robustness could still be *intrinsic*, (2) genetic canalization evolves by congruence associated with environmental canalization, or (3) there are alternative mechanisms for the evolution of genetic canalization. According to the above, it is natural to look for alternative scenarios where the genetic load is higher. For stabilizing selection, and the optimum phenotype coinciding with the mean, the load is essentially proportional to the amount of genetic variation maintained in the population. There are several situations where more genetic variation is available. Already in 1960, Ernst Mayr proposed that canalization should be favored in spatially structured populations with gene flow [29, p. 377]. Indeed there is more genetic variation maintained in populations with clines in a polygenetic trait [4, 45]. Preliminary results show that selection for canalizing modifiers is predicted to be strong [24]. The other alternative is to consider nonequilibrium situations in which genetic variation for a character under stabilizing selection can increase. For instance, this is the case when genes have pleiotropic effects on a character that is under directional selection [2]. Preliminary results show that selection for modifiers of pleiotropic effects can be quite strong [30].

5 CONCLUSIONS

1. The definition and the experimental demonstration of phenotypic robustness require the thorough determination of (a) the source of variation in order to define and measure variability and (b) a reference point with respect to which the robustness of a character state is measured.
2. Environmental robustness is experimentally well established and theoretical models show that it should be easy to evolve.
3. Genetic robustness is difficult to demonstrate. We conclude from a review of the literature that there is no convincing proof for the existence of genetic canalization (adaptive genetic robustness).
4. Most of the published models for the evolution of genetic canalization assume mutation selection balance. These models show that genetic canalization is unlikely to be selected for in the wild.
5. Studies of alternative scenarios for the evolution of genetic canalization are severely lacking. Particularly promising directions are models with population structure, nonequilibrium models, and models including plasto-genetic congruence.

ACKNOWLEDGMENTS

We thank Steve Stearns for valuable discussions and Homa Attar, Ashley Carter, Hans Larsson, Sean Rice, and Terri Williams for their helpful comments on the manuscript. Joachim Hermisson gratefully acknowledges financial support by an Emmy Noether fellowship of the *Deutsche Forschungsgesellschaft* (DFG).

REFERENCES

[1] Ancel, L. W., and W. Fontana. "Plasticity, Evolvability and Modularity in RNA." *J. Exp. Zool. (Mol. Dev. Evol.)* **288** (2000): 242–283.

[2] Baatz, M., and G. P. Wagner. "Adaptive Inertia Caused by Hidden Pleiotropic Effects." *Theor. Pop. Biol.* **51** (1997): 49–66.

[3] Bagheri-Chaichian, H. "Evolution of Mutational Effects in Metabolic Physiology." Ph.D. thesis, Department of Ecology and Evolutionary Biology, Yale University, New Haven, 2001.

[4] Barton, N. H. "Clines in Polygenic Traits." *Genet. Res. Camb.* **74** (1999): 223–236.

[5] Bornholdt, S., and K. Sneppen. "Error Thresholds on Correlated Fitness Landscapes." *Proc. Roy. Soc. Lond.* **B 267** (2000): 2281–2286.

[6] Bürger, R. *The Mathematical Theory of Selection, Recombination, and Mutation.* Wiley, Chichester, 2000.

[7] Clark, G. M., and J. A. McKenzie. "Developmental Stability of Insecticide Resistance in the Blowfly: A Result of Canalizing Natural Selection." *Nature* **325** (1987): 345–346.

[8] de Visser, J. A. G. M., J. Hermisson, G. P. Wagner, L. Ancel Meyers, H. Bagheri-Chaichian, J. L. Blanchard, L. Chao, J. M. Cheverud, S. F. Elena, W. Fontana, G. Gibson, T. F. Hansen, D. Krakauer, R. C. Lewontin, C. Ofria, S. H. Rice, G. von Dassow, A. Wagner, and M. C. Whitlock. "Perspective: Evolution and Detection of Genetic Robustness." *Evolution* **57** (2003): 1959–1972.

[9] Dunn, R. B., and A. S. Fraser. "Selection for an Invariant Character—'Vibrissae Number'—in the House Mouse." *Nature* **181** (1958): 1018–1019.

[10] Dykhuizen, D., and D. L. Hartl. "Selective Neutrality of 6pgd Allozymes in *E. coli* and the Effects of Genetic Background." *Genetics* **96** (1980): 801–817.

[11] Eigen, M., and P. Schuster. *The Hypercycle.* Berlin: Springer, 1979.

[12] Elena, S. F., and R. E. Lenski. "Epistasis between New Mutations and Genetic Background and a Test of Genetic Canalization." *Evolution* **55** (2001): 1746–1752.

[13] Elena, S. F., and R. E. Lenski. "Test of Synergistic Interactions among Deleterious Mutations in Bacteria." *Nature* **390** (1997): 395–398.

[14] Eshel, I., and C. Matessi. "Canalization, Genetic Assimilation and Preadaptation: A Quantitative Genetic Model." *Genetics* **149** (1998): 2119–2133.

[15] Feder, M. E., and G. E. Hofmann. "Heat-Shock Proteins, Molecular Chaperones, and the Stress Response: Evolutionary and Ecological Physiology." *Ann. Rev. Physiol.* **61** (1999): 243–282.

[16] Gärtner, K. "A Third Component Causing Random Variability Beside Environment and Genotype. A Reason for the Limited Success of a 30-Year-Long Effort to Standardize Laboratory Animals?" *Laboratory Animals* **24** (1990): 71–77.

[17] Gavrilets, S., and A. Hastings. "A Quantitative-Genetic Model for Selection on Developmental Noise." *Evolution* **48** (1994): 1478–1486.
[18] Gerhart, J., and M. Kirschner. *Cells, Embryos, and Evolution.* Oxford: Blackwell Scientific, 1997.
[19] Gibson, G., and S. van Helden. "Is the Function of the *Drosophila* Homeotic Gene *Ultrabithorax* Canalized?" *Genetics* **147** (1997): 1155–1168.
[20] Gibson, G., and G. P. Wagner. "Canalization in Evolutionary Genetics: A Stabilizing Theory?" *BioEssays* **22** (2000): 372–380.
[21] Hartman, J. L., IV, B. Garvik, and L. Hartwell. "Principles for the Buffering of Genetic Variation." *Science* **291** (2001): 1001–1004.
[22] Hermisson, J., O. Redner, H. Wagner, and E. Baake. "Mutation-Selection Balance: Ancestry, Load, and Maximum Principle." *Theor. Pop. Biol.* **62** (2002): 9–46.
[23] Hermisson, J., T. F. Hansen, and G. P. Wagner. "Epistasis in Polygenic Traits and the Evolution of Genetic Architecture under Stabilizing Selection." *Am. Natural.* **161** (2003): 708–734.
[24] Hermisson and Wagner, in prep.
[25] Houle, D. "How Should We Explain Variation in the Genetic Variance of Traits?" *Genetica* **102/103** (1998): 241–253.
[26] Kacser, H., and J. A. Burns. "The Molecular Basis of Dominance." *Genetics* **97** (1981): 6639–6666.
[27] Kawecki, T. J. "The Evolution of Genetic Canalization under Fluctuating Selection." *Evolution* **54** (2000): 1–12.
[28] Lynch, M., and J. B. Walsh. *Genetics and Analysis of Quantitative Traits.* Sinauer, Sunderland, 1998.
[29] Mayr, E. "The Emergence of Evolutionary Novelties." In *Evolution after Darwin*, edited by S. Tax, 349–380. Cambridge, MA: Harvard University Press, 1960.
[30] Mezey, J. "Pattern and Evolution of Pleiotropic Effects: Analysis of QTL Data and an Epistatic Model." Ph.D. thesis, Department of Ecology and Evolutionary Biology, Yale University, New Haven, 2000.
[31] Moreno, G. "Genetic Architecture, Genetic Behavior, and Character Evolution." *Ann. Rev. Ecol. Syst.* **25** (1995): 31–44.
[32] Nowak, M. A., M. C. Boerlijst, J. Crooke, and J. Maynard Smith. "Evolution of Genetic Redundancy." *Nature* **388** (1997): 167–171.
[33] Polacysk, P. J., R. Gasperini, and G. Gibson. "Naturally Occurring Genetic Variation Affects *Drosophila* Photoreceptor Determination." *Dev. Genes Evol.* **207** (1998): 462–470.
[34] Rendel, J. M. *Canalization and Gene Control.* New York: Logos Press, 1967.
[35] Rendel, J. M. "Canalization of the Scute Phenotype of *Drosophila*." *Evolution* **13** (1959): 425–439.
[36] Rice, S. H. "The Evolution of Canalization and the Breaking of von Baer's Laws: Modeling the Evolution of Development with Epistasis." *Evolution* **52** (1998): 647–656.

[37] Rutherford, S. L. "From Genotype to Phenotype: Buffering Mechanisms and the Storage of Genetic Information." *BioEssays* **22** (2000): 1095–1105.

[38] Rutherford, S. L., and S. Lindquist. "Hsp90 as a Capacitor for Morphological Evolution." *Nature* **396** (1998): 336–342.

[39] Scharloo, W. "Canalization: Genetic and Developmental Aspects." *Ann. Rev. Ecol. Syst.* **22** (1991): 65–93.

[40] Scharloo, W. *Selection on Morphological Patterns, in Population Genetics and Evolution*, edited by G. de Jong, 230–250. Berlin: Springer, 1988.

[41] Scheiner, S. M. "Genetics and The Evolution of Phenotypic Plasticity." *Ann. Rev. Ecol. Syst.* **24** (1993): 35–68.

[42] Schlichting, C., and M. Pigliucci. *Phenotypic Evolution: A Reaction Norm Perspective.* Sinauer, Sunderland, 1998.

[43] Schmalhausen, I. I. *Factors of Evolution: The Theory of Stabilizing Selection.* Chicago, IL: Chicago University Press, 1949. (Reprinted edition 1986.)

[44] Schuster, P., and J. Swetina. "Stationary Mutant Distributions and Evolutionary Optimization." *Bull. Math. Biol.* **50** (1988): 635–660.

[45] Slatkin, M. "Spatial Patterns in the Distribution of Polygenic Characters." *J. Theor. Biol.* **70** (1978): 213–228.

[46] Slotine, J.-J., and W. Li. *Applied Nonlinear Control.* Upper Saddle River, NJ: Prentice Hall, 1991.

[47] Sniegowski, P., P. Gerrish, T. Johnson, and A. Shaver. "The Evolution of Mutation Rates: Separating Causes from Consequences." *BioEssays* **22** (2000): 1057–1066.

[48] Stadler, B. M. R., P. Stadler, G. P. Wagner, and W. Fontana. "The Topology of the Possible: Formal Spaces Underlying Patterns of Evolutionary Change." *J. Theor. Biol.* (2001).

[49] Stearns, S. C. "The Evolutionary Links between Fixed and Variable Traits." *Acta Pal. Pol.* **38** (1994): 215–232.

[50] Stearns, S. C., and T. J. Kawecki. "Fitness Sensitivity and the Canalization of Life-History Traits." *Evolution* **48** (1994): 1438–1450.

[51] Stearns, S. C., M. Kaiser, and T. J. Kawecki. "The Differential Genetic and Environmental Canalization of Fitness Components in *Drosophila melanogaster*." *J. Evol. Biol.* **8** (1995): 539–557.

[52] van Nimwegen, E., J. P. Crutchfield, and M. Huynen. "Neutral Evolution of Mutational Robustness." *Proc. Natl. Acad. Sci. USA* **96** (1999): 9716–9720.

[53] von Dassow, G., E. Meir, E. M. Munro, and G. M. Odell. "The Segment Polarity Network is a Robust Developmental Module." *Nature* **406** (2000): 188–192.

[54] Waddington, C. H. "The Genetic Assimilation of an Aquired Character." *Evolution* **7** (1953): 118–126.

[55] Waddington, C. H. *The Strategy of the Genes.* New York: MacMillan, 1957.

[56] Wagner, A. "Does Evolutionary Plasticity Evolve?" *Evolution* **50** (1996): 1008–1023.

[57] Wagner, A. "Redundant Gene Functions and Natural Selection." *J. Evol. Biol.* **12** (1999): 1–16.
[58] Wagner, A., and P. F. Stadler. "Viral RNA and Evolved Mutational Robustness." *J. Exp. Zool. (Mol. Dev. Evol.)* **285** (1999): 119–127.
[59] Wagner, G. P., and L. Altenberg. "Complex Adaptations and the Evolution of Evolvability." *Evolution* **50** (1996): 967–976.
[60] Wagner, G. P., and J. Mezey. "Modeling the Evolution of Genetic Architecture: A Continuum of Alleles Model with Pairwise $A \times A$ Epistasis." *J. Theor. Biol.* **203** (2000): 163–175.
[61] Wagner, G. P., G. Booth, and H. Bagheri-Chaichian. "A Population Genetic Theory of Canalization." *Evolution* **51** (1997): 329–347.
[62] Wilke, C. O., J. L. Wang, C. Ofria, R. E. Lenski, and C. Adami. "Evolution of Digital Organisms at High Mutation Rates Lead to Survival of the Flattest." *Nature* **412** (2001): 331–333.
[63] Wilkins, A. S. "Canalization: A Molecular Genetic Perspective." *BioEssays* **19** (1997): 257–262.
[64] Wright, S. "The Evolution of Dominance." *Amer. Nat.* **63** (1929): 556–561.
[65] Wuchty, S., W. Fontana, I. L. Hofacker, and P. Schuster. "Complete Suboptimal Folding of RNA and the Stability of Secondary Structures." *Biopolymeres* **49** (1999): 145–165.

Principles and Parameters of Molecular Robustness

David C. Krakauer
Joshua B. Plotkin

1 THREE PRINCIPLES OF ROBUSTNESS

1.1 CANALIZATION, NEUTRALITY, AND REDUNDANCY

During the course of replication of RNA or DNA, genomes incorporate large numbers of mutations. These mutations can be small, such as the modification of a single nucleotide (point mutations), or large, involving repeating motifs of nucleotides (micro-satellites), damage to whole chromosomes (genetic instability), or even duplication or loss of whole chromosomes (aneuploidy). The influence of mutation on the evolutionary process is two-fold. On the one hand, mutation leads to phenotypic variance mediated by developmental dynamics, thereby providing the variability required by selection to fix new variants in a population. On the other hand, mutation undermines preadapted phenotypes by perturbing development in such a way as to lead to poorly adapted variants. The tension between the advantage of a few novel phenotypes, and the disadvantage of the majority of novel phenotypes is reflected in those mechanisms controlling the rate of mutation, and in those strategies influencing the impact of mutation on phenotypes. In an effort to understand the evolutionary response to mutations, we present three distinct, albeit closely related, principles. The principle of canalization, the principle of neutrality, and the principle of redundancy. These are contrasted with the parameters of robustness—the precise mechanisms by which these principles are realized. The principles and parameters metaphor is derived from linguistics [11] where the principles are the invariant properties of universal grammar and the parameters the local rules and practices of language.

The principle of canalization was introduced by Waddington [68] as a means of explaining the constancy of tissues and organ types during development. Canalization refers to those mechanisms that suppress phenotypic variation during development and thereby reduce the cumulative cost of deviations from a locally optimal trajectory. Waddington conceived of deviations as the result of mutations or environmental insults. Mutational canalization and environmental canalization are the terms Waddington applied to the unspecified mechanisms buffering these effects. Evidence for both types of mechanism has been observed [59, 74]. While Waddington conceived of canalization as an ensemble of mechanisms, work bearing directly on the concept has been largely functional, either of a theoretical nature [3, 71, 72], or related to experimental evolution [17] and quantitative genetics [62]. Fewer studies have directly identified the mechanisms of canalization. The term canalization exists as a general evolutionary principle describing adaptive suppression of phenotypic variation during the process of development and as a catch-all phrase referring to those mechanisms buffering deleterious mutations or environmental insults.

The principle of neutrality is best known to biologists in relation to the selective neutrality of alleles in populations. The neutral theory [35, 36] rose to prominence as a means of explaining the higher than expected level of variation in electrophoretic data [46]. Neither the traditional theory of rare wildtypes combined with common deleterious mutations [51], nor the balance theory [73] was adequately able to explain the observed diversity. Neutrality refers to the selective equivalence of different phenotypes. The idea does not require that mutations to a wildtype sequence leave the phenotype untouched (although this can be the case), but that the phenotypic differences are beyond the detection limit for selection. As populations become smaller, drift effects (random sampling of gametes) amplify [80], implying that selection coefficients must increase in order for selection to dominate drift. Neutrality, unlike canalization, is not assumed to be an adaptive means of suppressing variability. Nor is neutrality concerned with the development of phenotypes. Neutrality is simply a measure of the selective equivalence of phenotypes, and is primarily concerned with finite population effects. Through canalization, phenotypes can manifest neutrality, but it is meaningless to say that neutrality is a cause of canalization. More recently, interest has turned to the discussion of neutral networks [21, 22]. These networks are sets of selectively equivalent genotypes, connected via single mutational steps. These networks span large volumes of genotype space and provide a natural buffering mechanism. They are often treated as highly epistatic fitness landscapes [66].

The principle of redundancy is more ancient and more widespread. We restrict ourselves to discussing its modern biological interpretation which is derived largely from molecular biology. A very common means of identifying the function of a gene is to perform a knockout experiment, removing or silencing a gene early in development. By carefully assaying the phenotype, the putative function of the absent gene may be revealed. However, in many such experiments, there is no scoreable phenotype: the knockout leaves the phenotype as it was in the wildtype. Biologists then refer to the gene as redundant. This is taken to mean that this gene is but one of two or more genes contributing to the phenotype, and where removal of one leaves the phenotype unchanged. Of course, what we might be observing is something like experimental neutrality, an effect below the experimental detection limit [6]. Assuming that we are able to detect small changes, redundancy describes the degree of correlation among genes contributing to a single function [63, 64]. As with canalization and neutrality, the mechanisms (parameters) giving rise to robustness through redundancy are not explicitly stated. Unlike canalization, redundancy does not include a developmental component. Redundancy is merely a measure of the degree to which a set of genes shares the burden of function. Moreover, unlike canalization, redundancy need not be adaptive—it can be accidental (although this is unlikely). Once again, one can say redundancy gives rise to neutrality, but not that neutrality gives rise to redundancy.

In conclusion, all three principles, canalization, neutrality, and redundancy, are associated with a reduction in selective variance. Canalization is the adaptive suppression of variance during development. Neutrality is the selective equivalence of phenotypes. Redundancy is the overlap of gene function. Only redundancy and canalization assume that phenotypic variance and selective variance are co-linear. Very different phenotypes can be selectively equivalent and hence neutral. Neither redundancy nor neutrality assumes adaptation, whereas canalization is always an evolved character (in the sense of Waddington).

2 BEYOND THE REDUNDANCY PRINCIPLE: THE PRINCIPLE OF ANTIREDUNDANCY

2.1 REDUNDANCY (R) AS ERROR BUFFERING, AND ANTIREDUNDANCY (AR) AS ERROR ELIMINATION

Perhaps the most obvious way in which genes that are correlated in function promote redundancy is through several copies of a single gene [64]. This situation arises through gene duplication and leads to what are referred to as paralogous copies. Redundancy attributed to paralogues has been found in homeotic genes [49], transcription factors [47], signal transduction proteins [29], metabolic pathway genes [53], and among the variable genes encoding antibody peptides [76]. It is thought that paralogues promote robustness by "backing-up" important functions. The idea is that if one copy should sustain damage, then the paralogue will be sufficient to generate the required protein. However, this line of reasoning can be problematic. Because one copy is, in principle, as fit as two copies, in time one copy is expected to be lost from the population through random mutation [69]. To preserve two or more redundant, paralogous genes, there should be some asymmetry in the contribution of each gene to their shared function [40, 55]. For example, those genes making the larger fitness contribution could experience higher rates of deleterious mutation than those making smaller fitness contributions. Without such an asymmetry, those genes with the higher mutation rates are lost by random drift. Duplicated genes constitute a redundant mechanism, as both genes contribute to an identical function, and in the wildtype condition, only one gene is required. In a later section we provide case studies for mechanisms demonstrated to give rise to redundancy at the phenotypic level. We have discussed this one example early in order to consolidate our intuition of redundancy before describing a further principle of robust design—antiredundancy.

In many organisms, redundancy is rare. In viruses and bacteria, for example, the need for rapid replication and translation leads to small genomes with few or no duplicate genes, a small number of controlling elements, and overlapping reading frames. As a result, a single mutation will often damage several distinct functions simultaneously [37]. Within multicellular eukaryotes checkpoint genes, such as p53, enhance the cell damage caused by mutations which might otherwise accumulate in a tissue [44]. The decline in telomerase enzyme during the development of a cell lineage effectively ensures that cells are unable to propagate mutations indefinitely [65]. Similarly, it has been conjectured that the loss of key error repair genes in mitochondria might reduce the rate of mildly deleterious mutation accumulation [48]. In each of these cases we observe the evolution of mechanisms that promote antiredundancy—that is, mechanisms which sensitize cells or individuals to single gene damage and thereby eliminate them preemptively from a tissue or from a population of individuals. Unlike redundancy in which genes act together to share the burden of function, antiredundant mechanisms amplify the damage to other genes. The effect of these mechanisms is to increase the selective cost of each mutation.

3 REDUNDANCY AND FITNESS LANDSCAPES

3.1 FITNESS LANDSCAPES AS STATISTICAL MEANS OF DESCRIBING REDUNDANCY

The concept of a fitness landscape, firstc articulated by Sewall Wright in 1932 [81], has proven to be an enormously useful tool for evolutionary biologists and population geneticists. A fitness landscape is simply an assignment of a fitness value—that is, an intrinsic growth rate in the absence of density limitation—to each possible genotype. In other words, the landscape encodes a mapping between genotypes and their Darwinian fitness.

We will use single-peak fitness landscapes to encode the degree of redundancy or antiredundancy of an organism. Multiple-peak landscapes are discussed in a following section. We shall study landscapes that are symmetric around a central "wildtype" genotype. We define the set of genotypes as the set of all L-bit strings of zeroes and ones. There are thus 2^L possible genotypes, where L is the size (in bits) of the genome. One of these genotypes is distinguished as the "wildtype" whose fitness equals or exceeds that of all others. We will consider single-peak landscapes: genotypes far from the wildtype have lower fitness, whereas genotypes near the wildtype have higher fitness.

The L "bits" of which a genome is comprised may be interpreted as L genes (each of which is functional or mutated), or alternatively as L nucleotide positions. We do not lose any generality by assuming that each bit assumes two states, zero or one. The gene-wise or nucleotide-wise interpretation of the bits will depend upon context.

In these terms, the amount of redundancy of an organism is described by the steepness of its fitness landscape. Loosely speaking (see below for a formal definition), an organism features a buffered or redundant genome if mutations away from the wildtype do not dramatically lower fitness. Conversely, the genome is characterized by antiredundancy if mutations precipitously reduce the fitness. In other words, the degree of redundancy reflects the rate of phenotypic penetrance of deleterious mutations, as described by the geometry of the fitness landscape.

3.2 HAMMING CLASSES AND MULTIPLICATIVE LANDSCAPES

The landscapes we consider are symmetric around a fixed wildtype genome. In other words, the fitness of a genotype depends only upon the number of mutations between that genotype and the wildtype. This implicitly assumes that mutations to any part of the genome are equally deleterious.

The Hamming distance between two genomes is defined as the number of positions, or bits, at which the two genomes differ. The Hamming distance between a given genotype and the wildtype is thus a number between zero and L. All genotypes which are exactly k mutations away from the wildtype comprise the kth *Hamming class*. By our symmetry assumption, the fitness of a genome depends only on the Hamming class in which it lies.

We will generally explore multiplicative fitness landscapes, wherein the fitness, w_k of the kth Hamming class is proportional to

$$w_k \propto (1-s)^k \,. \tag{1}$$

The wildtype sequence, $k = 0$, is maximally-fit, and each deleterious mutation reduces fitness by an amount $(1 - s)$, independent of the other loci. The parameter s measures the deleteriousness of each mutation. By varying the magnitude of s we vary the steepness of the landscape and hence the effective degree of redundancy in the genome. A large value of s yields a steep, or antiredundant, landscape. A small value of s results in a shallower, redundant landscape.

The fitness loss caused by a random mutation to a genome is generally very small in a wide range of organisms. In *Drosophila*, for example, measured values of s rarely exceed one percent—despite the fact that several individual mutations are known to be lethal. Throughout this chapter, we will generally assume that the deleteriousness s of each mutation is small. Such an assumption is supported by current evidence [67].

The fitness formulation in eq. (1) is the canonical example of a *non-epistatic* landscape. In other words, mutations at one locus have the same deleterious effect on fitness independent of the status of other loci. Non-epistatic landscapes are generally easier to analyze than epistatic landscapes, and they will be the focus of our attention. Nevertheless, we pause to introduce a simple formulation of epistatic landscapes:

$$w_k \propto (1-s)^{k^\alpha} . \tag{2}$$

When $\alpha = 1$, this formulation reduces to the non-epistatic case. When α exceeds one, however, the landscape features *antagonistic epistasis*: each additional mutation has an increasingly deleterious effect on fitness. When α is less than one, the landscape features *synergistic epistasis*: each additional mutation has a less deleterious effect on fitness.

For a single, fixed landscape, geneticists are accustomed to thinking of the degree of redundancy itself in terms of the degree of epistasis in the landscape. Yet in this chapter, we have chosen to model redundancy by comparing a *family* of landscapes with varying degrees of steepness. We feel that such a framework—whether the family of landscapes has no epistatis ($\alpha = 1$), antagonistic epistatis ($\alpha > 1$), or synergistic epistatis ($\alpha < 1$)—yields a more intuitive measure of redundancy.

3.3 LANDSCAPE NORMALIZATION

In the previous section, we have introduced a model of redundancy in terms of the steepness, s, of a fitness landscape. Below, we will compare a family of landscapes by varying the steepness s. We will investigate under what conditions an organism will prefer a steep, antiredundant landscape, and under what conditions it will prefer a shallow landscape. In other words, we will allow organisms to evolve the steepness of the landscape itself.

What are the constraints on the evolution of fitness landscapes, and thereby, on the evolution of developmental programs? If an organism were allowed to evolve its landscape steepness in a fixed environment, it would certainly always prefer the shallowest landscape possible ($s = 0$), so that all genotypes would have the maximal fitness. In other words, in a fixed environment, an organism will evolve towards maximum redundancy whereby mutations do not effect fitness. In reality, however, this solution is not allowable in light of physiological constraints—i.e., intrinsic molecular costs associated with the evolution of redundancy.

Genotypes cannot evolve towards both maximum fitness and maximum redundancy simultaneously. Of the molecular mechanisms of redundancy discussed below, all incur some cost to the organism—through increased genome size, increased metabolism, or reduced binding specificity. We will model this cost by enforcing a tradeoff between the maximal height of the fitness landscape and its steepness—that is, we normalize all landscapes to have total volume one:

$$w_k = \frac{(1-s)^k}{\sum_{j=0}^{L}(1-s)^j} = \frac{s(1-s)^k}{1+(1-s)^L(s-1)}\,. \tag{3}$$

Equation (3) enforces a tradeoff between redundancy and wildtype fitness, constraining the family of landscapes which we consider. Although the precise form of the tradeoff curve is arbitrary and relatively unimportant (other normalizations yield similar results) it is essential that we impose some tradeoff between maximum fitness and redundancy.

4 A QUASISPECIES DESCRIPTION OF ROBUST POPULATIONS

In order to investigate the evolution of a population on a fixed landscape—and, eventually, the evolution of the landscape itself—we will use the quasispecies formulation introduced by Eigen [15, 16]. The quasispecies equation provides a very general framework for exploring mutation and selection in a heterogeneous, infinite population [7].

Eigen's quasispecies framework considers a large population of L-bit genomes, x_i, reproducing with imperfect fidelity according to their assigned fitnesses, w_i, with fixed total concentration:

$$\dot{x}_i = \sum_{j=1}^{j=2^L} w_j x_j Q_{ij} \;\; - x_i W\,. \tag{4}$$

In this equation, $W(t) = \sum w_j x_j(t)$ denotes the mean population fitness. The mutation matrix Q_{ij} denotes the probability of genotype i mutating into genotype j during a replication event.

Although eq. (4) is nonlinear, the change of variables

$$y_i(t) = \frac{x_i(t)}{\exp\left(-\int_0^t W(s)\,ds\right)} \tag{5}$$

produces a linear system whose solution satisfies $x_i(t) = y_i(t)/\sum_j y_j(t)$.

In the present investigation, we assume that the fitness depends only upon the Hamming class of a genotype. If we define the kth Hamming class as the sum

$z_k = \sum_{H(i)=k} y_i$ over the abundances of all sequences i which are k bits away from a fixed wildtype sequence, then eq. (5) reduces to:

$$\dot{z_k} = \sum_{i=0}^{L} z_i w_i P_{ki} \,. \tag{6}$$

The value w_i denotes the fitness of a genome in the ith Hamming class, and P_{ki} is the probability of mutation from an i-error genotype to a k-error genotype during replication.v

We will assume that each locus has a fixed chance of mutating at each replication event, independent of the other loci. Thus the chance of mutation P_{ki} from Hamming class i to class k is determined by the per-base forward and backward mutation rates, p and b:

$$P_{ki} = \sum_{l} \binom{i}{l} \binom{L-i}{k-i+l} p^{k-i+l} (1-p)^{L-k-l} b^l (1-b)^{i-l} \tag{7}$$

where we sum from $l = \max(0, i-k)$ to $l = \min(i, L-k)$. In this sum l denotes the number of back-mutations. The backward mutation rate b may be less than or equal to the forward mutation rate p, depending upon the interpretation of the loci as genes or nucleotides.

4.1 POPULATION MEAN FITNESS AT EQUILIBRIUM

For a fixed landscape, we are interested in the mean fitness of a population that has reached mutation-selection equilibrium. For a purely multiplicative landscape ($\alpha = 1$) with equal forward and backward mutation rates ($p = b$) the mean fitness can be solved exactly [28]. The equilibrium mean fitness will depend upon the genome length, L, the mutation rate p, and the steepness of the landscape, s.

The dominant eigenvector of $(w_l P_{kl})$ provides the equilibrium relative abundances of the Hamming classes. Moreover, the corresponding dominant eigenvalue equals the equilibrium mean fitness. As suggested in Woodcock and Higgs [79], we look for an eigenvector of the binomial form $z_k = \binom{L}{k} a^k (1-a)^{L-k}$, where a must yet be determined. In order to compute a, we solve the discrete-time equivalent of eq. (4), which reduces to the same eigen system problem [79].

Consider the random variable V_k defined as the Hamming class after one generation of replication (with mutation) of a genotype starting in Hamming class k. The generating function of V_k is defined as

$$G(V_k) = \sum_{i=0}^{\infty} \mathbf{P}(V_k = i) X^i \,.$$

where X is a formal variable.

For a one-bit genome ($L = 1$), we clearly have

$$G(V_1) = p + qX$$
$$G(V_0) = q + pX\,,$$

where $q = 1 - p$. For $L > 1$, V_k is sum of L independent random variables, one for each bit in the genome. Hence $G(V_k)$ is the product of generation functions:s

$$G(V_k) = (p + qX)^k (q + pX)^{L-k}.$$

Given the current abundance of each Hamming class, $\mathbf{z} = (z_0, z_1, \ldots z_L)$, then

$$G(V_{\mathbf{z}}) = \sum_k z_k (q + pX)^{L-k} (p + qX)^k$$

is the generating function for the Hamming class after mutation of a randomly chosen individual in the population. Similarly,

$$G(V_{\mathbf{z}}^{\mathrm{s}}) = \sum_k (1 - s)^k z_k (q + pX)^{L-k} (p + qX)^k$$

is the generating function for the Hamming class after mutation of an individual chosen according to its fitness.

In equilibrium, the eigenvector $\hat{\mathbf{z}}$ satisfies $G(V_{\hat{\mathbf{z}}}^{\mathrm{s}}) = \lambda \sum_k \hat{z_k} X^k$, or

$$\sum_k t^k \hat{z_k} (q + pX)^{L-k} (p + qX)^k = \lambda \sum_k \hat{z_k} X^k\,,$$

where we have defined $t = 1 - s$. Substituting our binomial assumption for the equilibrium eigenvector $\hat{\mathbf{z}}$, we may solve the following system to find the value of a:

$$\sum_k \binom{L}{k} a^k (1 - a)^{L-k} t^k (q + pX)^{L-k} (p + qX)^k =$$
$$\lambda \sum_k \binom{L}{k} a^k (1 - a)^{L-k} X^k\,.$$

Equivalently, we solve

$$[ta(p + qX) + (q + pX)(1 - a)]^L = \lambda [aX + 1 - a]^L$$
$$ta(p + qX) + (q + pX)(1 - a) = \lambda^{-L} (aX + 1 - a)\,.$$

Setting coefficients of X^0 and X^1 equal, we find that

$$tap + q(1 - a) = \lambda^{-L} (1 - a)$$
$$taq + p(1 - a) = \lambda^{-L} a$$
$$= \frac{tap + q(1 - a)}{1 - a} \cdot a\,,$$

and so

$$a = \frac{1}{2}\left(1 - p + \frac{2p}{s} - \sqrt{\left(1 - p + \frac{2p}{s}\right)^2 - \frac{4p}{s}}\right). \tag{8}$$

In equilibrium, the mean Hamming distance from the wildtype is $\bar{k} = aL$.

The mean population fitness in equilibrium, $\bar{w}$, is easy to calculate once we know the complete (binomial) distribution of equilibrium Hamming classes: $\bar{w} = \sum w_k \hat{z}_k$. Note that $\bar{w}$ depends on genome length L, the mutation rate p, and the landscape steepness, s.

AN APPROXIMATION FOR MEAN FITNESS

In some situations, it may be difficult to find the full distribution of equilibrium Hamming classes, $(\hat{z}_0, \hat{z}_1, \ldots, \hat{z}_k)$, as we did above for a multiplicative landscape. Nevertheless we may often be able to find the first two moments $\bar{k}$ and var(k) of the equilibrium Hamming distribution. In such cases we can recover a good approximation of the mean fitness according to the Taylor expansion of k around $\bar{k}$:

$$\bar{w} = w(\bar{k}) + \frac{1}{2}\mathrm{var(k)w''(\bar{k})} + \ldots$$

where we use the notations w_k and $w(k)$ interchangeably. In our case $w(k) \propto (1-s)^k$ and $w''(k) \propto (1-s)^k \log(1-s)^2 \approx (1-s)^k(s^2 + s^3 + O(s^4))$. Since we are assuming a small selective value s throughout, $w''(k)$ is negligible compared to s. Therefore, we may use

$$\bar{w} \approx w(\bar{k}) \tag{9}$$

as a good approximation of equilibrium mean fitness in terms of the fitness of the equilibrium mean Hamming class.

4.2 STOCHASTIC DYNAMICS AND THE INFLUENCE OF POPULATION SIZE

The quasispecies framework introduced by Eigen applies only to infinite populations of replicating individuals. But we are primarily interested in the effects of redundancy and antiredundancy in finite and even very small populations. For a constant, finite population size, the population mean fitness does not steadily approach a fixed equilibrium value. Instead, the stochastic process of mutation and selection produces variation in the mean population fitness over time. Nevertheless, the stochastic process approaches a steady state—that is, the expected population fitness (where expectation here denotes ensemble average) assumes a fixed value for large times.

Assuming a small mutation rate p, moment equations [79] allow us to compute the steady state population mean Hamming class in terms of the population size N:

$$\langle \bar{k} \rangle = \frac{L}{2}\left(1 + \frac{2p}{s} + \frac{1}{2sN} - \sqrt{\left(1 + \frac{2p}{s} + \frac{1}{2sN}\right)^2 - \frac{4p}{s} - \frac{1}{sN} - 2p}\right). \quad (10)$$

Note that the ensemble average, $\langle \cdot \rangle$, is taken after the population average. Substitution into Eqs. 3 and 9 yields the expected equilibrium mean fitness of a finite population in steady state. These equations determine the relationship between mean population fitness, the strength of selection, the rate of mutation, the genome length, and the size of the population.

Figure 1 shows the relationship between the level of redundancy, s, and the expected mean population fitness for several different population sizes. Both in theory (fig. 1(a)) and in individual-based stochastic simulations (fig. 1(b)) we see that redundancy increases the mean fitness in small populations, while it decreases fitness in large populations. This result has an intuitive explanation. In small populations, mutational drift contributes disproportionately to the population fitness. There is a large temporal variance in the mean Hamming class, and redundancy can effectively mask these mutations. Small populations are thus better served by shallow landscapes—i.e., by slightly decreasing the fitness of the wildtype, but increasing the fitness of its nearby neighbors. Large populations, however, are not at risk of being "swept off" the fitness peak by the stochastic fluctuations that afflict small populations; the temporal variance in the mean Hamming class is small. It is better, therefore, for large populations to amplify the phenotypic penetrance of deleterious genes via sharp landscapes.

4.3 EVOLUTIONARY ACCESSIBILITY OF ROBUST LANDSCAPES

Our results on equilibrium mean population fitness (fig. 1) constitute a population-based argument for the evolution of redundancy in small populations and antiredundancy in large populations. These results do not, in themselves, demonstrate that such strategies are evolutionarily stable or achievable. In other words, we must yet demonstrate that individual replicators subject to individual-level selection evolve degrees of redundancy consistent with the optimal population mean fitness. If we allow individuals to modify the heritable steepness of their own individual landscapes through mutation, however, we have previously shown that small populations do, indeed, evolve toward redundancy, and large populations towards antiredundancy via individual-level selection (fig. 2).

The evolutionary stability of these two strategies—sensitivity in large populations and redundancy in small populations—has an intuitive explanation. The stability rests on the fact that flatter landscapes have lower fitness peaks. A large population on a steep landscape is highly localized near the wildtype (low $\bar{k}$). Mutants with different s-values are thus most often generated near the wildtype—precisely where a more shallow landscape would be disadvantageous to them. Conversely, small populations with shallow landscapes are de-localized

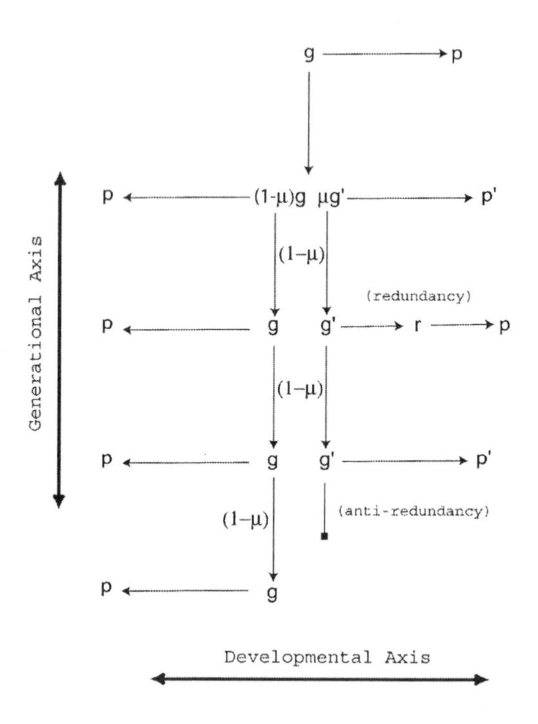

FIGURE 1 The principles of redundancy and antiredundancy. A genome (g) develops into a phenotype (p). Genomes are transmitted across generations with perfect fidelity with a probability $1 - \mu$. Genomes experience mutations with a probability μ causing the wildtype g to transform into a mutant g'. The mutant genome g' develops into the mutant phenotype p'. Mechanisms of redundancy occur during development, buffering the ability of heritable mutations to produce wildtype phenotypes. However, redundancy does not influence the mutant genotype. Mechanisms of antiredundancy purge mutant genomes from the population, leaving only wildtypes to replicate. The generational axis refers to iterations of genome replication, whereas the developmental axis refers to the non-replicative production of phenotype from genotype.

(high $\bar{k}$). In this case, landscape mutants tend to arise far from the wildtype—precisely where a steeper landscape would decrease their fitness. Thus, the landscape itself acts as a mechanism for ensuring the robustness of the incumbent strategy, in each population size.

4.4 THE IMPORTANCE OF BACK MUTATION

Unlike many treatments of the quasispecies equation, we allow for back mutations. According to two well-known principles of population genetics, the neglect of back mutations introduces pathologies into the equilibrium state of both infinite and finite populations. If back-mutations are neglected, then, according the Haldane-Muller principle, the mean equilibrium fitness of an infinite haploid population is independent of the landscape's steepness (provided the wildtype is maintained). Hence, without some rate of back mutation, we cannot detect a preference for one landscape over another in an infinite population.

Similar problems apply if we ignore back mutations in small, finite populations. In this case, the dynamics will proceed by the gradual accumulation of mutations. Once a mutation is shared by all the members of the finite population, then (ignoring back mutations) the mutation is fixed for all time thereafter. This phenomenon, called Muller's ratchet [18, 25, 51], implies that (for a finite genome length L) the equilibrium mean fitness will equal the minimum fitness, regardless of the steepness of the landscape. Therefore, in order to detect adaptive benefits or costs of redundancy—in finite and infinite populations—we cannot ignore back mutations.

Despite the importance of not ignoring back mutation, it is important to allow for differences between the forward mutation rate p and the back-mutation rate b. If we interpret each bit of the genome as a base-pair, then certainly $p \approx b$; if we interpret each bit as indicating whether or not a given gene is functional, then the forward mutation rate will exceed the backward rate. Fortunately, our qualitative results (figs. 1(b), 2) remain essentially unchanged; for forward mutations occurring at twice the rate of backward mutations, small populations still favor redundancy, and large populations still favor antiredundancy. However, as the backward mutation rate becomes proportionately smaller than the forward rate, larger populations are required to favor steep landscapes. This makes intuitive sense as, when back mutations are rare, drift away from the wildtype is more problematic and requires more buffering.

4.5 EPISTATIC EFFECTS

The analytical results on mean equilibrium fitness derived in sections 4.1 and 4.2 apply to families of non-epistatic landscapes: $w_k \propto (1-s)^k$, normalized to unit volume. We used such landscapes because they are analytically tractable and because epistatic effects can often be confounding (although more so for diploid models). In this section we briefly discuss the consequences of epistasis, $w_k \propto (1-s)^{k^\alpha}$, on the tendency to evolve redundancy or antiredundancy.

For small to moderate degrees of synergistic or antagonistic epistasis, $0.8 < \alpha < 1.2$, all of our qualitative results remain essentially unchanged: redundancy is preferred in small populations and antiredundancy in large populations. However, it is interesting to note that antagonistic epistasis ($\alpha > 1$) accentuates this general trend. When comparing a family of antagonistic landscapes with varying degrees of steepness, small populations have an even stronger preference for redundancy and large populations for antiredundancy. Similarly, synergistic epistatis ($\alpha < 1$) mitigates these preferences (see figs. 3 and 4).

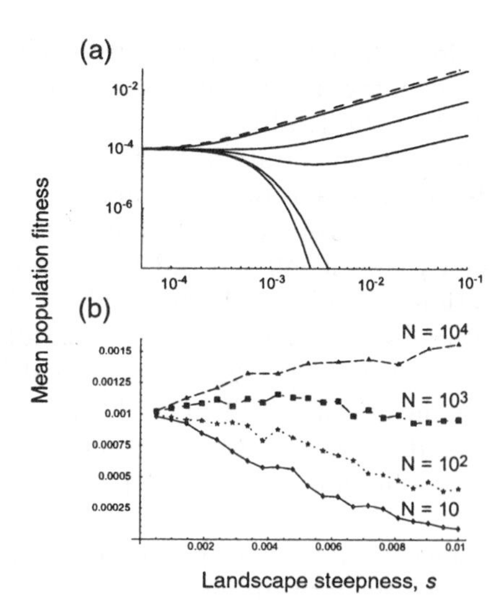

FIGURE 2 The theoretical relationship between redundancy and equilibrium mean fitness for populations of various sizes ($N = 10$, 100, 500, 1000, 10000, and N infinite). Small populations benefit from redundant (i.e., flatter) landscapes, but large population prefer antiredundant (i.e., steeper) landscapes. The curves in the figure, given by eqs. 8, 9, and 10, correspond to genome length $L = 10^4$ and mutation rate $u = 5 \cdot 10^{-5}$. (b) The relationship between redundancy and equilibrium mean fitness as observed from individual-based computer simulations of the quasispecies equation ($L = 1000$, $u = 5 \cdot 10^{-4}$). Each individual is characterized by its Hamming distance from wildtype. The mean population fitness is computed by averaging the last 20% of 10,000 generations with selection and mutation. In each discrete generation, N parents are chosen probabilistically from the previous generation according to their relative fitnesses. The offspring of a parent is mutated according to eq. (7). The numerical studies confirm the theoretical prediction: small populations prefer shallow landscapes, while large ones prefer steep landscapes.

4.6 SELECTIVE VALUES

We must emphasize that our results on redundancy, antiredundancy, and population size assume (i) haploid asexual reproduction, (ii) constant population size, (iii) roughly equal forward and backward mutation rates, (iv) finite genome length, and (v) small selective values s. These are all reasonable assumptions for a broad range of biological circumstances. In particular, the assumption that a random mutation in the genome has a small deleterious effect, s [67]. There are few extant organisms known for which, on average, a single random mutation decreases replicative ability more than one percent.

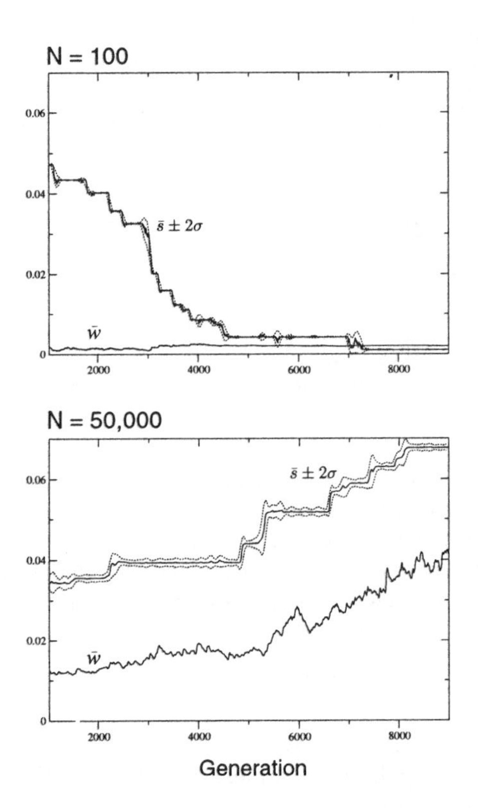

FIGURE 3 The evolution of redundancy in a small population (top) and of antiredundancy in a large population (bottom). We perform quasispecies simulations in which individuals are characterized both by their Hamming class k and their individual landscape steepness s. We choose genome size $L = 500$ and per-base mutation rate $p = .001$. On a slow timescale (probability .0005 per replication), an individual's landscape is heritably mutated (uniformally within $[s - .005, s + .005]$). For the small population, all individuals begin as wildtypes with a fairly steep landscape $s = 0.05$. For the large population, all individuals begin as wildtypes with $s = .025$. After an initial transient, in both cases the population's mean landscape steepness $\bar{s}$ evolves towards its preferred level. Simultaneously, the mean population fitness $\bar{f}$ increases; although it increases less dramatically for the small population. Throughout the evolutionary timecourse, the within-population variance in redundancy (σ^2) is small. In addition, over time the small population becomes de-localized ($\bar{k}$ increases), whereas the large population becomes increasingly localized (data not shown).

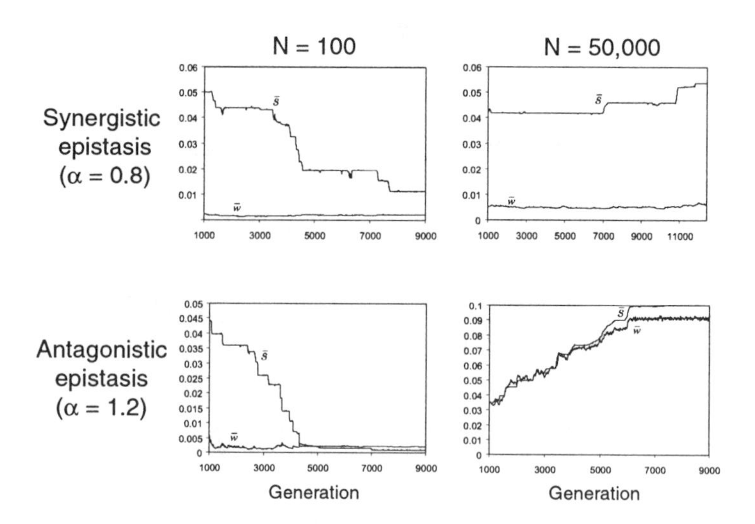

FIGURE 4 The relationship between redundancy and equilibrium mean fitness for epistatic landscapes. Results are derived from individual-based computer simulations as in figure 1(b). Neither synergistic (top) nor antagonistic (bottom) epistasis alters the preference for shallow landscapes in small populations and steep landscapes in large populations. Note that the effect of population size on a preferred landscape is more dramatic under antagonistic epistasis.

However, it is important to be aware of the equilibrium behavior of landscape families which allow very large s values, $s > 0.1$. In these cases, even small populations which start, comprising mostly wildtypes, can preserve the wildtype and fail to evolve flatter landscapes. In other words, if the landscape is *sufficiently* steep, then even a small, constant-sized population of wildtypes evolves to keep the landscape steep. Mutants are rapidly removed before they have a chance to evolve towards shallow landscapes. This phenomenon is perhaps not so significant in multicellular organisms, but becomes important for replicating RNA molecules under prebiotic conditions [41].

4.7 MULTIPEAK LANDSCAPES

In the preceding analysis and discussion we have assumed a symmetrical single-peak landscape model.

Our interest has been the preservation or maintenance of an optimal genotype configuration—designated the wildtype. In biology, landscapes are rarely single peaked, and it is frequently the case that different genomes map onto identical fitness values. In other words, we have neglected neutral networks. We have done so for several reasons: (1) multipeak landscape models are largely

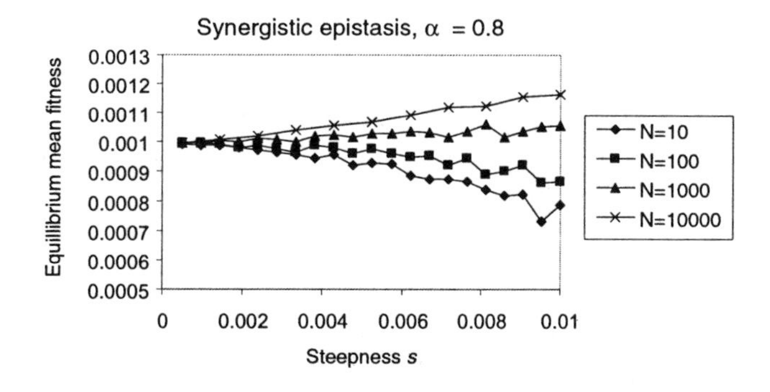

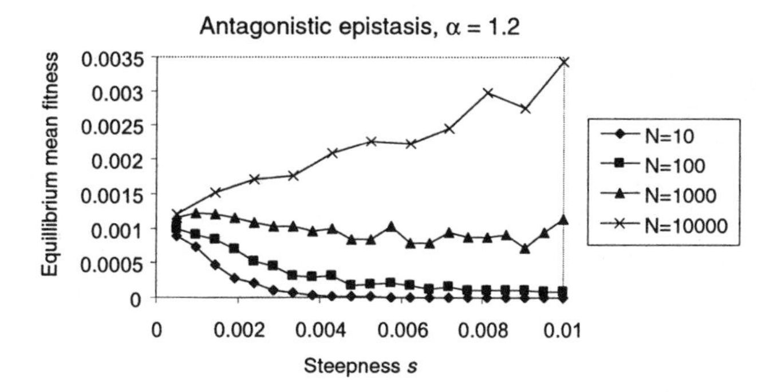

FIGURE 5 The evolution of redundancy in a small population (left) and of antiredundancy in a large population (right) under synergistic (top) and antagonistic (bottom) epistasis. As in figure 2, we perform quasispecies simulations in which individuals are characterized both by their Hamming class and their individual landscape steepness s. As in figure 2, large populations evolve steep landscapes and small populations evolve flatter landscapes. The effect of population size is more dramatic under antagonistic epistasis.

concerned with optimization, (2) neutral network models are frequently single-peaked, and (3) for increased model tractability.

A recurrent question in the study of multipeak landscapes is the fixation probability of reaching a maximum peak or the number of mutations required to reach from one peak to another [66]. In landscapes with multiple equivalent maxima, optimization leads to a symmetry breaking event in which a single peak is selected according to the initial system configuration. In rugged landscapes with irregular local optima, the problem is not so much a question of symmetry breaking, but a question of the preferred mechanism for hill-climbing when numerous suboptimal solutions exist (in other words, combinatorial optimization problems). Neither of these questions is concerned with the robustness of the final solution—the stability of the optimal system configuration—the wildtype. A robustness problem in a multipeak landscape would address the question of mean-population fitness as a function of the spatial separation of iso-fit maximum peaks. Thus, in relation to our present concerns, this includes transitions between alternative functional wildtypes for different population sizes and mutation rates. This problem has not yet been attempted and is beyond the scope of the present chapter.

Existing analytical studies of neutral networks assume single-peak landscapes. These are frequently plateau landscapes in which one neutral network occupies the high ground and another the low ground [66, 75]. While many different genotype configurations map onto a single fitness class, there are only two fitness classes. Our single-peak landscape model with high positive epistasis reduces to this case.

One problem with the robustness of rugged landscapes involves adopting a model formalism that remains tractable. A variety of models have been used to study optimization on rugged landscapes including spin glasses, genetic algorithms [66], NK-models [34] and stochastic additive scale population genetics models [24]. These are all powerful formalisms, but they result in few transparent results when assuming many heterogeneous peaks.

5 THE PARAMETERS OF REDUNDANT AND ANTIREDUNDANT DESIGN: CASE STUDIES

As a result of our modeling efforts we have discovered that population size influences the degree of redundancy we expect to be expressed by a genome. In large populations of microorganisms, such as viruses and bacteria, and in large populations of rapidly dividing cells within multicellular organisms, we predict an evolution towards antiredundant mechanisms. For small populations, on the other hand, we expect a tendency towards redundancy. In biological systems we find a large variety of molecular mechanisms capable of producing redundancy and antiredundancy. While our quasispecies formulation does not incorporate the explicit details of these mechanisms, it does provide a statistical treatment of the parameter s, that is assumed to be the developmental end-point of all these processes. In this section we provide brief case studies for six adaptive mechanisms of redundancy and the same number for antiredundancy. The principal purpose of this section is to demonstrate the utility of dividing molecular mechanisms into two groups according to the classification suggested by our model.

Those mechanisms described as redundant or antiredunt all influence development and somatic processes by modifying the effective degree of deleteriousness, s, of mutations. Redundant mechanisms (low s values) preserve mutations by masking their influence. These mechanisms are termed redundant as a consequence of their neutrality in the wildtype and their buffering capacity in the mutant. Antiredundant mechanisms (high s values) remove mutant genomes from populations (either of individuals or cells). The term antiredundancy derives from a capacity to amplify mutational damage.

When available, both sets of strategies can be exploited, even simultaneously, by a single organism according to population size constraints. Where data on the incidence of a mechanism in relation to population size are available we report results. However, these are usually fairly qualitative, as population size estimates are hard to come by, and the incidence of mechanisms often reflects sampling bias rather than genuine absence. It is also the case that a plurality of selection pressures impinge on each of these mechanisms making evaluation of data from literature sources particularly difficult and urging caution in interpretation!

5.1 REDUNDANCY THROUGH DOMINANCE MODIFIERS

Perhaps the best known example of the buffering of mutation occurs upon the mutation of a single allele in a diploid organism. Fisher [20] noted that a great number of these mutations leave the phenotype unchanged. In other words, the wildtype is almost completely dominant over the deleterious mutation. Fisher proposed that this observation was the outcome of a protracted selective process, in which modifier genes evolved to increase the recessivity of mutations. Wright [80] stood in opposition to this view, proposing that dominance was a nonselected (neutral) consequence of the kinetic structure of metabolic pathways. Wright's idea can be reinterpreted as stating that enzymes have little influence on the flux through a pathway unless they are rate limiting. Thereby the reduction in enzyme concentration by a half upon mutation into the hemizygote, will have little effect on the pathway. This interpretation has been verified through metabolic control theory [31]. However, it is now thought that the robustness of kinetic pathways can themselves be evolved properties. Rather than thinking in terms of modifiers dampening the expression of a deleterious allele (sensu Fisher), modifiers are now thought to act on the kinetic parameters of enzymatic pathways [5].

By definition, dominance is a property of diploid genomes, and thus dominance is rarely observed in short-lived populous microorganisms. However, there are populous diploid organisms. In *Drosophila* there are estimates of the average coefficients of dominance (h) for deleterious spontaneous mutations. A parameter value of $h = 1$ signifies complete dominance, $h = 0.5$ signifies co-dominance. In *Drosophila* this is estimated at $h = 0.1$ [23]. In Daphnia the estimated average value of h is 0.3 [13], that is, a range between approximately co-dominant and dominant.

5.2 REDUNDANCY THROUGH EPIGENETICS AND IMPRINTING

Epigenetics describes heritable changes in gene expression without changes in underlying nucleic acid sequences. Imprinting is a special case of epigenetic inheritance, and is often thought of as the expression of only one allele at a locus, dependent on the parental origin of the allele. Mutations to DNA, giving rise to repeated runs of nucleotides containing nonsense and missense mutations, arise relatively frequently through recombinational slippage. DNA-DNA pairing can detect these repeats and induce MIP (methylation induced premeiotically), by which duplicated sequences are extensively methylated leading to transcriptional gene silencing (TGS) [78]. Defective genes are no longer expressed. Imprinting plays an important role in guarding against the transformation of healthy cells into cancerous cells. Loss of imprinting is often an early step in cancer progression. Once imprinting is lost, it becomes very difficult to distinguish homologous from nonhomologous chromosomes. In an attempt to repair large mutant sequences arising through "microsattelite instability" during homologous recombination, there is inappropriate recognition, and this increases the incidence of mutation [12]. Epigenetic silencing can therefore promote redundancy by hiding the effects of mutation.

Imprinting has been reported in a number of non-mammalian groups, and by implication, in groups that tend to live in large population sizes. These groups include the yeasts, the dipterans and even in the plants: rye and maize [30]. In most of these cases imprinting is as crude as the elimination of chromosomes from one parent. However, imprinting is far more common among mammals (smaller population sizes) and also more elaborate.

5.3 REDUNDANCY THROUGH AUTOPHAGY

Autophagy is a cell membrane trafficking process that occurs in response to changes in cell nutrient concentrations or specific kinds of mutation. During autophagy, cytoplasmic material is sequestered into double membrane compartments or vesicles, known as autophagosomes. These vesicles then fuse with the lysosome, which releases hydrolases breaking down vesicle contents, allowing them to be recycled [1]. The importance of autophagy in relation to redundancy becomes apparent in cancer. Transformed cells dependent on hormones for growth, are killed through autophagous processes upon the removal of hormones. Moreover, overexpression of the autophagy-related gene beclin is capable of reversing the transformed state of cancerous cells and inhibiting their ability to grow [60]. In other words, autophagy is capable of breaking down and recycling the translated protein products of oncogenes.

5.4 REDUNDANCY THROUGH mRNA SURVEILLANCE

A sizeable fraction of mRNAs of eukaryotes contain premature termination codons. These arise through misincorporation errors during transcription, or derive from mutations within the DNA template. The result is the production of mRNAs encoding nonsense. These mRNAs are found to be less stable than the wildtypes as a result of "nonsense-mediated mRNA decay" (NMD) or mRNA surveillance [57]. NMD is thought to protect cells from the deleterious effects of high concentrations of truncated proteins, reducing the number of defective mRNA transcripts prior to translation. To date at least seven different genes have been discovered involved in NMD [9]. NMD is thus a mechanism of redundancy as it discovers and eliminates errors in the mRNA but is unable to remove defective DNA.

5.5 REDUNDANCY THROUGH tRNA SUPPRESSORS

Whereas mRNA surveillance intercepts error prior to translation into protein, tRNA suppressors intercept errors during translation. This is achieved through a special class of modified tRNA, in which the anticodon is modified to be able to recognize a nonsense codon [14]. In the absence of tRNA suppressor molecules, termination codons within the mRNA are recognized by proteins known as releasing factors, terminating translation. In the presence of tRNA suppressors, the termination codon is bound by a tRNA suppressor and an amino acid is inserted into the growing polypeptide. In this way a nonsense mutation is transformed into a missense mutation. There exist tRNA suppressors for each type of termination codon. An interesting consequence of suppression is that the suppressor tRNAs must compete with the protein release factors for the termination codons. If suppression is too effective, then there will be extensive read-through of the true termination codon, producing an excess of C-terminal product. This problem is overcome by making sure that the suppressors are much less than 100% effective (from 50% for amber to 10% for ochre). Redundancy is promoted through suppression by masking many of the nonsense mutations to the DNA sequence.

5.6 REDUNDANCY THROUGH CHAPERONES

Chaperones are proteins that facilitate the folding of nascent polypeptides. The majority of chaperones reside within the endoplasmic reticulum (ER), through which most polypeptides pass after translation in order to be folded for export from the cell or for recirculation in the cytoplasm. Within the ER, chaperones play an essential role in protein quality control by retaining misfolded or misassembled proteins, eliminating proteins through ER-associated protein degradation, or preventing the accumulation of unfolded proteins through the unfolded protein response (UPR) [19]. Retention is facilitated by an excess of glycosylation on slowly folding defective proteins acting as a signal for retention by chaperones. Degradation requires first the recognition of aberrant polypepetides, second, the retrotranslocation (export of the proteins from the ER back to the cytoplasm), and third, the degradation of the polypeptide by proteosomes.

The yeast chaperones, BiP and calnexin have been implicated in each of these pathways. Chaperones can promote redundancy by reducing the impact of mutations on protein structure. Since chaperones operate at the level of translation upwards, they can only buffer the effect of mutation, and are unable to purge mutations.

5.7 ANTIREDUNDANCY THROUGH OVERLAPPING READING FRAMES

Single sequences of DNA or RNA encoding parts of more than one polypeptide are said to possess overlapping reading frames. In principle, three different amino acid sequences can be obtained by initiating transcription from each of the three nucleotides constituting a single codon. This gives rise to three different readings of a genetic message, all of them out of phase with one another. Alternatively, transcription might begin in phase, but from a codon further downstream than the traditional initiation codon. Overlapping reading frames are a preferred strategy of genomic compression found among viruses, bacteria, and even some eukaryotic genes [37, 54]. Overlapping genes are of interest as they can increase mutational load: a single mutation can result in damage to more than one protein. In section 2 we introduced the multiplicative fitness landscape of the form, $w_k = (1 - s_1)^k$. If we now consider a single sequence of length $2N - M$ containing two genes of length N and with an overlapping region of length M, the multiplicative fitness landscape for this sequence is rendered as, $w_k^{(o)} = \sum_i \binom{k}{i} p^i q^{k-i} (1 - s_1)^i (1 - s_2)^{k-i}$, where $p = \frac{M}{2N-M}$, $q = 1 - p$ and s_1 and s_2 are the selection coefficients for mutation to one gene and two genes respectively, where $s_2 > s_1$. Note that $w_k^{(o)} < w_k$ and hence, on average, mutations to genomes with overlapping reading frames will tend to be more deleterious than mutations to genomes without overlapping reading frames. This gives rise to an enhancement of point mutations consistent with antiredundancy.

Overlapping reading frames are primarily a mechanism of genomic compression, and hence are rarely, if ever, observed in eukaryotes [37]. Thus this is a mechanism that produces antiredundancy incidentally and always in large population sizes.

5.8 ANTIREDUNDANCY THROUGH NONCONSERVATIVE CODON BIAS

The genetic code gives rise to high levels of synonym redundancy. There are four nucleotides and a triplet code, whereas there are only 20 amino acids. This produces a ratio of 16:5 codons to amino acids. Assuming an equal abundance of each of the codons, and a selective equivalence or "neutrality" of each codon, then we would expect equal frequencies of nucleotides in the genome. This is not observed, different species often have consistent and characteristic codon biases [2]. The possible causes of codon biases are numerous, including translation selection for increased gene expression [32, 42], translation selection for parasite immunity [38]), and structural stability [33], weak selection, and drift.

One potential consequence of codon bias is to increase the rate of amino acid substitution in proteins. As an example we can consider a GC-rich genome in which, for each of the amino acids G or C, nucleotides are used preferentially. If we consider the four codons for Serine we have TCT, TCC, TCA, and TCG. In a GC-rich genome, TCG will be most common. In this genome random mutations are most likely to introduce Gs or Cs at each site. Given this assumption, around 100% of mutations to the first site, and around 50% of mutations to the second and third sites, will lead to a different amino acid in each case. Thus, around two thirds of mutations to the serine codon are deleterious. With equal frequencies of G, C, A, and T, around one half of mutations are deleterious. Thus codon bias can lead to a greater chance of a non-synonymous amino acid substitution following a point mutation, promoting antiredundancy in the genome.

Variation in codon usage is greater among microorganisms than in mammals, birds, amphibians, and reptiles (codon usage database: http://www.kazusa.or.jp/codon/). This could reflect any number of different independent variables, including population size. In large populations, greater variation is expected simply as a consequence of random mutation. This indirect effect could still affect variation in gene expression rates, and thus the level of redundancy.

5.9 ANTIREDUNDANCY THROUGH APOPTOTIC CHECKPOINT GENES

Apoptosis or programmed cell death describes a series of adaptive phases cells undergo, including mitochondrial breakdown, blebbing, degradation of chromatin, and membrane fragmentation, before being engulfed by phagocytic cells. The genetic and cellular cues initiating the apoptotic pathway are numerous. These include infection, developmental signals, the removal of trophic factors, heat stress, and mutation. Without apoptosis, deleterious mutations that leave cells capable of proliferation or increase the rate of proliferation, can lead to an increase in the frequency of mutant genes contained within body tissues [27]. Tumor suppressor genes or checkpoint genes, such as the transcription factor p53, respond to mutation by inducing apoptosis. The range of mutations p53 is capable of responding to includes double strand breaks in DNA, chemical damage to DNA, and DNA repair intermediates [44]. Apoptotic check point genes enhance the deleterious effects of mutation so as to increase their likelihood of being purged through local selection pressures. These are, therefore, mechanisms for increasing the selective cost of mutation, and are mechanisms of antiredundancy as they remove defective cells and genomes.

Most of the apoptosis genes are confined to multicellular eukaryotes. Moreover, within the metazoa, apoptosis is more frequently observed in large populations of cells. In populations of cells with effective population sizes approaching zero (such as oocytes and neurons), apoptosis is almost completely inhibited.

5.10 ANTIREDUNDANCY THROUGH GENETIC BOTTLENECKS

Genetic bottlenecks arise when the effective population size of a gene, or set of genes, experiences a dramatic reduction. The transmission of mitochondria through the germline, the transmission of bacterial or viral pathogens between hosts, and the alternation of diploid and haploid generations, all lead to severe bottlenecks in the genetic variation of founder populations. The consequence of this reduction of variation is the exposure of formerly masked mutations [4]. One of the clearest examples is provided by mitochondria. Cells containing in excess of 10% wildtype mitochondrial genomes are capable of almost perfect aerobic metabolism [10]. Without a bottleneck, the binomial sampling of mitochondria to provision daughter cells, leads to few daughters with less than 10% wildtypes. Hence most daughter cells are equally fit. By imposing a bottleneck, heterogeneity in daughter cells is increased, increasing variation in metabolism, and allowing mitochondrial genomes to compete for survival [39]. Genetic bottlenecks are a mechanism for antiredundancy as they expose and purge deleterious mutations from a population.

The extent of the genetic bottleneck is very nicely correlated with effective population sizes. In those species producing many offspring the bottleneck is minimal, whereas in species producing few offspring the bottleneck is most severe. Thus, species producing few offspring make use of the abundance of their gametes to increase the efficiency of local selection pressures [39].

5.11 ANTIREDUNDANCY THROUGH INACTIVATION OF TELOMERASE

The ends of chromosomes are capped by protective, nucleoprotein structures known as telomeres. At each cell division there is a loss of part of the noncoding repeat sequence constituting the telomeres. When telomeres are allowed to erode beyond a certain critical threshold, this leads to the proliferative arrest of mitotic cells. The erosion of the telomeres can be reversed through the action of the telomerase enzyme. In early embryonic development telomerase is active, favoring the steady proliferation of cells and the growth of tissues and organ systems. At maturity, telomerase is inactivated, imposing an upper limit on the lifetime of somatic cells [65]. In cancer cells, telomerase is very often over-expressed, allowing transformed cells to propagate chromosomes for an almost indefinite number of generations. The expression of telomerase effectively allows mutant lineages to increase in frequency. The loss of telomerase leads to the purging of mutant cells, and is, therefore, a mechanism of antiredundancy.

The repression of telomerase appears to be confined to humans and other long-lived mammals. Telomerase repression is, therefore, a feature of long-lived individuals comprising large populations of actively replicating cells. Rodents do not possess the same stringent controls on telomerase inactivation [13].

5.12 ANTIREDUNDANCY THROUGH LOSS OF DNA ERROR REPAIR

Mutational damage to DNA is minimized through the actions of mechanisms that recognize changes to the genome and repair them. The mechanisms of DNA repair include excision-repair (removal of damaged regions and replacement), mismatch repair (replacing noncomplementary bases in opposite strands of a double helix), and direct repair (the reversal of damage to nucleotides). The loss of one or more of these classes of repair mechanism can lead to the lethal accumulation of mutations and genetic instability of the genome [45]. In finite populations, the accumulation of mildly deleterious mutations can lead to the eventual extinction of a lineage through Muller's ratchet [18]. The rate of the ratchet can be reduced either by increasing the efficacy of repair, or paradoxically, by eliminating some forms of repair altogether. This latter strategy is a mechanism of antiredundancy, as it increases the deleterious effects of mutations. It has been suggested that the absence of direct DNA repair in mitochondria enhances the efficiency of selection.

6 LEVELS OF SELECTION AND THE ROBUST EVOLUTIONARY INDIVIDUAL

Natural selection operates on any entity capable of replication, assuming a reasonably stable pattern of heredity. Selected entities exist at many levels of biological organization, from short ribonucleotide sequences to genes, genomes, cells, organisms, and populations. These levels are called the levels of selection, and they are of interest to biology, as selection at lower levels is often incompatible with stable heredity at more inclusive levels of organization. The ways in which selection acting at lower levels creates higher levels, and the ways in which higher levels feed back to influence lower levels, have been called the "fundamental problem of biology" [43].

The dynamics of redundancy and antiredundancy reveal reciprocal relations among the "levels of selection." In multicellular organisms, rapidly dividing cells experience selection as members of a large quasispecies—much like viruses or bacteria. Each cell, bacterium, or virus is in immediate competition with its neighbors for survival factors and nutrients. There is a premium on fast replication, with the attendant loss of many daughters. In smaller populations comprising more slowly dividing cells, robust replication is more important, and fewer cells can be sacrificed in favor of haste.

There is a potential conflict between the organismal and cellular levels of selection. Multicellular organisms living in small populations would benefit from a redundant (flatter) fitness landscape, whereas the abundant cells from which the organism is composed would benefit from an antiredundant (steeper) landscape. In some cases this conflict has a synergistic resolution: antiredundancy at the cellular level is an effective means of ensuring redundancy and robustness at the organismal level. Antiredundant mechanisms activated in mutant or damaged cells cause their removal, thereby ensuring stability (redundancy) against mutation in tissues. This coordination of interest will not always be observed. Considering only two levels of selection, there are four combinations of strategies available: (1) redundancy at the cellular level promoting redundancy at the organismal level (for example, polyploidy), (2) redundancy at the cellular level promoting antiredundancy at the organismal level (loss of molecular checkpoints), (3) antiredundancy at the cellular level promoting redundancy at the organismal level (checkpoint genes inducing apoptosis), and (4) antiredundancy at the cellular level promoting antiredundancy at the organismal level (bottlenecks in organelle transmission within and between generations).

The preferred strategy will depend upon the local population size experienced by the cell and by the organism, and any further constraints placed on their ability to replicate. Whereas bacteria and viruses replicate their genomes over a potentially indefinite number of generations, the somatic cells of many animals are only able to replicate over a small number of generations (this is often the result of the loss of telomerase). The division of cell lines into "somatic" and "germline" was an innovation of enormous importance for the evolution of organization [30]. It effectively handicapped selection acting at the level of cells in favor of selection acting on the germline and somatic cell aggregate. This aggregate has come to be known as the evolutionary individual [8]. Individuals are characterized by an evolved common interest among levels (for example cell and organism). The emergence of individuals at more inclusive levels of organization has come to be known as the "major transitions" [50].

Because our model does not separate germline from soma, we cannot directly address the evolutionary conflicts of interest between cells and organism. In other words, we do not consider the developmental dynamics of the individual. A thorough treatment of cancer progression would require such an approach. In cancer, mutant cells strive to increase cellular redundancy in order to mitigate the deleterious effects of their mutations. These mutations impose a cost on cells by damaging the individual organism. The "parliament of genes," coming under selection from the whole organism, seeks to promote cellular antiredundancy so as to remove mutant cells and increase organismal redundancy (case 3). The fact that so many antiredundant mechanisms are found to respond to cancer at the cellular level should be viewed as a victory for the muticellular individuals.

7 EVOLVING LANDSCAPE PARAMETERS

7.1 MODIFIER MODELS

We have spoken of redundancy and antiredundancy in very general terms as principles of robustness. We have also spoken of the parameters of robustness in terms of a diverse set of molecular adaptations. The steepness parameter has been assumed to represent the net contribution of these mechanisms to the final plasticity of the phenotype. Selection acting on the individual must be able to modify the degree of redundancy, through modification of these or similar mechanisms. Mutations do not only alter fitness, they also alter these mechanisms, and thereby alter the genotype to phenotype map [71]. In order to arrive at an approximately continuous change in landscape steepness, we should assume an approximately continuous degree of variation in the efficacy of mechanisms. From the list of mechanisms that we have provided this is not hard to imagine. Dominance is commonly described as varying between complete dominace through incomplete dominance to recessivity; autophagy, mRNA surveillance and tRNA suppression are stochastic phenomena, working on a variable proportion of products; reading frames can overlap to varying degrees; codon bias can be more or less extreme; and bottlenecks are of varying severity. These mechanisms are all compatible with an approximately continuous distribution of variation. We can, therefore, think of redundancy as the outcome of a multilocus system in which we have multiple modifiers of redundancy (chaperone genes, methylation genes, tRNA suppressors etc.), and where at each locus, there are several alleles. Working out the population genetics of such a system represents an open challenge. An even greater problem is ascertaining why there should be so many different redundancy modifiers.

7.2 EVIDENCE FOR DISTRIBUTED CONTROL

Recent work on the robustness of yeast development, and on the mechanisms of redundancy, has highlighted the distributed, multi-locus, multi-allele nature of redundancy. In yeast, the ability to buffer against the effects of gene knockout are largely independent from the genetic distance between paralogues [70]. This suggests that gene duplication is unable to produce significant functional redundancy in yeast, and is suggestive of more distributed mechanisms for buffering. Knockout of the chaperone gene HSP90 leads to the formerly undetectable expression (cryptic variation) of polymorphisms at multiple loci [58]. Hence a single gene can conceal variation in multiple different genetic pathways. A review of synthetic lethal mutations in yeast (genes for which a double knockout is lethal) finds that any given gene has a synthetic lethal relationship with, at most, 26 other genes in the genome[26]. While this would suggest that genes are buffered from the activity of the majority of genes in the genome, it also shows that genes are buffered at multiple loci. The accumulating evidence paints a picture of connected modules within which there is a great deal of dependence, and among which activity is fairly independent. Above these modules, there are shared processes playing essential buffering roles. These processes give rise to the principles of redundancy and antiredundancy. They also give us a crucial insight into biological complexity.

8 ROBUST OVERDESIGN AND BIOLOGICAL COMPLEXITY

The study of macroscopic organization has lead to an evolutionary worldview in which evolution produces carefully crafted, painstakingly parsimonious, and reliably robust structures. The role of evolutionary theory has been, traditionally, to explain these engineered properties of biological systems. A standard perspective on adaptation is the degree of agreement between a biological trait and a comparable engineered system, such as bird wings and those of aircraft; eyes and the lenses of cameras. While this approach has been extremely fruitful—not only because it enables us to understand how natural devices operate—it has lead us to neglect the baroque extravagance of natural designs. In other words, we have often ignored the important fact that so much of the natural world seems to be over-designed. When we look at the cell we do not so much think of a John Harrison clock [61] as a Rube Goldberg cartoon [77], with device mounted upon device to accomplish the simplest of tasks.

What is overdesign? In a deep sense this has been the subtext of most of our evolutionary questions: Why proteins when nucleic acids seem cable of so much function? Why diploidy when this requires twice as much resource as haploidy? Why multicellularity when unicellularity seem so efficient? Why cellular differentiation when coordinated control is so uncertain? Why sexuality when asexuality has a two-fold advantage? Why cooperation when selfishness provides higher payoffs? These are all questions in which a reasonable design solution is discarded in favor of an apparently unreasonable solution. Evolutionary biology seems to be the science of unreasonable solutions, whereas engineering is the science of reasonable solutions. This does not mean that biological systems can not be studied in terms of engineering, but that biological problems have different properties than classical engineering problems, and the issues of stability and robustness play a very central role in this difference.

Considering the list of dichotomous solutions listed in the previous paragraph and reviewing some standard explanations for the observed solutions, (1) proteins over nucleic acids—amino acids act as cofactors increasing binding specificity; (2) diploidy over haploidy—diploidy facilitates DNA repair; (3) multicellularity over unicellularity—multicellularity increases control over selfish cytoplasmic elements; (4) sexuality over asexuality—sex reduces the mutational load or helps evade parasitism; (5) cooperation over selfishness—cooperation increases the fitness of aggregates. In each of these examples one factor is repeatedly in evidence: the need for stability or robustness. In other words, our canonical theories for biological organization seem to be implicitly formulated in terms of robust designs. While this comes as little surprise upon reflection, it highlights some very important notions of adaptation that have been neglected by the classical engineering schools of life.

1. Biology is deeply stochastic—no absolute zero.
2. Biology is deeply historical—no tabula rasa.

3. Biology is deeply conflictual—no garden of Eden.
4. Biology is deeply connected—no free agents.

Robustness provides a unifying theme running through all of these ideas. Consider only genetics as an example. Stochasticity (1) presenting itself as mutation, leads to the evolution of DNA repair enzymes, mRNA surveillance, tRNA suppression, and checkpoints. Historicity (2) present in the canonical genetic code leads to diverse translational strategies, preferred codon frequencies, and biased amino acid usage. Conflict (3) arising from selfish, parasitic elements, leads to diploidy and parliaments of genes. Finally, the fact of networks of interdependence (4), leads to modularity and distributed control. In each of these cases, and having considered only the genetic level, we have observed how the notion of robustness is deeply related to the uniquely biological property of over-design, and how over-design reflects a need to incorporate redundancy and canalization. In other words, the remarkable diversity and complexity of living things arise in part as robust and redundant solutions to instabilities that evolve alongside and above primary design goals.

REFERENCES

[1] Abeliovich, H., and D. J. Klionsky. "Autophagy in Yeast: Mechanistic Insights and Physiological Function." *Microbiol. Mol. Biol. Rev.* **65(3)** (2001): 463–479.

[2] Akashi, H. "Inferring Weak Selection from Patterns of Polymorphism and Divergence at 'Silent' Sites in *Drosophila* DNA." *Genetics* **139(2)** (1995): 1067–1076.

[3] Ancel, L. W., and W. Fontana. "Plasticity, Evolvability, and Modularity in RNA." *J. Exp. Zool.* **288(3)** (2000): 242–283.

[4] Bergstrom, C. T., and J. Pritchard. "Germline Bottlenecks and the Evolutionary Maintenance of Mitochondrial Genomes." *Genetics* **149(4)** (1998): 2135–2146.

[5] Bourguet, D. "The Evolution of Dominance." *Heredity* **83(Pt. 1)** (1999): 1–4.

[6] Brookfield, J. F. "Genetic Redundancy." *Adv. Genet.* **36** (1997): 137–155.

[7] Burger, R. *The Mathematical Theory of Selection, Recombination, and Mutation.* New York: Wiley, 2000.

[8] Buss, L. W. *The Evolution of Individuality.* Princeton, NJ: Princeton University Press, 1987.

[9] Cali, B. M., S. L. Kuchma, J. Latham, and P. Anderson. "smg-7 is Required for mRNA Surveillance in *Caenorhabditis elegans*." *Genetics* **151(2)** (1999): 605–616.

[10] Chomyn, A., A. Martinuzzi, M. Yoneda, A. Daga, O. Hurko, S. T. Johns, I. Lai, C. Nonaka, G. Angelini, and D. Attardi. "MELAS Mutation in mtDNA

Binding Site for Transcription Termination Factor Causes Defects in Protein Synthesis and in Respiration but No Change in Levels of Upstream and Downstream Mature Transcripts." *Proc. Natl. Acad. Sci. USA* **89(10)** (1992): 4221–4225.

[11] Chosmky, N. "Principles and Parameters in Syntactic Theory." In *Explanations in Linguistics*, edited by N. Hornstein and D. Lightfoot. London: Longman, 1981.

[12] Cui, J., F. Shen, F. Jiang, Y. Wang, J. Bian, and Z. Shen. "Loss of Heterozygosity and Microsatellite Instability in the Region Including BRCA1 of Breast Cancer in Chinese." *Zhonghua Yi Xue Yi Chuan Xue Za Zhi* **15(6)** (1998): 348–350.

[13] Deng, H. W., and M. Lynch. "Inbreeding Depression and Inferred Deleterious-Mutation Parameters in Daphnia." *Genetics* **147(1)** (1998): 147–155.

[14] Eggertsson, G., and D. Soll. "Transfer Ribonucleic Acid-Mediated Suppression of Termination Codons in *Escherichia coli*." *Microbiol. Rev.* **52(3)** (1988): 354–374.

[15] Eigen, M. "Self-Organization of Matter and the Evolution of Biological Macromolecules." *Naturwissenschaften* **58(10)** (1971): 465–523.

[16] Eigen, M., W. Gardiner, P. Schuster, and R. Winkler-Oswatitsch. "The Origin of Genetic Information." *Sci. Am.* **244(4)** (1981): 88–92, 96, et passim.

[17] Elena, S. F., and R. E. Lenski. "Epistasis between New Mutations and Genetic Background and a Test of Genetic Canalization." *Evol. Int. J. Org. Evol.* **55(9)** (2001): 1746–1752.

[18] Felsenstein, J., and S. Yokoyama. "The Evolutionary Advantage of Recombination. II. Individual Selection for Recombination." *Genetics* **83(4)** (1976): 845–859.

[19] Fewell, S. W., K. J. Travers, J. S. Weissman, and J. L. Brodsky. "The Action of Molecular Chaperones in the Early Secretory Pathway." *Ann. Rev. Genet.* **35** (2001): 149–191.

[20] Fisher, R. A. "The Possible Modification of the Response of the Wildtype to Recurrent Mutations." *Am. Nat.* **62** (1928): 115–126.

[21] Fontana, W., and P. Schuster. "Continuity in Evolution: On the Nature of Transitions." *Science* **280(5368)** (1998): 1451–1455.

[22] Fontana, W., P. F. Stadler, E. G. Bornberg-Bauer, T. Griesmacher, I. L. Hofacker, M. Tacker, P. Tarazona, E. D. Weinberger, P. Schuster. "RNA Folding and Combinatory Landscapes." *Phys. Rev. E. Statistical Physics, Plasmas, Fluids, and Related Interdisciplinary Topics* **47(3)** (1993): 2083–2099.

[23] Garcia-Dorado, A., and A. Caballero. "On the Average Coefficient of Dominance of Deleterious Spontaneous Mutations." *Genetics* **155(4)** (2000): 1991–2001.

[24] Gillespie, J. H. *The Causes of Molecular Evolution.* Oxford, Oxford University Press, 1991.

[25] Haigh, J. "The Accumulation of Deleterious Genes in a Population—Muller's Ratchet." *Theor. Pop. Biol.* **14(2)** (1978): 251–267.
[26] Hartman, J. L. T., B. Garvik, and L. Hartwell. "Principles for the Buffering of Genetic Variation." *Science* **291(5506)** (2001): 1001–1004.
[27] Hartwell, L. H., and M. B. Kastan. "Cell Cycle Control and Cancer." *Science* **266(5192)** (1994): 1821-1828.
[28] Higgs, P. G. "Error Thresholds and Stationary Mutant Distributions in Multi-locus Diploid Genetics Models." *Genet. Res. Cam.* **63** (1994): 63–78.
[29] Hoffmann, F. M. "*Drosophila* Abl and Genetic Redundancy in Signal Transduction." *Trends Genet.* **7(11-12)** (1991): 351–355.
[30] Jablonka, E., and M. J. Lamb. "Epigenetic Inheritance and Evolution." In *The Lamarckian Dimension.* Oxford: Oxford University Press, 1995.
[31] Kacser, H., and J. A. Burns. "The Molecular Basis of Dominance." *Genetics* **97(3-4)** (1981): 639–666.
[32] Karlin, S., and J. Mrazek. "Predicted Highly Expressed Genes of Diverse Prokaryotic Genomes." *J. Bacteriol.* **182(18)** (2000): 5238–5250.
[33] Karlin, S., and J. Mrazek. "What Drives Codon Choices in Human Genes?" *J. Mol. Biol.* **262(4)** (1996): 459–472.
[34] Kauffman, S. A. *The Origins of Order.* Oxford, Oxford University Press, 1993.
[35] Kimura, M. "DNA and the Neutral Theory." *Phil. Trans. Roy. Soc. Lond. B Biol. Sci.* **312(1154)** (1986): 343–354.
[36] Kimura, M. "The Neutral Theory of Molecular Evolution." *Sci. Am.* **241(5)** (1979): 98–100, 102, 108 passim.
[37] Krakauer, D. C. "Stability and Evolution of Overlapping Genes." *Evolution* **54** (2000): 731–739.
[38] Krakauer, D. C., and V. Jansen. "Red Queen Dynamics of Protein Translation." *J. Theor. Biol.* (2002): in press.
[39] Krakauer, D. C., and A. Mira. "Mitochondria and Germ-Cell Death." *Nature* **400(6740)** (1999): 125–126.
[40] Krakauer, D. C., and M. A. Nowak. "Evolutionary Preservation of Redundant Duplicated Genes." *Semin. Cell Dev. Biol.* **10(5)** (1999): 555–559.
[41] Krakauer, D. C., and A. Sasaki. "Noisy Clues to the Origin of Life." *Proc. Roy. Soc. Lond. B Biol. Sci* **269(1508)** (2002): 2423–2428.
[42] Kurland, C. G. "Codon Bias and Gene Expression." *FEBS Lett.* **285(2)** (1991): 165–169.
[43] Leigh, E. G., Jr. "Levels of Selection, Potential Conflicts, and Their Resolution: The Role of the 'Common Good.'" In *Levels of Selection in Evolution,* edited by L. Keller, 15–31. Princeton, NJ: Princeton University Press, 1999.
[44] Levine, A. J. "The Cellular Gatekeeper for Growth and Division." *Cell* **88(3)** (1997): 323–331.
[45] Lewin, B. *Genes V.* Oxford: Oxford University Press, 1994.
[46] Lewontin, R. C. "The Problem of Genetic Diversity." *Harvey Lect.* **70(Series)** (1975): 1–20.

[47] Li, X., and M. Noll. "Evolution of Distinct Developmental Functions of Three *Drosophila* Genes by Acquisition of Different cis-regulatory Regions." *Nature* **367(6458)** (1994): 83–87.
[48] Lynch, M., R. Burger, D. Butcher, and W. Gabriel. "The Mutation Meltdown in Small Asexual Population." *J. Heredity* **84** (1993): 339–344.
[49] Maconochie, M., S. Nonchev, A. Morrison,and R. Krumlauf. "Paralogous Hox Genes: Function and Regulation." *Ann. Rev. Genet.* **30** (1996): 529–556.
[50] Maynard Smith, J., and E. Szathmary. *The Major Transitions in Evolution.* San Francisco, CA: W. H. Freeman, 1995.
[51] Muller, H. J. "Our Loads of Mutations." *Am. J. Hum. Genet.* **2** (1950): 111–176.
[52] Newbold, R. F. "The Significance of Telomerase Activation and Cellular Immortalization in Human Cancer." *Mutagenesis* **17(6)** (2002): 539–550.
[53] Normanly, J., and Bartel. "Redundancy as A Way of Life—IAA Metabolism." *Curr. Opin. Plant Biol.* **2(3)** (1999): 207–213.
[54] Normark, S., S. Bergstrom, T. Edlund, T. Grundstrom, B. Jaurin, F. P. Lindberg, and O. Olsson. "Overlapping Genes." *Ann. Rev. Genet.* **17** (1983): 499–525.
[55] Nowak, M. A., M. C. Boerlijst, J. Cooke, and J. M. Smith. "Evolution of Genetic Redundancy." *Nature* **388(6638)** (1997): 167–171.
[56] Palmer, R. "Optimization on Rugged Landscapes." In *Molecular Evolution on Rugged Landscapes: Proteins, RNA and the Immune System*, edited by A. S. Pereleson and S. A. Kauffman. Santa Fe Institute Studies in the Sciences of Complexity. Reading, MA: Addison Wesley, 1991.
[57] Pulak, R., and P. Anderson. "mRNA Surveillance by the *Caenorhabditis elegans* smg Genes." *Genes Dev.* **7(10)** (1993): 1885–1897.
[58] Rutherford, S. L., and S. Lindquist. "Hsp90 as a Capacitor for Morphological Evolution." *Nature* **396(6709)** (1998): 336–342.
[59] Scheiner, S. M. "Genetics and the Evolution of Phenotypic Plasticity." *Ann. Rev. Eco. Syst.* **24** (1993): 35–68.
[60] Schulte-Hermann, R., W. Bursch, B. Grasl-Kraupp, B. Marian, L. Torok, P. Kahl-Rainer, and A. Ellinger. "Concepts of Cell Death and Application to Carcinogenesis." *Toxicol Pathol.* **25(1)** (1997): 89–93.
[61] Sobel, D. *Longitude: The True Story of a Lone Genius Who Solved the Greatest Scientific Problem of His Time.* New York, Penguin Books, 1995.
[62] Stearns, S. C., and T. J. Kawecki. "Fitness Sensitivity and the Canalization of Life History Traits." *Evolution* **48** (1994): 438–1450.
[63] Tautz, D. "A Genetic Uncertainty Problem." *Trends Genet.* **16(11)** (2000): 475–477.
[64] Tautz, D. "Redundancies, Development and the Flow of Information." *Bioessays* **14(4)** (1992): 263–266.
[65] Urquidi, V., D. Tarin, and S. Goodison. "Role of Telomerase in Cell Senescence and Oncogenesis." *Ann. Rev. Med.* **51** (2000): 65–79.

[66] van Nimwegen, E., J. P. Crutchfield, and M. Huynen. "Neutral Evolution of Mutational Robustness." *Proc. Natl. Acad. Sci. USA* **96(17)** (1999): 9716–9720.

[67] Voelker, R. A., C. H. Langley, A. J. Leigh Brown, S. Ohnishi, B. Dickson, E. Montgomery, and S. C. Smith. "Enzyme Null Alleles in Natural Populations of *Drosophila melanogaster*: Frequencies in North Carolina Populations." *Proc. Natl. Acad. Sci. USA* **77** (1980): 1091–1095.

[68] Waddington, C. H. "Canalization of Development and the Inheritance of Acquired Characters." *Nature* **150** (1942): 563–565.

[69] Wagner, A. "Redundant Gene Functions and Natural Selection." *J. Evol. Biol.* **12** (1999): 1–16.

[70] Wagner, A. "Robustness Against Mutations in Genetic Networks of Yeast." *Nat. Genet.* **25(1)** (2000): 3–4.

[71] Wagner, G., and L. Alternberg. "Complex Adaptations and the Evolution of Evolvability." *Evolution* **50** (1996): 967–976.

[72] Wagner, G. P., G. Booth, and H. Bagheri-Chaichian. "A Population Genetic Theory of Canalization." *Evolution* **51** (1997): 329–347.

[73] Wallace, B. *Fifty Years of Genetic Load.* Ithaca: Cornell University Press, 1991.

[74] Whitlock, M. C., P. C. Phillips, G. Moore, and S. J. Tonsor. "Multiple Fitness Peaks and Epistasis." *Ann. Rev. Ecol. Syst.* **26** (1995): 601–629.

[75] Wilke, C. O. "Adaptive Evolution on Neutral Networks." *Bull. Math. Biol.* **63** (2001): 715–730.

[76] Williamson, A. R., E. Premkumar, and M. Shoyab. "Germline Basis for Antibody Diversity." *Fed Proc* **34(1)** (1975): 28–32.

[77] Wolfe, M. F. *Rube Goldberg: Inventions.* Simon and Shuster, 2000.

[78] Wolffe, A. P., and M. A. Matzke. "Epigenetics: Regulation through Repression." *Science* **286(5439)** (1999): 481–486.

[79] Woodcock, G., and P. G. Higgs. "Population Evolution on a Multiplicative Single-Peak Fitness Landscape." *J. Theor. Biol.* **179(1)** (1996): 61–73.

[80] Wright, S. "Evolution in Mendelian Populations." *Genetics* **16** (1931): 97–159.

[81] Wright, S. "Surfaces of Adaptive Value Revisited." *Am. Nat.* **131** (1988): 115–123.

Directing the Evolvable: Utilizing Robustness in Evolution

Christopher A. Voigt
Stephen L. Mayo
Zhen-Gang Wang
Frances Arnold

1 INTRODUCTION

The use of directed evolution techniques has greatly accelerated the discovery of new and useful biological molecules and systems [10]. Through iterative cycles of diversity creation (e.g., mutation or recombination) and selection, proteins, antibodies, pathways, viruses, and organisms have been evolved to perform tasks optimized for pharmaceutical and industrial applications. Before directed evolution became established, it was unclear how successful such an approach would be. It was not obvious that randomized mutagenesis and selection would find improvements, due to a combinatorial explosion in the number of possible offspring and the observation that few of these are functional, much less have improvements in desired properties.

Directed evolution is successful, in part, because prior to being evolved in vitro, these systems have a long history of evolution in vivo. As a result of this history, they have properties that make them amenable to both natural and laboratory evolution. This *evolvability* represents the ability of a system to produce fit offspring in a dynamic environment. This chapter will review some of the features that make a system evolvable. A particular emphasis will be made on the contribution of *robustness*, or the ability for a system to survive perturbations of its internal parameters. Robustness enhances the ability of a population to sample parameter space, thus enabling the discovery of novel phenotypes.

Understanding the basis for evolvability will aid the design of efficient evolutionary algorithms that accelerate the in vitro discovery process. Achieving this goal will require the combination of computational models with data from in vitro evolution experiments. In this review, we describe the initial steps of this effort. First, we provide a general definition of robustness and explain its relationship to evolvability. In the following sections, we apply these ideas: first to the evolution of proteins through mutagenesis and recombination, and then to the evolution of genetic networks.

1.1 ROBUSTNESS

The behavior of a system can be defined by a set of internal *parameters*. In the case of a metabolic network, the parameters are the kinetic constants and concentrations of the component enzymes, which determine the products and their rates of production. Similarly, the activity of an enzyme is defined by its amino acid sequence, solvent conditions, and temperature. A system is robust if it can absorb variations in these parameters without disrupting its behavior [111, 114]. The parameters can be perturbed by various insults, for example, kinetic constants can be altered by mutations, temperature variations, or exposure to different environments. Because we are describing robustness from the perspective of directed evolution, the insults to these systems are defined by the experimental technique, such as point mutagenesis or recombination.

It is important to contrast parameter robustness with *variable* stability, which describes resilience to perturbations in the inputs of a system. Stability implies that the state of the system before and after the perturbation remains unchanged. As an illustration, a bridge is stable regarding variables such as car weight. The bridge is considered stable if it reliably returns to the same state after cars have passed over it. In contrast, robustness describes the collection of bridge design parameters, such as the cable strength for a suspension bridge. If the Golden Gate Bridge in San Francisco can be reproduced in Alaska without disturbing its function, then the design is robust. In this example, the strength of the materials is the parameter, and this parameter is perturbed by a change in environment.

In determining the robustness of a system, the behavior and the parameters need to be defined. A system may be robust with regards to one behavior while being sensitive with regards to a different behavior [14]. There are several convenient metrics for measuring robustness. For parameters that are continuous, the sensitivity S is defined as the change in a behavior b with respect to a parameter p [114],

$$S(b,p) = \frac{db}{dp}. \tag{1}$$

When the parameter is discontinuous (e.g., amino acid sequence), then it is useful to define an entropy which captures the number of states that are consistent with retaining the system behavior. The entropy of parameter i is defined as,

$$s_i = \sum_{j=1}^{20} p_i(a) \ln p_i(a) \tag{2}$$

where $p_i(s)$ is the probability that parameter i is in state s. These probabilities can be derived from several sources. They can be calculated explicitly if the energetic consequence of each state is known. They can also be obtained from a list of possible states, obtained either through a simulation or by experimental observation of variation in the system (e.g., examining amino acid variability through a sequence alignment). A useful application of eq. (2) is to quantify the variability of a protein residue i with respect to amino acid substitutions s [115, 130].

1.2 EVOLVABILITY

Evolvability is the capacity of a system to react at the genetic level to changing requirements for survival [72]. Upon environmental change, an evolvable system will produce offspring whose perturbed parameters improve the new fitness. Those systems that require the least dramatic parameter changes (e.g., the fewest mutations) have the smallest entropic barrier to being discovered [72, 129]. Architectures that minimize the entropic barriers are the ones that are likely to find improvements first and, therefore, survive. In directed evolution, the evolutionary constraints in vitro differ from those properties that were selected for by nature. With this "environmental change," an evolvable architecture is more likely to result in a successful directed evolution experiment.

Robustness can reduce the entropic barrier by separating the parameters that define the various behaviors of a system. For example, an enzyme may improve the evolvability of its activity by separating those residues that maintain its stability from those that tune its activity. If a residue contributes to both properties, then it would be more difficult to make substitutions to improve one property without degrading the other. To further reduce the entropic barrier, it is important that small changes in the evolvable parameter lead to large changes in the behavior of the system. Evolvability will be reduced if regions of parameter space that are devoid of any behavior have to be traversed (fig. 1). Robustness can also improve evolvability by facilitating the exploration of parameter space through neutral drift [14, 42, 61, 62, 72]. Neutral drift drastically increases the

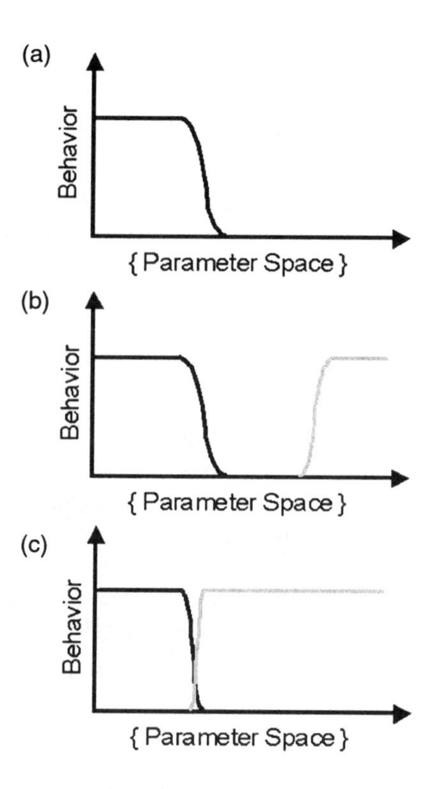

FIGURE 1 The behavior of a biological system is plotted as the function of some internal parameter. When the behavior is robust it is insensitive to large variations in parameter space (a). It is possible that variation in a parameter can sample new behaviors (gray line b). If this behavior is attainable without having to go through dead parameter space, the behavior is evolvable (c).

fraction of parameter space that can be sampled, thus increasing the likelihood of discovering novel behaviors.

Measuring robustness, either by calculating sensitivity or entropy, is fairly straightforward. In contrast, the various ways in which evolvability can manifest itself makes its quantification a more challenging task. For example, an evolvable parameter does not have to be robust. Some measures of evolvability include the number of behaviors that can be sampled and the nature of the transitions between these behaviors. A highly cooperative transition and small separations in parameter space are indicators of evolvability (fig. 1).

2 UTILIZING ROBUSTNESS TO OPTIMIZE MUTANT LIBRARIES

In directed evolution, genetic diversity can be tolerated due to the intrinsic robustness of proteins. The ability to predict how and where a protein is robust has led to the design of evolutionary algorithms where point mutations or recombination is targeted [9, 131, 132]. In describing the robustness of a protein, the behavior describes the combination of properties that needs to be retained for function. This is mainly the stability of the three-dimensional structure, but can be a more complex combination of properties. For example, in evolving an antibody for use as a pharmaceutical, besides maintaining the stability and affinity for the target, it may also be important to evade the human immune response. When diversity is generated using random point mutagenesis, the relevant parameter space is the amino acid state of each residue. A system that is robust can absorb variation in the parameters (amino acid substitutions) without altering a defined behavior (e.g., stability). The capacity to discover new behaviors via amino acid substitutions is the protein's evolvability. In this section, we describe the realization of robustness and evolvability in protein structures. This provides a basis for the introduction of strategies that accelerate evolutionary searches.

2.1 ROBUST PROTEIN ARCHITECTURES

The total number and order of interactions between amino acids affect the average robustness of a protein (fig. 2). When there are many interactions, there are more constraints that need to be satisfied, thus increasing the probability that a mutation is deleterious. This effect worsens as the order of the interactions increases. For example, a system that is dominated by two-body interactions is more robust on average than one that is dominated by three-body interactions [69, 71]. Robustness is also affected by the distribution of interactions in the protein structure. A scale-free distribution of interactions has been demonstrated to be particularly robust [64]. A property of scale-free distributions is that there are a few, highly interacting residues and many weakly interacting residues. A protein can achieve a scale-free-like distribution of interaction by increasing the ratio of surface area to volume. Reducing the number of interactions at many residues increases the average robustness of the protein at the cost of making a few residues highly sensitive to perturbation. If a mutation occurs at such a residue, this has a catastrophic effect on the stability. In other words, increasing the sensitivity of a few parameters makes the overall system robust, a scenario described as "highly optimized tolerance" [31].

Modularity is also important for structural robustness. This effect was demonstrated using a simple protein lattice model to enumerate the number of sequences that fold into various two-dimensional structures [78]. Many sequences folded into a few highly robust structures, whereas most structures were fragile, with only a few or no sequences folding into them. The robust structures were

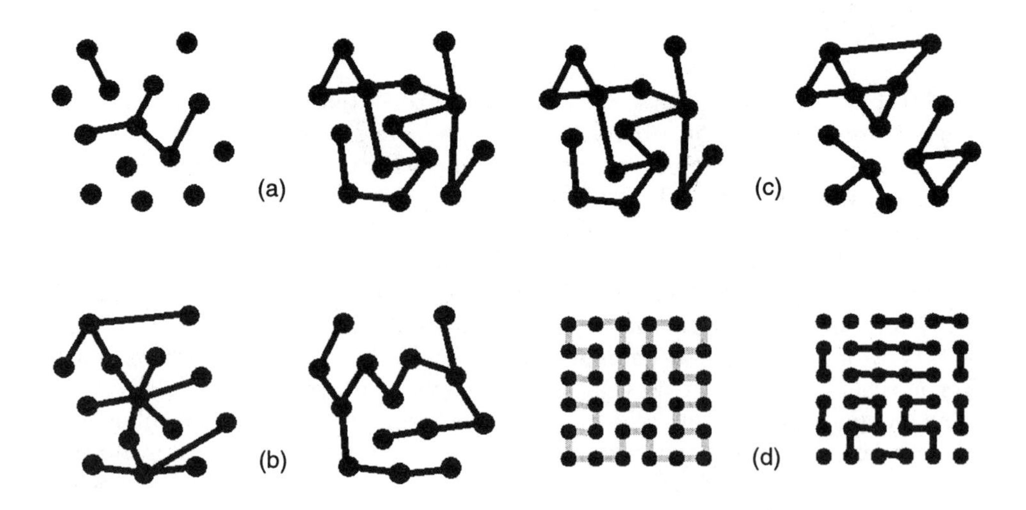

FIGURE 2 Some examples of robust and fragile architectures for a system of nodes and edges. If the system is a protein structure, the nodes represent residues and the edges are amino acid interactions. If the system is a protein network, then the nodes can be component proteins and the edges represent protein-protein interactions. More robust structures have fewer interactions (A — left), a scale-free-like distribution of interactions (B — left), and a modular structure (C — left). Modularity does not necessarily have to be on the level of interactions. The most robust two-dimensional protein structure, as determined by Wingreen and co-workers, is shown on the left, where gray lines mark the progression of the carbon backbone [78]. This structure represents the repetition of a smaller, robust motif that is not ascertainable from the interaction topology (black lines, right).

found to be modular, with a small robust motif copied throughout the larger structure (fig. 2) [140]. The repetition of modular peptide subunits is a common theme in protein structures [100].

Another mechanism to improve robustness is to increase the thermal stability of a structure. Many theoretical models have demonstrated that mutational and thermal stability are strongly correlated [7, 23, 25, 27, 88]. In understanding this correlation, it is important to note that the adjectives "mutational" and "thermal" describe insults rather than behaviors. The relevant behavior is existing in the state of a folded protein. This behavior can be rewritten as maintaining a large energy gap between the folded ground state and the ensemble of unfolded conformations. The energy gap can be perturbed either by increasing temperature or by the disruption of amino acid interactions via a mutation.

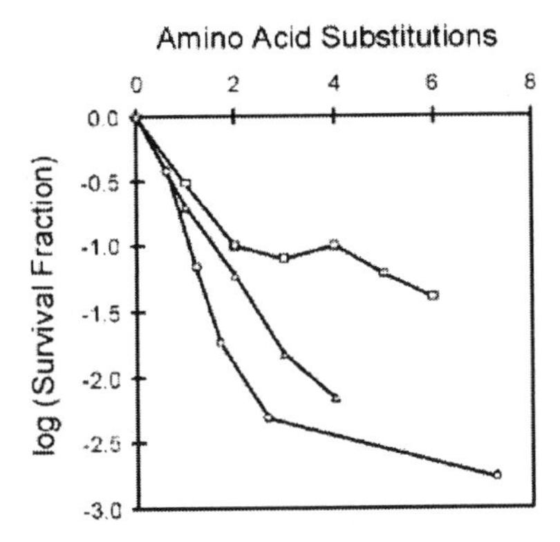

FIGURE 3 The knock-out graphs of HIV RT (○), T4 Polymerase (Δ), and an antibody (□) are shown [38, 128]. A steeper initial slope indicates the behavior is sensitive to mutations. Note that the same protein could have different slopes if different behaviors (or different stringencies) were measured. At high mutation rates, there is a transition in the slope, implying the emergence of compensating mutations.

2.2 MEASURING ROBUSTNESS

One feature of robustness is that the fitnesses of sequences close in sequence space are highly correlated so information is retained upon mutation [141]. Fitness correlation provides an experimentally attainable measure of robustness. This information is typically in the form of a plot of the mutation rate versus the percent of offspring that retain some function (fig. 3) [38, 128]. A small initial slope indicates that more mutations can be accumulated without degrading the function, indicating that this function is more robust [143].

We have developed a computational algorithm that calculates the structural robustness of a protein [130]. This algorithm calculates the stabilization energy of all amino acid sequences folded onto a specified three-dimensional structure using the ORBIT protein design software to calculate the amino acid interactions [37] and mean-field theory to accelerate the calculation [115]. The energetic information is condensed into a residue entropy (eq. (2)), where a high entropy indicates that a residue is tolerant to amino acid substitution. Using this algorithm, those residues that can be mutated while preserving the structural stability can be identified.

2.3 ROBUSTNESS IMPROVES FUNCTIONAL PLASTICITY

Proteins are particularly plastic with regard to tuning function and exploring novel function space. This is evidenced by the observation that some common structural motifs are able perform a wide variety of functions [22, 100]. Directed evolution experiments have demonstrated that significant functional variability can be obtained with few mutations [119]. One way to achieve functional evolvability is to improve the robustness of the behaviors that are essential for function but are not being optimized. By improving the robustness of a structure, mutations are less likely to be destabilizing and sequence space can be more readily explored for new properties through neutral drift [11, 54]. Indeed, mutagenesis experiments have repeatedly demonstrated that protein structures are amazingly robust regarding mutations [11, 12, 13, 60, 81, 109]. Besides improving the overall robustness of the structure, there are several additional mechanisms for improving evolvability, including the separation of parameters and the presence of suppressor mutations.

Evolvability can be improved through the separation of parameters that control different behaviors [72]. This allows one behavior to be optimized without negatively affecting the remaining behaviors. For example, a protein is more evolvable when those residues that maintain stability are isolated from those that control activity. This isolation can be observed when independent mutations that improve different behaviors have additive effects when combined. Additivity has been observed frequently in mutagenesis data and it has been proposed to take advantage of this property in protein engineering [110, 123, 142].

When several homologous parents have different properties and high sequence identity, functional additivity can be utilized to produce a library of offspring with combinations of properties from the parents. In one such study, 26 parental subtilisin genes were recombined to produce a library of offspring [94]. The offspring were screened for activity and stability in various conditions, such as acidic or basic environments or in organic solvents. The hybrid proteins in the library demonstrated a broad range of combined properties. Additivity can also be on the level of parameters that control activity, such as individual components of substrate specificity. By separating the effects of the parameters that confer specificity from the requirements imposed by the catalytic mechanism and structural maintenance, these parameters can be perturbed individually to produce offspring with diverse specificities. Additive parameters that confer specificity are apparent in the recombination of two triazine hydrolases [108]. These two enzymes only differ at 9 residues out of 475, but have very different activities. Recombined offspring were found to catalyze reactions on a variety of triazine compounds with chemically distinct R-groups. Sequence analysis revealed that different residues were important in controlling different physical components of specificity. For example, residues 84 and 92 determine the size of the R-group that could be bound. Mutations at these residues are free to alter aspects of specificity independently without disturbing the catalytic mechanism or stability.

Separating the residues that control activity and stability can be achieved by minimizing the number of stabilizing interactions at functionally important residues. One way this is manifested is through the prevalence of loops and exposed regions near the active site. These loops often control substrate specificity and have been the frequent target of mutagenesis [5, 28, 43, 57, 83, 101]. Further, the complementarity-determining regions (CDRs) of antibodies are composed of loops and have been shown to be robust [26, 28]. Structural tolerance can be achieved without resorting to loop structures. Patel and Loeb demonstrated that the active site of DNA polymerase I, an antiparallel β-strand, is very tolerant to mutagenesis [104].

Deleterious mutations can sometimes be overcome by additional, compensating mutations that act to suppress the negative effect [13, 66]. The appearance of compensating mutations is apparent in diagrams that plot the percent of a library that is functional versus the mutation rate (fig. 3). Typically, these plots demonstrate an exponential decay proportional to the robustness. However, some of these curves recover at high mutation rates, implying the existence of compensating mutations [38, 128]. When a single mutation compensates for an extraordinary number of deleterious mutations it is referred to as a "global suppressor." An example is the M182T mutation in beta-lactamase, which was found to generally compensate for locally destabilizing mutations in a loop near the active site [59, 99, 121]. When this mutation is present, it becomes possible to make additional mutations that rearrange the active site without degrading the stability. The trend of initially accumulating mutations that improve the evolvability has been observed in directed evolution experiments. In the directed evolution of the substrate specificity of beta-glucoronidase, the intermediate mutants first broadened the substrate specificity [83]. After the activity was made more plastic, additional synergistic mutations tuned activity towards the new substrate, and the broad specificity was lost. In another study, the crystal structures of wildtype and evolved esterases were compared [125]. It was found that several loops that form the entrance to the active site cavity are ordered into a specific conformation by mutations distant from the active site. This initial fixation provided the basis for additional mutations in later generations.

2.4 TARGETING DIVERSITY

Algorithms that have been proposed to optimize directed evolution can be separated into two general categories [131]. Several methods have been proposed that optimize the mutation rate as a function of the number of mutants that can be screened. In addition, the effectiveness of a screening algorithm, such as pooling, can be explored. Another approach has been to target specific residues for mutagenesis, either by comparing sequence alignments or using computational methods to identify those residues that are structurally tolerant. Each of these methods is fundamentally reliant on underlying assumptions regarding the robustness and evolvability of the enzyme.

Several theoretical models have been used to study the optimal mutation rate as a function of the size of the screening library and the ruggedness of the fitness landscape [84, 131]. As the number of interactions increases, the probability that a mutation is deleterious also increases. When multiple mutations

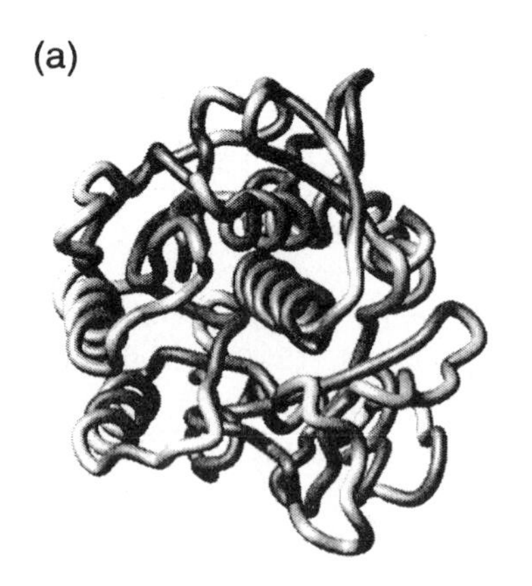

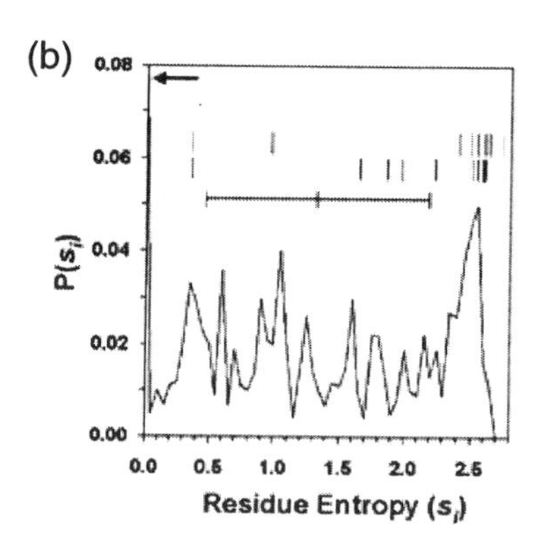

FIGURE 4 (a) Calculated site entropies are shown mapped onto the structure of subtilisin E [130]. High entropy residues, where many amino acid substitutions are allowed, are yellow, medium entropy residues are red and low entropy residues are gray. Solvent accessible residues tend to have higher entropies, but there is little correlation ($R^2 = 0.55$). (b) The probability distribution of site entropies in subtilisin E is shown, where $P(s_i)$ is the probability that a residue has entropy si (e.g., 5% of the residues of subtilisin E have an entropy of 2.5). The black bars mark the mean and standard deviation of this distribution and the arrow indicates that 7.5% of the residues had zero entropy. Residues where mutations were found to improve thermostability and activity in an organic solvent are marked with the red and blue bars, respectively. The majority of the mutations that improve these properties occur at residues predicted to have high entropies (mutations to the right of the extreme black bar have entropies greater than one standard deviation above the mean).

are accumulated on a gene, a larger fraction of these mutations will decrease the fitness. This effect quickly erodes the beneficial effect of any positive mutations. Therefore, to search rugged landscapes, a smaller mutation rate is optimal.

We used a computational model to demonstrate that the directed evolution algorithm preferentially discovers beneficial mutations at structurally tolerant residues [130]. For a given structure, the energetic effect of each amino acid was calculated using mean-field theory and condensed into a residue entropy (eq. (2)). Seventeen out of the twenty-two mutations found by directed evolution to improve the activity of subtilisin E and T4 lysozyme were found to occur at structurally robust residues (fig. 4). Targeting those residues that are structurally robust increases the fraction of the library that is folded and stable. This should also increase the probability of discovering functional improvements.

There is evidence that the immune system targets the generation of diversity to structurally tolerant residues during somatic mutagenesis. There are residue hot spots where mutations are concentrated by various cellular mechanisms during the affinity maturation process [17, 18, 35, 95, 120]. Through structural studies of germline and affinity-matured antibodies, it has been observed that somatic mutations generally preserve the structure of the binding site and antibody-antigen binding interactions [98, 126]. The somatic mutations could be targeted towards structurally robust residues to accelerate the discovery of higher affinity mutant antibodies. Antibodies are frequently the target of directed evolution, to improve antigen binding as well as to improve the activity of catalytic antibodies [21, 79, 118, 145]. Targeting those residues that have been identified as hot spots or those residues that are calculated to be structurally tolerant may improve the diversity of an in vitro library [144].

Additivity is essential in the success of pooling algorithms and recombination strategies. A pooling algorithm involves the screening of multiple mutations simultaneously and then recombining the best mutants from each pool. If all of the mutations are additive, then a pooling strategy drastically reduces the screening requirements to discover the optimal combination of mutations. As the number of nonadditive mutations increases, then pooling strategies become less reliable [70]. Similarly, the success of recombining several mutations onto a single offspring is dependent on the strength of interactions between the mutants [93]. If the mutations do not interact, then simply combining all of the mutations onto a single offspring is optimal. A theoretical method to identify the optimal strategy for combining the mutations has been proposed for the case when the mutants are interacting and the number of mutants that can be screened is limited [2].

Consensus design has been proposed as a method to improve the thermostability of enzymes [65, 77]. A sequence alignment of naturally divergent sequences is used to create a consensus sequence that contains the most common amino acid at each location. This method has been used successfully to improve the thermostability (and the mutational robustness) of several enzymes [77]. It is unclear why the consensus sequence improves thermostability, rather than just accumulating neutral mutations. One possibility is that if natural evolution behaves like a random walk, then it is expected that the time spent in an amino acid state is proportional to the energy of that state and more stable amino acids will reside longer. It is possible that the consensus amino acids reflect large residence times, and, therefore, low energies.

The success of each of these optimization strategies depends on the robustness of the enzyme. These strategies can be improved through the development of algorithms that can predict the effect of mutations on the structure [37, 130]. Those mutations that are additive are more likely to be combined successfully by pooling, consensus, and recombination strategies. Further, the ability to predict the overall robustness of a system, either computationally or through the analysis of an experimentally generated knock-out graph, will be useful. Besides calculating the additivity of some properties, there are currently no computational methods that can predict the evolvability of specific residues. Understanding how to identify residues that contribute to various properties will lead to powerful design tools.

3 ROBUSTNESS TO RECOMBINATION

Recombination is a powerful tool in directed evolution as it can combine traits from multiple parents onto a single offspring [36, 127]. Recombination plays a key role in the natural evolution of proteins, notably in the generation of diverse libraries of antibodies, synthases, and proteases [53]. These proteins have well-defined domain boundaries and recombination shuffles domains into different configurations. The beadlike or loop topologies of these structures make them robust to recombination events [30]. When there is no obvious domain topology, mechanisms, such as introns, can focus crossovers towards specific regions of the protein structure. In terms of in vitro recombination, the ability to focus the diversity towards regions that are robust with regards to recombination will improve the quality of the library and reduce the number of hybrids that need to be screened. In this section, we first describe the observed correlation between intron locations and protein structures and then demonstrate how exon shuffling can achieve functional diversity. Finally, an algorithm based on identifying compact structural units will be used to demonstrate that successful recombination events occur in regions separating structural modules.

3.1 EVOLUTION OF INTRON LOCATIONS

Many eukaryotic genes are composed of pieces of coding DNA (exons), separated in the genome by regions of non-coding DNA (introns). After transcription, introns are removed from the mRNA through a splicing mechanism. Of the many proposed functions of introns, one is that they facilitate the swapping of exons [20, 49, 53]. When two genes are recombined, the crossovers in the mature gene will be biased towards the interface between exons. Longer introns will increase the crossover probability at that location under the assumption that crossovers can occur at each nucleotide with equal probability. If exons correspond to structural or functional subunits of protein structure, then the reconstructed gene would have a higher probability of being stable and functional. Indeed, this correlation has been demonstrated for a large number of genes [39, 51, 52, 53, 102].

There are several possible routes by which introns could have emerged in eukaryotic genes [40, 50, 53]. The "introns-early" theory states that exons correspond to structural motifs that were discovered early in evolutionary history. These exons were pieced together by recombination and gene duplication to build the genes that are now observed. This view asserts that prokaryotes lost their introns due to the strong selection on genome size. In contrast, the "introns-late" theory states that introns were inserted in genes late in evolutionary history, thus explaining their existence in eukaryotes. The early versus late debate is ongoing and it is likely that some introns emerged early and were lost and others emerged late.

If an intron emerged due to the early mechanism, then it is clearly going to correspond to a structural subunit. Arriving late, it could appear anywhere throughout the structure equally, without any structural preference. This idea has led to the argument that the observed correlation between introns and structural units is evidence of an early mechanism [50]. However, if introns were to appear at random locations in a population of genes, then selection could drive the introns towards regions separating structural modules, if the existence of an intron increases an organism's fitness by promoting successful recombination events on a reasonably fast time scale. In other words, selection drives the creation of a robust gene structure.

Theory that has been developed to optimize genetic algorithms provides insight into the relationship between recombination and protein structure. In this literature, the concept of a schema, or a cluster of interacting bits, is useful in predicting the success, or failure, of recombination [58]. When crossovers frequently divide a schema, then these interactions are disrupted and the offspring are more likely to have inferior fitnesses. When schema disruption is not controlled, genetic algorithms will often fail to converge on an optimal solution [91, 90]. In a particularly interesting study, the success of a genetic algorithm was improved by recording where past crossovers resulted in fit offspring [116]. This information was used to bias crossovers in future generations. In this way, selection automatically biased the recombination markers towards regions that separated schemas. Extending these results to biology, this simulation demonstrates the advantage of shifting introns towards the regions that separate structural schemas.

3.2 EXONS AS FUNCTIONAL SWITCHES

Exon swapping can occur on evolutionary timescales or on the timescale of gene splicing in the cell [53]. The ability to swap exons without disrupting the structure improves the evolvability of the gene by promoting functional switches between different molecular properties [20, 49]. These switches have been found to alter the substrate specificity, the tissue distribution, and the association properties of the translated proteins. It has been suggested that performing exon swapping in vitro will produce functionally diverse libraries [46, 73].

There have been several examples of achieving functional diversity through the in vitro swapping of exons that correspond to structural modules of enzymes. Go and co-workers altered the coenzyme specificity of isocitrate dehydrogenase by calculating the structural module corresponding to the NADP-binding site [146]. When this module was swapped with a NAD-binding site, the reaction was shown to proceed with the new coenzyme. In another experiment, a module of the β-subunit of hemoglobin was swapped with the corresponding module of the α-subunit [63, 139]. This hybrid protein folded into the correct tertiary structure, but the association of different subunits was altered, suggesting that the function of the fourth module is to regulate subunit association. This substitution did not affect other properties of hemoglobin, including oxygen binding. In a particularly dramatic experiment, the catalytic activities of α-lactalbumin and lysozyme were swapped by shuffling the exon corresponding to the amino acids that surround their active sites [75]. The success of this experiment hinged on the observation that the two enzymes share the same structure and distribution of exons. Swapping exons can also alter specificity. For example, the swapping of alternate exons in a human cytochrome P450 changed its substrate specificity and tissue distribution [32]. This implies that the gene structure of P450 promotes the swapping of functional modules such that this enzyme can participate in different biological functions. In the dehydrogenase, hemoglobin, and lysozyme experiments, subportions of the structural module were swapped as controls. In each case, swapping a portion rather than the whole module resulted in an unstable or nonfunctional enzyme.

The immune system effectively uses exon shuffling to create antibody variants that can bind a broad range of antigens. Mimicking in vivo antibody selection, Borrebaeck and co-workers used recombination techniques to shuffle the naturally-occurring human exons that encode the CDR regions to generate a large binding repertoire [124]. The library containing 10^9 antibodies was screened against a wide array of hapten and protein targets, and antibodies with nanomolar binding affinities were reliably found. This stunning work represents the ability to create a full antibody repertoire in the test tube. When combined with directed-evolution-like somatic mutagenesis, a nearly complete artificial immune system will be created.

3.3 IN VITRO RECOMBINATION PRESERVES STRUCTURAL SCHEMA

The success of in vitro recombination is based on the assumption that the parents share similar structures. For a hybrid protein to demonstrate new or improved properties, a prerequisite is that it fold into a well-defined (and presumably similar) structure. Therefore, crossovers are more likely to be successful when they occur in regions that lie between schemas [132]. In this context, schema are defined by the pattern of stabilizing interactions between amino acids and recombination is most successful when the crossovers break the fewest interactions. The hybrids with the minimum schema disruption are the most likely to retain the structure of the parents.

Using a computational algorithm that predicts the location of schemas, data were analyzed from five independent directed evolution experiments where several parents were shuffled to create random libraries of recombinant offspring [132]. Crossovers in the offspring that survived selection were strongly biased towards regions that minimize the schema disruption. To further demonstrate the requirement that schema be preserved, two β-lactamases were recombined that have similar structures, but share little sequence identity. The three-

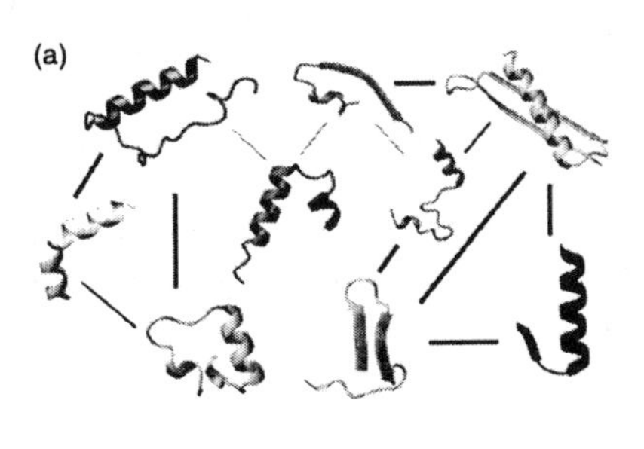

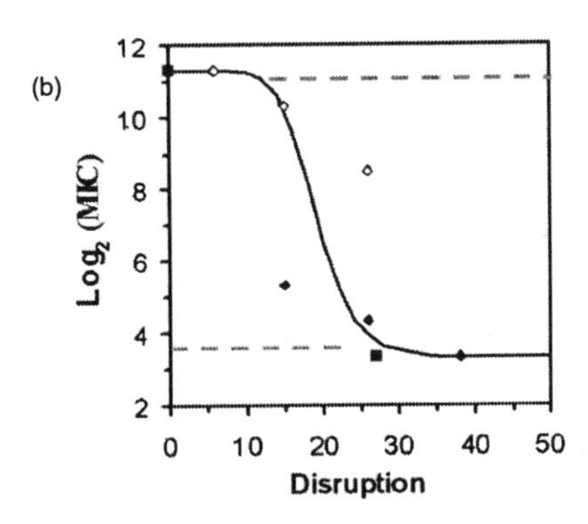

FIGURE 5 The division of beta-lactamase TEM-1 into shuffleable domains (a) and the results of experiments recombining TEM-1 with PSE-4 (b) [132]. The domains are color-coded and the width of the bars is proportional to the number of interactions between domains. The domains were swapped experimentally and the activity of beta-lactamase (measured as the minimum inhibitory concentration, MIC) was recorded as a function of the number of disrupted interactions. The dashed line at the top of the graph is the wildtype activity and the lower line is the MIC of the cells, corresponding to no measurable beta-lactamase activity. A very sharp transition in activity is observed as the disruption increases past a threshold. Libraries can be rich with functional hybrid proteins when the disruption is maintained below this transition point.

dimensional structure was divided into schemas and the interaction strengths between the different schemas were calculated. Experimentally, hybrid proteins were constructed where the schemas were exchanged between structures and each hybrid was tested for activity. A sharp transition was found in the activity as the disruption of the hybrid increased (fig. 5). Recombination events that cause disruption above this threshold resulted in nonfunctional hybrids. These experiments demonstrate in real-time how selection for folded, function offspring can bias crossover-focusing mechanisms towards regions separating structural schemas. In addition, this algorithm will improve directed evolution as it enables the design of libraries with an enriched fraction of properly folded hybrid proteins.

4 EVOLUTION OF GENETIC CIRCUITS AND METABOLIC PATHWAYS

Metabolic pathways and genetic circuits have recently become targets for in vitro evolution [85, 117]. The robustness of a network describes the resilience of a behavior to perturbations in parameters. In the case of a metabolic pathway, the behavior may be the chemicals generated and the rate at which they are produced. For a genetic circuit, the behavior is the integrity of the computation, for example, the ability to behave like an oscillator, toggle switch, logic gate, or memory [19, 44, 47, 149]. A network is evolvable if it can change behaviors through the perturbation of its internal parameters.

There are several means by which a diverse library of networks can be created. One method involves the randomization of the component genes through mutagenesis or recombination. Mutations can change the behavior of a network by altering the kinetic constants for an activity, the substrate specificity, and the products produced. If the mutated DNA encodes a repressor or activator protein or a related DNA binding site, then mutations will vary the strength of repression or activation [16]. Further, mutations can stabilize or destabilize a protein, which affects the dynamics of the network by changing the protein's residence time. The library of offspring is more likely to contain the desired properties if the diversity is applied to those parameters that are evolvable or those parameters that are robust regarding properties required for the desired behavior, but not being explicitly optimized.

A combinatorial library of networks can also be created by randomly combining modules of preconstructed combinations of transcription units. For example, the combination of different genes with different repressor sites can produce a library of many possible dynamic circuits. In this way, different interaction topologies can be created and tested for various behaviors. For this combinatorial approach to be successful, it is necessary that the modules are robust regarding the form of the inputs and the network and cellular environments in which they are inserted [113, 133].

4.1 ROBUST NETWORK TOPOLOGIES

The topology of a network describes the architecture of interactions between the network components. For a metabolic pathway, an interaction may represent the enzymatic conversion of a substrate into a product or the effect of a species on the control of a reaction. An interaction may also describe the effect of a repressor or activator on the expression of another gene. Network topologies can be visualized as a set of nodes and edges, where a node represents a component and an edge represents an interaction (fig. 1). The topologies of large networks can be very complex and it is difficult to predict those features that are robust from those that are fragile (fig. 6). Some topological motifs that confer robustness have been identified, including feedback loops, a modular architecture, and a scale-free distribution of nodes and edges. Alternative topologies that contained multiple, small feedback loops in sequence or nested, proved to be less robust.

Feedback control buffers the variables of the system towards external perturbations; in other words, it improves the stability of the system. Beyond improving stability, feedback control also improves robustness with respect to internal parameters [111]. Feedback control can occur with various topologies. The simplest,

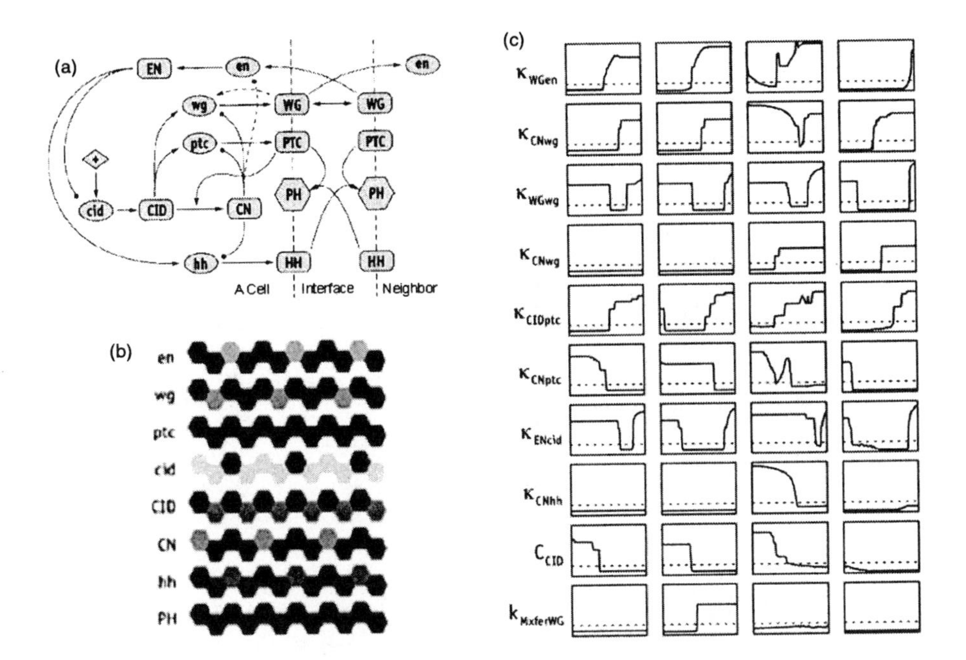

FIGURE 6 The segment polarization network of *Drosophila*, as modeled by Odell and co-workers [133]. (a) The topology of the network showing the interactions between proteins, including intercellular interactions. The behavior of this network is to produce a specific pattern of expression for a group of cells (b). The wildtype behavior was found to be remarkably robust. Often, the parameters controlling the network could be varied by several orders of magnitude (c). In addition to the robustness, it was found that varying some parameters could change the behavior of the network (different expression patterns), indicating that the network is evolvable.

autoregulation, where the immediate products inhibit the reaction, has been shown to improve robustness [15, 112]. Larger feedback loops, when downstream products control the first steps of metabolism, proved to be more robust [111].

Barkai and Leibler proposed that robustness is manifested in signal transduction networks through the topology of feedback control [14]. They constructed a model of bacterial chemotaxis and found a set of parameters that reproduced the desired network behavior, in this case, to react properly to an attractant gradient. This network turned out to be remarkably robust as nearly 80% of networks generated by randomly varying all of the parameters two-fold from the starting point were found to demonstrate the correct behavior. When varied individually, each parameter could be varied by several orders of magnitude. This robust-

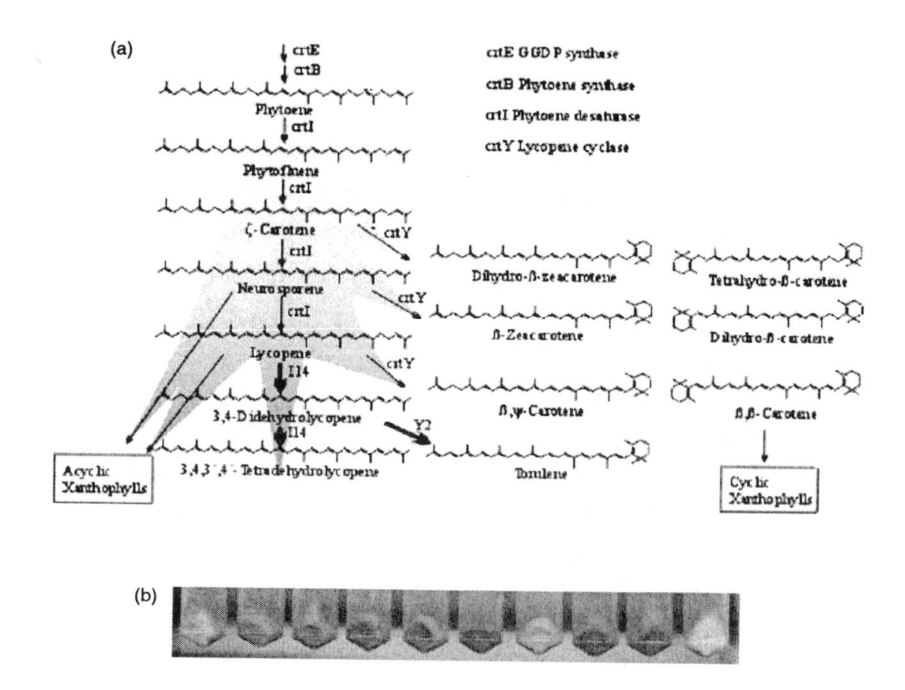

FIGURE 7 The range of carotenoids (a) and their colors (b) that can be created through directed evolution [117]. The *crtE* and *crtB* enzymes create the initial C40 backbone (top, left). This can then be transformed into different carotenoids by desaturating bonds with *crtI* and cyclizing the ends with *crtY*. By evolving the properties or *crtI* and *crtY*, the production of a variety of carotenoids was possible.

ness was later verified experimentally [4] and proved to result from an integral feedback control loop [147]. This result supports the concept that robustness is inherent to the topology of the network and does not emerge through a specific combination of parameter values.

The robustness of a network topology towards the removal of nodes or edges, as the result of knock-out or specificity-altering mutations, has been studied using a number of theoretical models [3, 33, 34, 64]. A general result from these studies is that a scale-free topology of interactions tends to be robust. In a scale-free network, the distribution of nodes and edges follows a power law, $P(k) \sim k^{-\alpha}$, where $P(k)$ is the probability of a node having connectivity k, and a is a power-log exponent. Considering robustness, there is an optimal value of $\alpha \approx 2.5$, which is shared by the structure of the internet [33] and metabolic networks [137]. In a comparison across metabolic networks of 43 species, it was found that the most highly connected nodes were the most conserved, whereas the least connected varied considerably [64]. In other words, the least connected nodes are highly tolerant to changes, similar to what was found for protein structures [130].

4.2 EVOLVABILITY OF NETWORKS

An evolvable network has the ability to sample many behaviors by varying the internal parameters. To be robust, a network has to retain some behavior under parameter variation. In contrast, evolvability requires that other behaviors are attainable without traversing regions of parameter space that are devoid of behavior. Very little is understood as to how the topology of a network affects evolvability. There is some evidence that the evolvability of a network can be improved through modularity, switch points, and broad substrate specificities. Understanding these properties will facilitate the creation of in vitro strategies to create libraries of networks with diverse properties.

When shuffling portions of gene networks, it is essential that the subcomponents are modular. A portion of a genetic network forms a module if it can be substituted in other environments or other networks and performs the same qualitative behavior. If a network is modular, evolution can reuse motifs by rewiring the network inputs and outputs [56]. In terms of the distribution of nodes and edges, a modular topology can be divided into subsets of nodes, where the nodes are highly interacting within the subset and not interacting between subsets. In addition, the network should generate the same behavior for a variety of input stimuli [133]. This reduces the demands on the form of the inputs, making it easier to combine with other subnetworks. Using kinetic models, Bhalla and Iyengar found that a wide range of complex dynamic behaviors was attainable by coupling multiple, independent signaling pathways [19]. The behaviors that could be obtained included memory, timers, switching regulatory wiring, and time and concentration thresholds for response.

Stochastic mechanisms can alter the behavior of a network [86, 87]. The random component can be introduced by low concentrations of reacting species, slow reaction rates, or limited availability of catalytic centers. Biological systems exploit this randomness as a switch that can determine the behavior of an individual cell or create diversity in a population in cells. For a network to be able to utilize stochastic effects, it is necessary to stabilize the effect of fluctuations on the remainder of the network through redundancy and feedback loops. In other words, maintaining the overall robustness of the network promotes evolvability through the variation of other parameters. Stochastic switches that select between two alternate pathways can occur [8, 86]. Identifying a switch point and then targeting it using directed evolution could be used to create a diverse library of network behaviors.

Gene duplication creates functional redundancy. The robustness conferred by a duplication event is likely to deteriorate rapidly as there is no disadvantage for the function of one of the copied genes to be destroyed [96, 134, 135]. However, the evolutionary stability of the duplication can be ensured initially by increasing regulatory reliability or by changing the function of the duplicated gene. This creates the opportunity for the duplicated gene to add or optimize functions, possibly in unrelated pathways [86]. An ancestral gene with broad substrate specificity can potentially participate in more pathways and is, therefore, more evolvable. This mechanism has been duplicated in vitro by demonstrating that a metabolic enzyme with broad substrate specificity could be evolved to specifically participate in two different metabolic pathways [67].

Secondary metabolites are produced by various organisms and are typically not essential for survival. These chemicals may confer properties such as the color or fragrance or they may have therapeutic properties, such as antibiotic or tumor-suppressing activities, or serve in biological warfare or defense. While many of the chemicals produced by an organism may not confer a selective advantage, the existence of a secondary metabolism is advantageous because it gives the organism the potential to discover a few potent chemicals [45]. To optimize the discovery process, secondary metabolisms appear to have evolved to maximize the diversity of chemicals that are attainable when the components of the pathway are perturbed. The evolvability of secondary metabolic enzymes is enhanced by low substrate specificities, thus increasing the number of potential downstream products. These specificities can be tuned and fixed by evolution as beneficial metabolites are discovered. The components of secondary metabolism have been observed to be robust to environmental changes. This facilitates the transfer of a network from one organism to another.

The robustness and evolvability of secondary pathways makes them particularly well-suited targets for in vitro evolution to tune the production of desired chemicals. New biosynthetic pathways can be created by combining genes from different sources and subjecting these genes to mutagenesis and recombination. Selection is then applied for the production of specific compounds or for the generation of diverse chemicals. Directed evolution has been applied to the carotenoid biosynthetic pathway to alter the diversity of the chemicals produced [117]. Four genes that encode different component enzymes were combined. The basic C40 carotenoid building block is constructed by two synthases. The inclusion of a phytoene desaturase (*crtI*) and lycopene cyclase (*crtY*) allowed the C_{40} backbone to be modified to produce distinct carotenoids. These two modifying enzymes control the number of double bonds and the formation of cyclic rings, respectively. Libraries of *crtI* and *crtY* genes were created through the recombination of homologous genes. The production of different carotenoids could be visualized by the change in color caused by these modifications. Mutants were found that produced carotenoids with different degrees of desaturation (fewer double bonds results in a yellowish color, more double bonds results in a pinkish color) and different degrees of cyclization (resulting in an orange-red to purple-red color). One clone produced torulene, a carotenoid that had not been previously observed in organisms that contain these parental biosynthetic genes. Organisms that produce torulene in nature do so by a different metabolic route. These experiments demonstrate that applying directed evolution to a secondary metabolic pathway can alter the range of chemicals produced by the network and extend the product diversity beyond what is produced by the parents.

5 CONCLUSIONS

Understanding the robustness and evolvability of biological systems is essential for constructing algorithms that guide evolutionary design strategies. To quantify the potential for evolutionary improvement, stability theory is useless as it is concerned with the resilience of a single state of a system when variables are perturbed. In contrast, evolution samples many states by perturbing the system's internal parameters, such as kinetic constants, interaction topologies, and component stabilities. A robust system has the ability to sample many of these states while preserving the fundamental behaviors of the system. During this process of drifting through parameter space, new behaviors can be sampled. Some systems have a higher capacity for such change and are, therefore, evolvable.

The robust and evolvable properties of three hierarchies in biology have been reviewed: the mutation of individual genes, the recombination of clusters of amino acids from different genes, and the evolution of biological networks. While these systems have some fundamental differences, there are many common motifs that confer robustness and evolvability. In all of these systems, there are similar motifs that improve robustness, such as the distribution of interactions, modularity, the separation of parameters (additivity), and redundancy. These themes are common to biological systems not reviewed here, such as RNA structures, viruses, genomes, and whole ecologies. An understanding of the basis for robust and evolvable systems in biology will facilitate the design of a new generation of techniques in directed evolution.

REFERENCES

[1] Adami, C., C. Ofria, and T. C. Collier. "Evolution of Biological Complexity." *Proc. Natl. Acad. Sci. USA* **97** (2000): 4463–4468.

[2] Aita, T., and Y. Husimi. "Theory of Evolutionary Molecular Engineering through Simultaneous Accumulation of Advantageous Mutations." *J. Theor. Biol.* **207** (2000): 543–556.

[3] Albert, R., H. Jeong, and A.-L.Barabasi. "Error and Attack Tolerance of Complex Networks." *Nature* **406** (2000): 378-382.

[4] Alon, U., M. G. Surette, N. Barkai, and S. Leibler. "Robustness in Bacterial Chemotaxis." *Nature* **397** (1999): 168–171.

[5] Altamirano, M. M., J. M. Blackburn, C. Aguayo, and A. R. Ferscht. "Directed Evolution of New Catalytic Activity using the a/b-barrel Scaffold." *Nature* **403** (2000): 617–622.

[6] Alves, R., and M. A. Savageau. "Comparing Systemic Properties of Ensembles of Biological Networks by Graphical and Statistical Methods." *Bioinformatics* **16** (2000): 527–533.
[7] Ancel, L. W., and W. Fontana. "Plasticity, Evolvability, and Modularity in RNA." *J. Exp. Zool.* **288** (2000): 242–283.
[8] Arkin, A., J. Ross, and H. H. McAdams. "Stochastic Kinetic Analysis of Developmental Pathway Bifurcation in Phage l-Infected *Escherichia coli* Cells." *Genetics* **149** (1998): 1633–1648.
[9] Arnold, F. H. "Combinatorial and Computational Challenges for Biocatalyst Design." *Nature* **409** (2001): 253–257.
[10] Arnold, F. H., ed. *Evolutionary Protein Design*. San Diego, CA: Academic Press, 2001.
[11] Aronsson, H. G., W. E. Royer, Jr., and W. A. Hendrickson. "Quantification of Tertiary Structural Conservation Despite Primary Sequence Drift in the Globin Fold." *Protein Science* **3** (1994): 1706–1711.
[12] Axe, D. D., N. W. Foster, and A. R. Fersht. "Active Barnase Variants with Completely Random Hydrophobic Cores." *Proc. Natl. Acad. Sci. USA* **93** (1996): 5590–5594.
[13] Baase, W. A., N. C. Gassner, X.-J. Zhang, R. Kuroki, L. H. Weaver, D. E. Tronrud, and B. W. Matthews. "How Much Sequence Variation Can the Functions of Biological Molecules Tolerate?" In *Simplicity and Complexity in Proteins and Nucleic Acids*, edited by H. Frauenfelder, J. Deisenhofer, and P. G. Wolynes, 297–311. Dahlem University Press, 1999.
[14] Barkai, N., and S. Leibler. "Robustness in Simple Biochemical Networks." *Nature* **387** (1997): 913–917.
[15] Becskei, A., and L. Serrano. 1998
[16] Becskei, A., and L. Serrano. "Engineering Stability in Gene Networks by Autoregulation." *Nature* **405** (2000): 590–593.
[17] Berek, C., and C. Milstein. "Mutation Drift and Repertoire Shift in the Maturation of the Immune Response." *Immunol. Rev.* **96** (1987): 23–41.
[18] Betz, A. G., C. Rada, R. Pannell, R. Milstein, and M. S. Neuberger. "Passenger Transgenes Reveal Intrinsic Specificity of the Antibody Hypermutation Mechanism: Clustering, Polarity, and Specific Hot Spots." *Proc. Natl. Acad. Sci. USA* **90** (1993): 2385–2388.
[19] Bhalla, U. S., and R. Iyengar. "Emergent Properties of Networks of Biological Signal Pathways." *Science* **283** (1999): 381–387.
[20] Blake, C. C. F. "Exons Encode Protein Functional Units." *Nature* **277** (1979): 598–598.
[21] Boder et al. 2000
[22] Bolon, D. N., and S. L. Mayo. "Enzyme-like Proteins by Computational Design." *Proc. Natl. Acad. Sci. USA* **98** (2001): 14274–14279.
[23] Bornberg-Bauer, E., and H. S. Chan. "Modeling Evolutionary Landscapes: Mutational Stability, Topology, and Superfunnels in Sequence Space." *Proc. Natl. Acad. Sci. USA* **96** (1999): 10689–10694.

[24] Bornholdt, S., and T. Rohlf. "Topological Evolution of Dynamical Networks: Global Criticality from Local Dynamics." *Phys. Rev. Lett.* **84** (2000): 6114–6117.

[25] Broglia, R. A., G. Tiana, H. E. Roman, E. Vigezzi, and E. Shakhnovich. "Stability of Designed Proteins Against Mutations." *Phys. Rev. Lett.* **82** (1999): 4727–4730.

[26] Brown, M., M. B. Rittenberg, C. Cheng, and V. A. Roberts. "Tolerance to Single, but not Multiple, Amino Acid Replacements in Antibody V_H CDR2." *J. Immunol.* **156** (1996): 3285–3291.

[27] Buchler, N. E. G., and R. A. Goldstein. "Universal Correlation between Energy Gap and Foldability for the Random Energy Model and Lattice Proteins." *J. Chem. Phys.* **111** (1999): 6599–6609.

[28] Burks, E. A., G., Chen, G. Georgiou, and B. L. Iverson. "In vitro Scanning Saturation Mutagenesis of an Antibody Binding Pocket." *Proc. Natl. Acad. Sci. USA* **94** (1997): 412–417.

[29] Callaway, D. S., M. E. J. Newman, S. H. Strogatz, and D. J. Watts. "Network Robustness and Fragility: Percolation on Random Graphs." *Phys. Rev. Lett.* **85** (2000): 5468–5471.

[30] Campbell, I. D., and M. Baron. "The Structure and Function of Protein Modules." *Phil. Trans. Roy. Soc. Lond. B.* **332** (1991): 165–170.

[31] Carlson, J. M., and J. Doyle. "Highly Optimized Tolerance: Robustness and Design in Complex Systems." *Phys. Rev. Lett.* **84** (2000): 2529–2532.

[32] Christmas, P., J. P. Jones, C. J. Patten, D. A. Rock, Y. Zheng, S.-M. Cheng, B. M. Weber, N. Carlesso, D. T. Scadden, A. E. Rettie, and R. J. Soberman. "Alternative Splicing Determines the Function of CYP4F3 by Switching Substrate Specificity." *J. Biol. Chem.* **41** (2001): 38166–38172.

[33] Cohen, R., K. Erez, D. Ben-Avraham, and S. Havlin "Resilience of the Internet to Random Breakdowns." *Phys. Rev. Lett.* **85** (2000): 4626–4628.

[34] Cohen, R., K. Erez, D. Ben-Avraham, and S. Havlin "Breakdown of the Internet under Intentional Attack." *Phys. Rev. Lett.* **86** (2001): 3682–3685.

[35] Cowell, L. G., H.-J. Kim, T. Humaljoki, C. Berek, and T. B. Kepler. "Enhanced Evolvability in Immunoglobin V Genes under Somatic Hypermutation." *J. Mol. Evol.* **49** (1999): 23–26.

[36] Crameri, A., S.-A. Raillard, E. Bermudez, and W. P. C. Stemmer. "DNA Shuffling of a Family of Genes from Diverse Species Accelerates Directed Evolution." *Nature* **391** (1998): 288–291.

[37] Dahiyat, B. I., and S. L. Mayo. "De Novo Protein Design: Fully Automated Sequence Selection." *Science* **278** (1997): 82–87.

[38] Daugherty, P. S., G. Chen, B. I. Iverson, and G. Georgiou. "Quantitative Analysis of the Effect of the Mutation Frequency on the Affinity Maturation of Single Chain Fv Antibodies." *Proc. Natl. Acad. Sci. USA* **97** (2000): 2029–2034.

[39] de Souza, S. J., M. Long, L. Schoenbach, S. W. Roy, and W. Gilbert. "Intron Positions Correlate with Module Boundaries in Ancient Proteins." *Proc. Natl. Acad. Sci. USA* **93** (1996): 14632–14636.

[40] de Souza, S. J., M. Long, R. J. Klein, S. Roy, S. Lin, and W. Gilbert. "Toward a Resolution of the Introns Early/Late Debate: Only Phase Zero Introns are Correlated with the Structure of Ancient Proteins." *Proc. Natl. Acad. Sci. USA* **95** (1998): 5094–5099.

[41] Edwards, R., and L. Glass. "Combinatorial Explosion in Model Gene Networks." *Chaos* **10** (2000): 691–704.

[42] Eigen, M., and J. McCaskill. "The Molecular Quasi-Species." *Adv. Chem. Phys.* **75** (1989): 149–263.

[43] El Hawrani, A. S., K. M. Moreton, R. B. Sessions, A. R. Clarke, and J. J. Holbrook. "Engineering Surface Loops of Proteins—A Preferred Strategy for Obtaining New Enzyme Function." *TIBTECH* **12** (1994): 207–211.

[44] Elowitz, M. B., and S. Leibler. "A Synthetic Oscillatory Network of Transcriptional Regulators." *Nature* **403** (2000): 335–338.

[45] Firn, R. D., and C. G. Jones. "The Evaluation of Secondary Metabolism—A Unifying Model." *Mol. Microbiol.* **37** (2000): 989–994.

[46] Fisch, I., R. E. Kontermann, R. Finnern, O. Hartley, A. S. Soler-Gonzalez, A. D. Griffiths, and G. Winter. "A Strategy of Exon Shuffling for Making Large Peptide Repertoires Displayed on Filamentous Bacteriophage." *Proc. Natl. Acad. Sci. USA* **93** (1996): 7761–7766.

[47] Gardner, T. S., C. R. Cantor, and J. J. Collins. "Construction of a Genetic Toggle Switch in *Escherichia coli*." *Nature* **403** (2000): 339–342.

[48] Germain, R. N. "The Art of the Probable: System Control in the Adaptive Immune System." *Science* **293** (2001): 240–245.

[49] Gilbert, W. "Why Genes in Pieces?" *Nature* **271** (1978): 501–501.

[50] Gilbert, W., S. J. de Souza, and M. Long. "Origin of Genes." *Proc. Natl. Acad. Sci. USA* **94** (1997): 7698–7703.

[51] Go, M. "Correlation of DNA Exonic Regions with Protein Structural Units in Haemoglobin." *Nature* **291** (1981): 90–92.

[52] Go, M. "Modular Structural Units, Exons, and Function in Chicken Lysozyme." *Proc. Natl. Acad. Sci. USA* **80** (1983): 1964–1968.

[53] Go, M. "Protein Structures and Split Genes." *Adv. Biophys.* **19** (1985): 91–131.

[54] Govindarajan, S., and R. A. Goldstein. "Evolution of Model Proteins on a Foldability Landscape." *Proteins* **29** (1997): 461–466.

[55] Hartman, J. L., IV, B. Garvik, and L. Hartwell. "Principles for the Buffering of Genetic Variation." *Science* **291** (2001): 1001–1004.

[56] Hartwell, L. H., J. J. Hopfield, S. Leibler, and A. W. Murray. "From Molecular to Modular Biology." *Nature* **402** (1999): C47–C52.

[57] Hedstrom, L., L. Szilagyi, and W. J. Rutter. "Converting Trypsin to Chymotrypsin: The Role of Surface Loops." *Science* **255** (1992): 1249–1253.

[58] Holland, J. H. *Adaptation in Natural and Artificial Systems.* Cambridge, MA: MIT Press, Cambridge, 1975.

[59] Huang, W., and T. Palzkill. "A Natural Polymorphism in β-lactamase is a Global Suppressor." *Proc. Natl. Acad. Sci. USA* **94** (1997): 8801–8806.

[60] Huang, W., J. Petrosino, M. Hirsch, P. S. Shenkin, and T. Palzkill. "Amino Acid Sequence Determinants of Beta-lactamase Structure and Activity." *J. Mol. Biol.* **258** (1996): 688–703.

[61] Huynen, M. A. "Exploring Phenotype Space through Neutral Evolution." *J. Mol. Evol.* **43** (1996): 165–169.

[62] Huynen, M. A. "Smoothness within Ruggedness: The Role of Neutrality in Adaptation." *Proc. Natl. Acad. Sci. USA* **93** (1998): 397–401.

[63] Inaba, K., K. Wakasugi, K. Ishimori, T. Konno, M. Kataoka, and I. Morishima. "Structural and Functional Roles of Modules in Hemoglobin." *J. Biol. Chem.* **272** (1997): 30054–30060.

[64] Jeong, H., B. Tombor, R. Albert, Z. N. Oltvai, and A.-L. Barabasi. "The Large-Scale Organization of Metabolic Networks." *Nature* **407** (2000): 651–654.

[65] Jermutus, L., M. Tessier, L. Pasamontes, A. P. G. M. van Loon, and M. Lehmann. "Structure-based Chimeric Enzymes as an Alternative to Directed Evolution: Phytase as a Test Case." *J. Biotechnology* **85** (2001): 15–24.

[66] Jucovic, M., and A. R. Poteete. "Protein Salvage by Directed Evolution: Functional Restoration of a Defective Lysozyme Mutant." *Ann. NY Acad. Sci.* **870** (1999): 404–407.

[67] Jürgens, C., A. Strom, D. Wegener, S. Hettwer, M. Wilmanns, and R. Sterner. "Directed Evolution of a $(\beta\alpha)_8$-barrel Enzyme to Catalyze Related Reactions in Two Different Metabolic Pathways." *Proc. Natl. Acad. Sci. USA* **97** (2000): 9925–9930.

[68] Kaneko, S., S. Iwamatsu, A. Kuno, Z. Fujimoto, Y. Sato, K. Yura, M. Go, H. Mizuno, K. Taira, T. Hasegawa, I. Kusakabe, and K. Hayashi. "Module Shuffling of a Family F/10 Xylanase: Replacement of Modules M4 and M5 of the FXYN of *Streptomyces olivaceoviridis* E-86 with those of the Cex of *Cellomonas fimi.*" *Protein Eng.* **13** (2000): 873–879.

[69] Kauffman, S. A., and S. Levin. "Towards a General Theory of Adaptive Walks on Rugged Landscapes." *J. Theor. Biol.* **128** (1987): 11–45.

[70] Kauffman, S. A., and W. G. Macready. "Search Strategies for Applied Molecular Evolution." *J. Theor. Biol.* **173** (1995): 427–440.

[71] Kauffman, S. A., and E. D. Weinberger. "The NK Model of Rugged Fitness Landscapes and Its Application to Maturation of the Immune Response." *J. Theor. Biol.* **141** (1989): 211–245.

[72] Kirschner, M., and J. Gerhart. "Evolvability." *Proc. Natl. Acad. Sci. USA* **95** (1998): 8420–8427.

[73] Kolkman, J. A., and W. P. C. Stemmer. "Directed Evolution of Proteins by Exon Shuffling." *Nature Biotechnology* **19** (2001): 423–428.

[74] Kono, H., and J. G. Saven. "Statistical Theory for Protein Conformational Libraries. Packing Interactions, Backbone Flexibility, and the Sequence Variability of a Main-Chain Structure." *J. Mol. Biol.* **306** (2001): 607–628.

[75] Kumagai, I., S. Takeda, and K.-I. Miura. "Functional Conversion of the Homologous Proteins a-lactalbumin and Lysozyme by Exon Exchange." *Proc. Natl. Acad. Sci. USA* **89** (1992): 5887–5891.

[76] Lauffenberger, D. A. "Cell Signaling Pathways as Control Modules: Complexity for Simplicity?" *Proc. Natl. Acad. Sci. USA* **97** (2000): 5031–5033.

[77] Lehman, M., L. Pasamontes, S. F. Lassan, and M. Wyss. "The Consensus Concept for Thermostability Engineering of Proteins." *Biochemica et Biophysica Acta* **1543** (2000): 408–415.

[78] Li, H., R. Helling, C. Tang, and N. Wingreen. "Emergence of Preferred Structures in a Simple Model of Protein Folding." *Science* **273** (1996): 666–669.

[79] Lo, N. M., P. Holliger, and G. Winter. "Mimicking Somatic Hypermutation: Affinity Maturation of Antibodies Displayed on Bacteriophage using a Bacterial Mutator Strain." *J. Mol. Biol.* **260** (1996): 359–368.

[80] Lockless, S. W., and R. Ranganathan. "Evolutionary Conserved Pathways of Energetic Connectivity in Protein Families." *Science* **286** (1999): 295–299.

[81] Loeb, D. D., R. Swanstrom, L. E. Everitt, M. Manchester, S. E. Stamper, and C. A. Hutchson, III. "Complete Mutagenesis of the HIV-1 Protease." *Nature* **340** (1989): 397–400.

[82] Macken, C. A., P. S. Hagan, and A. S. Perelson. "Evolutionary Walks on Rugged Landscapes." *SIAM J. Appl. Math.* **51** (1991): 799–827.

[83] Matsumura, I., and A. D. Ellington. "In vitro Evolution of beta-glucuronidase into a beta-galactosidase Proceeds through Non-specific Intermediates." *J. Mol. Biol.* **305** (2001): 331–339.

[84] Matsuura, T., T. Yomo, S. Trakulnaleamsai, Y. Ohashi, K. Yamamoto, and I. Urabe. "Nonadditivity of Mutational Effects on the Properties of Catalase I and Its Application to Efficient Directed Evolution." *Protein Engineering* **11** (1998): 789–795.

[85] McAdams, H. H., and A. Arkin. "Gene Regulation: Towards a Circuit Engineering Discipline." *Curr. Biol.* **10** (2000): R318–R320.

[86] McAdams, H. H., and A. Arkin. "It's a Noisy Business! Genetic Regulation at the Nanomolar Scale." *TIG* **15** (1999): 65–69.

[87] McAdams, H. H., and A. Arkin. "Stochastic Mechanisms in Gene Expression." *Proc. Natl. Acad. Sci. USA* **94** (1997): 814–819.

[88] Mélin, R., H. Li, N. S. Wingreen, and C. Tang. "Designability, Thermodynamic Stability, and Dynamics in Protein Folding." *J. Chem. Phys.* **110** (1999): 1252–1262.

[89] Mezey, J. G., J. M. Cheverud, and G. P. Wagner. "Is the Genotype-Phenotype Map Modular?: A Statistical Approach using Mouse Quantitative Trait Loci Data." *Genetics* **156** (2000): 305–311.

[90] Mitchell, M. *An Introduction to Genetic Algorithms.* Cambridge, MA: MIT Press, 1998.

[91] Mitchell, M., J. H. Holland, and S. Forrest. "When Will a Genetic Algorithm Outperform Hill Climbing?" In *Advances in Neural Information Processing Systems 6*, edited by J. Cowan, G. Tesauro, and J. Alspector. San Francisco, CA: Morgan Kauffman, 1994.

[92] Miyazaki, K., and F. H. Arnold. "Exploring Nonnatural Evolutionary Pathways by Saturation Mutagenesis: Rapid Improvement of Protein Function." *J. Mol. Evol.* **49** (1999): 716–720.

[93] Moore, J. C., H.-M. Jin, O. Kuchner, and F. H. Arnold. "Strategies for the in vitro Evolution of Protein Function: Enzyme Evolution by Random Recombination of Improved Sequences." *J. Mol. Biol.* **272** (1997): 336–347.

[94] Ness, J. E., M. Welch, L. Giver, M. Bueno, J. R. Cherry, T. V. Borchert, W. P. C. Stemmer, and J. Minshull. "DNA Shuffling of Subgenomic Sequences of Subtilisin." *Nature Biotech.* **17** (1999): 893–896.

[95] Neuberger, M. S., and C. Milstein. "Somatic Hypermutation." *Curr. Opin. Immun.* **7** (1995): 248–254.

[96] Nowak, M. A., M. C. Boerlijst, J. Cooke, and J. M. Smith. "Evolution of Genetic Redundancy." *Nature* **388** (1997): 167–171.

[97] Oprea, M., and T. B. Kepler. "Genetic Plasticity of V Genes under Somatic Hypermutation: Statistical Analysis using a New Resampling-Based Methodology." *Genome Res.* **9** (1999): 1294–1304.

[98] Orencia, C., M. A. Hanson, and R. C. Stevens. "Structural Analysis of Affinity Matured Antibodies and Laboratory-Evolved Enzymes." *Advances in Protein Chemistry* **55** (2000): 227–259.

[99] Orencia, M. C., J. S. Yoon, J. E. Ness, W. P. C. Stemmer, and R. C. Stevens. "Predicting the Emergence of Antibiotic Resistance by Directed Evolution and Structural Analysis." *Nature Struct. Biol.* **8** (2001): 238–242.

[100] Orengo, C. A., D. T. Jones, and J. M. Thornton. "Protein Superfamilies and Domain Superfolds, +." *Nature* **372** (1994): 631–634.

[101] Palzkill T., Q.-Q. Le, K. V. Venkatachalam, M. LaRocco, and H. Ocera. "Evolution of Antibiotic Resistance: Several Different Amino Acid Substitutions in an Active Site Loop Alter the Substrate Profile of β-lactamase." *Mol. Microbiology* **12** (1994): 217–229.

[102] Panchenko, A. R., Z. Luthey-Schulten, and P. G. Wolynes. "Foldons, Protein Structural Units, and Exons." *Proc. Natl. Acad. Sci. USA* **93** (1996): 2008–2013.

[103] Parisi, G., and J. Enchave. "Structural Constraints and Emergence of Sequence Patterns in Protein Evolution." *Mol. Biol. Evol.* **18** (2001): 750–756.

[104] Patel, P. H., and L. A. Loeb. "DNA Polymerase Active Site is Highly Mutable: Evolutionary Consequences." *Proc. Natl. Acad. Sci. USA* **97** (2000): 5095–5100.

[105] Petit, A., L. Maveyraud, F. Lenfant, J.-P. Samama, R. Labia, and J.-M. Masson. "Multiple Substitutions at Position 104 of β-lactamase TEM-1: Assessing the Role of this Residue in Substrate Specificity." *Biochem J.* **305** (1995): 33–40.

[106] Petrosino, J., C. Cantu, III, and T. Palzkill. "β-lactamases: Protein Evolution in Real Time." *Trends in Microbiology* **6** (1998): 323–327.

[107] Powell, S. K., M. A. Kaloss, A. Pinkstaff, R. McKee, I. Burimski, M. Pensiero, E. Otto, W. P. C. Stemmer, and N.-W. Soong. "Breeding of Retroviruses by DNA Shuffling for Improved Stability and Processing Yields." *Nature Biotech.* **18** (2000): 1279–1282.

[108] Raillard, S., A. Krebber, Y. Chen, J. E. Ness, E. Bermudez, R. Trinidad, R. Fullem, C. Davis, M. Welch, J. Seffernick, L. P. Wackett, W. P. C. Stemmer, and J. Minshull. "Novel Enzyme Activities and Functional Plasticity Revealed by Recombining Highly Homologous Enzymes." *Chemistry & Biology* **8** (2001): 891-898.

[109] Rennell, D., S. E. Bouvier, L. W. Hardy, and A. R. Poteete. "Systematic Mutation of Bacteriophage T4 Lysozyme." *J. Mol. Biol.* **222** (1991): 67–87.

[110] Sandberg, W. S., and T. C. Terwilliger. "Engineering Multiple Properties of a Protein by Combinatorial Mutagenesis." *Proc. Natl. Acad. Sci. USA* **90** (1993): 8367–8371.

[111] Savageau, M. "The Behavior of Intact Biochemical Systems." *Curr. Top. Cell. Regul.* **6** (1972): 63–130.

[112] Savageau, M. "Comparison of Classical and Autogenous Systems of Regulation in Inducible Operons." *Nature* **252** (1974): 546–549.

[113] Savageau, M. "Design Principles for Elementary Gene Circuits: Elements, Methods, and Examples." *Chaos* **11** (2001): 142–159.

[114] Savageau, M. "Parameter Sensitivity as a Criterion for Evaluating and Comparing the Performance of Biological Systems." *Nature* **229** (1971): 542–544.

[115] Savan, J. G., and P. G. Wolynes. "Statistical Mechanics of the Combinatorial Synthesis and Analysis of Folding Macromolecules." *J. Phys. Chem. B* **101** (1997): 8375–8389.

[116] Schaffer, J. D., and A. Morishima. "An Adaptive Crossover Distribution Mechanism for Genetic Algorithms." In *Genetic Algorithms and Their Applications: Proceedings of the Second International Conference on Genetic Algorithms*, 36–40. 1987.

[117] Schmidt-Dannert, C., D. Umeno, and F. H. Arnold. "Molecular Breeding of Carotenoid Biosynthetic Pathways." *Nature Biotech.* **18** (2000): 750–753.

[118] Schultz, P. G., and R. A. Lerner. "From Molecular Diversity to Catalysis: Lessons from the Immune System." *Science* **269** (1995): 1835–1842.

[119] Shao, Z., and F. H. Arnold. "Engineering New Functions and Altering Existing Functions." *Curr. Opin. Struct. Biol.* **6** (1996): 513–518.

[120] Sharon, J., M. L. Gefter, L. J. Wyosocki, and M. N. Margolies. "Recurrent Somatic Mutations in Mouse Antibodies to p-azophenylarsonate Increase Affinity for Hapten." *J. Immunol.* **142** (1989): 596–601.

[121] Sideraki, V., W. Huang, T. Palzkill, and H. F. Gilbert. "A Secondary Drug Resistance Mutation of TEM-1 b-lactamase that Suppresses Misfolding and Aggregation." *Proc. Natl. Acad. Sci. USA* **98** (2001): 283–288.

[122] Sinha, N., and R. Nussinov. "Point Mutations and Sequence Variability in Proteins: Redistributions of Preexisting Populations." *Proc. Natl. Acad. Sci. USA* **98** (2001): 3139–3144.

[123] Skinner, M. M., and T. C. Terwilliger. "Potential Use of Additivity of Mutational Effects in Simplifying Protein Engineering." *Proc. Natl. Acad. Sci. USA* **93** (1996): 10753–10757.

[124] Soderlind, E., L. Strandberg, P. Jirholt, N. Kobayashi, V. Alexeiva, A. M. Aberg, A. Nilsson, B. Jansson, M. Ohlin, C. Wingren, L. Danielsson, R. Carlsson, and C. A. K. Borrebaeck. "Recombining Germline-Derived CDR Sequences for Creating Diverse Singe-framework Antibody Libraries." *Nature Biotech.* **19** (2000): 852–856.

[125] Spiller, B., A. Gershenson, F. H. Arnold, and R. C. Stevens. "A Structural View of Evolutionary Divergence." *Proc. Natl. Acad. Sci. USA* **96** (1999): 12305–12310.

[126] Spinelli, S., and P. M. Alzari. "Structural Implications of Somatic Mutations during the Immune Response to 2-phenyloxazolone." *Res. Immunol.* **145** (1994): 41–45.

[127] Stemmer, W. P. C. "DNA Shuffling by Random Fragmentation and Reassembly: In vitro Recombination for Molecular Evolution." *Proc. Natl. Acad. Sci. USA* **91** (1994): 10747–10751.

[128] Suzuki, M., F. C. Christians, B. Kim, A. Skandalis, M. E. Black, and L. A. Loeb. "Tolerance of Different Proteins for Amino Acid Diversity." *Molecular Diversity* **2** (1996): 111–118.

[129] van Nimwegen, E. "The Statistical Dynamics of Epochal Evolution." Ph.D thesis, Utrecht University, The Netherlands, 1999.

[130] Voigt, C. A., S. L. Mayo, F. H. Arnold, and Z.-G. Wang. "Computational Method to Reduce the Search Space for Directed Evolution." *Proc. Natl. Acad. Sci. USA* **98** (2001): 3778–3783.

[131] Voigt, C. A., S. Kauffman, and Z.-G. Wang. "Rational Evolutionary Design: The Theory of in vitro Protein Evolution." *Advances in Protein Chemistry* **55** (2001): 79–160.

[132] Voigt, C. A., C. Martinez, Z.-G.Wang, S. L. Mayo, and F. H. Arnold. "Protein Building Blocks Preserved by Recombination." *Nat. Struct. Biol.* **9(7)** (2002): 553–558.

[133] von Dassow, G., E. Meir, E. M. Munro, and G. M. Odell. "The Segment Polarity Network is a Robust Developmental Module." *Nature* **406** (2000): 188–192.

[134] Wagner, A. "Robustness Against Mutations in Genetic Networks of Yeast." *Nature Genetics* **24** (2000): 355–361.

[135] Wagner, A. "The Yeast Protein Interaction Network Evolves Rapidly and Contains Few Redundant Duplicate Genes." *Mol. Biol. Evol.* **18** (2001): 1283–1292.

[136] Wagner, G., and L. Altenberg. "Complex Adaptations and the Evolution of Evolvability." *Evolution* **50** (1996): 967–976.

[137] Wagner, G. Fell 2001.

[138] Wagner, G., C.-H. Chiu, and T. F. Hansen. "Is Hsp90 a Regulator of Evolvability?" *J. Exp. Zool.* **285** (1999): 116–118.

[139] Wakasugi, K., K. Ishimori, K. Imai, Y. Wada, and I. Morishima. "'Module' Substitution in Hemoglobin Subunits." *J. Biol. Chem.* **269** (1994): 18750–18756.

[140] Wang, T., J. Miller, N. S. Wingreen, C. Tang, and K. A. Dill. "Symmetry and Designability for Lattice Protein Models." *J. Chem. Phys.* **113** (2000): 8329–8336.

[141] Weinberger, E., "Correlated and Uncorrelated Fitness Landscapes and How to Tell the Difference." *Biol. Cybern.* **63** (1990): 325–336.

[142] Wells, J. A. "Additivity of Mutational Effects in Proteins." *Biochemistry* **29** (1990): 8509–8517.

[143] Wilke, C. O., and C. Adami. "Interaction between Directional Epistasis and Average Mutational Effects." *Proc. Roy. Soc. Lond. B* **268** (2001): 1469–1474.

[144] Wittrup, D. Personal communication.

[145] Xu, J., Q. Deng, J. Chen, K. N. Houk, J. Bartek, D. Hilvert and I. A. Wilson. "Evolution of Shape Complementarity and Catalytic Efficiency from a Primordial Antibody Template." *Science* **286** (1999): 2345–2348.

[146] Yaoi, T., K. Miyazaki, T. Oshima, Y. Komukai, and M. Go. "Conversion of the Coenzyme Specificity of Isocitrate Dehydrogenase by Module Replacement." *J. Biochem,* **119** (1996): 1014–1018.

[147] Yi, T-M., Y. Huang, M. I. Simon, and J. Doyle. "Robust Perfect Adaptation in Bacterial Chemotaxis through Integral Feedback Control." *Proc. Natl. Acad. Sci. USA* **9** (2000): 4649–4653.

[148] Yuh, C. H., H. Bolouri, and E. H. Davidson. "Genomic cis-regulatory Logic: Experimental and Computational Analysis of a Sea Urchin Gene." *Science* **279** (1998): 1896–1902.

[149] Yuh, C. H. et al. 2000.

Robustness in Neuronal Systems: The Balance Between Homeostasis, Plasticity, and Modulation

Eve Marder
Dirk Bucher

1 INTRODUCTION

Animals live long lives. Humans, sea turtles, elephants, and lobsters all may live for a good portion of a century. In these animals the nervous system that generates appropriate behavior consists, for the most part, of neurons that are born early in the animal's life. During early development the complex neuronal circuits that mediate behavior are formed, and these circuits retain their functional integrity for most of the healthy animal's lifetime. Several features of nervous systems make this particularly remarkable. (1) The nervous system must retain its essential flexibility and plasticity so that the animal can adapt to changes in its environment and learn. (2) Although individual neurons can live for tens or even one hundred years, all of the ion channels, receptors, and cell signaling proteins that give each neuron its characteristic structure and electrical properties turn over in minutes, hours, days, or weeks. Indeed, unlike electronic circuits which are built of static components, the nervous system consists of a relatively stable structure that is constantly rebuilding itself, as all of its constituent molecules are replaced. Moreover, during normal behavior the properties of individual neurons and their synaptic connections are constantly varying, but do so in a manner that ensures functional changes in network output, and rarely leads to loss of network stability.

There are three fundamental and linked questions that we must address in understanding the nervous system:

1. How do short-term modifications of synaptic strength and neuronal properties lead to adaptive changes in network dynamics while avoiding regimes in which network stability is lost?
2. How is behaviorally relevant neuronal function maintained during growth of the animal and nervous system?
3. How does the nervous system maintain a relatively constant structure and function while constantly replacing its molecular structure?

As an example, in this chapter we will use a small model nervous system, the crustacean stomatogastric nervous system, that allows us to address these issues experimentally.

2 NETWORK DYNAMICS DEPEND ON THE INTERACTION BETWEEN SYNAPTIC AND INTRINSIC PROPERTIES

Years of both experimental and computational studies have demonstrated that the dynamics of networks depend both on the properties of the synapses connecting neurons and on the intrinsic properties of the network neurons. It is well accepted that synaptic connections vary in sign, strength, and time course [34], and that modifications of any of these properties can alter network dynamics. In contrast, it is less well-appreciated that modifications of neuronal intrinsic properties can also be important for alterations in circuit dynamics.

We call a neuron's intrinsic properties its electrical properties when studied in isolation of all synaptic inputs (fig. 1). Some neurons are silent, others fire slowly, still others fire bursts of action potentials in isolation. Some neurons respond to strong inhibition with post-inhibitory rebound firing [35, 36]. Others show bistability and generate plateau potentials [29]. Still others show spike-frequency adaptation. These disparate firing properties depend on how many ion channels of different types an individual neuron has [12]. Because there are numerous forms of K^+, Na^+, and Ca^{2+} channels that differ in their voltage and time dependence [24] as well as density and location in the neuronal membrane, wide variations in the dynamics of firing produced by individual neurons are possible. Therefore, as will be seen below, the control of the number, kind, and distribution of the ion channels that give neurons their characteristic intrinsic properties is important. Because intrinsic neuronal properties determine how a neuron responds to a given synaptic input, modifications of any of the membrane currents

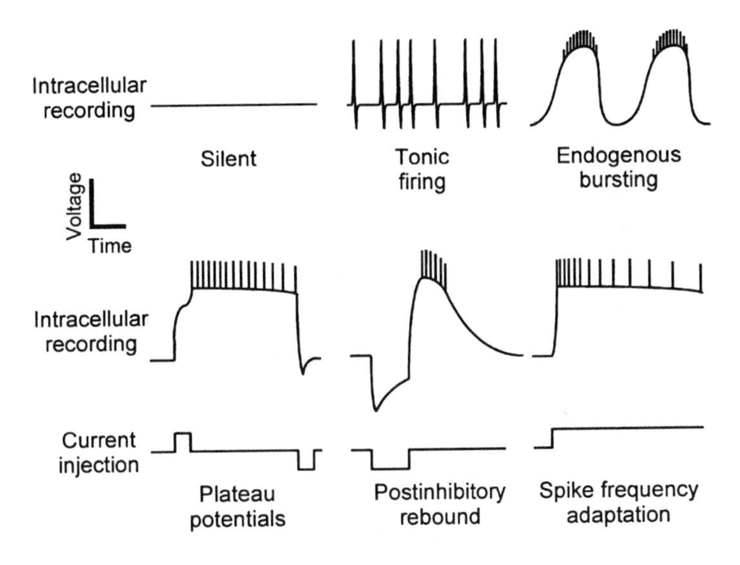

FIGURE 1 Intrinsic membrane properties of neurons. Upper panel: Neurons show different spontaneous activity in isolation from synaptic input. Lower panel: Voltage-dependent currents lead to nonlinear responses to input.

that shape a neuron's intrinsic properties can alter network dynamics [17]. That said, it is important to recognize that there are many different mixtures of ion conductance densities that can produce similar activity patterns [12, 16]. Therefore, individual neurons with similar activity patterns may do so by utilizing different biophysical mechanisms.

3 THE STOMATOGASTRIC NERVOUS SYSTEM

The stomatogastricc nervous system governs the movements of the crustacean foregut [18, 49]. It consists of a group of four ganglia that are linked by nerves (fig. 2(a)). The bilateral commissural ganglia (CoGs) each have 400–500 neurons including a number that project down the connecting nerves to modulate the stomatogastric ganglion (STG). The single esophageal ganglion (OG) consists of 14–18 neurons, several of which also project into the STG. The STG consists of about 30 neurons, and contains the motor neurons for two different motor patterns, the rapid pyloric rhythm and the slower gastric mill rhythm. These motor patterns can be recorded extracellularly from the motor nerves or intracellularly from the somata of the STG neurons themselves (fig. 2(a)). Figure 2(b) illustrates the triphasic pyloric rhythm. Shown are three simultaneous intracellular

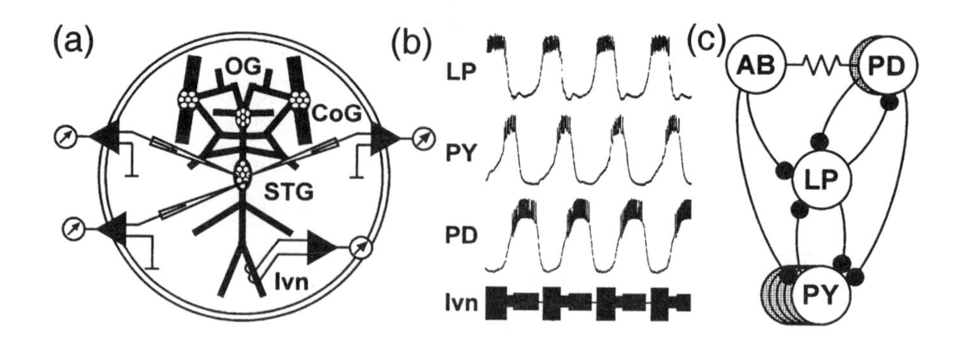

FIGURE 2 The crustacean stomatogastric system. (a) The isolated stomatogastric nervous system placed in a dish. (b) Intracellular and extracellular recordings from cells of the pyloric system. (c) Simplified connectivity diagram for the pyloric rhythm. OG: oesophageal ganglion; CoG: comissural ganglion; STG: stomatogastric ganglion; LP: lateral pyloric (motor neuron); PY: pyloric; PD: pyloric dilator; AB: anterior burster; lvn: lateral ventricular nerve.

recordings from the lateral pyloric (LP), pyloric (PY), and pyloric dilator (PD) neurons and an extracellular recording from the lateral ventricular nerve (lvn) which shows the activity of all three.

Figure 2(c) is a simplified connectivity diagram for the pyloric rhythm. The anterior burster (AB) neuron is an interneuron that is intrinsically oscillatory when isolated from other pyloric neurons [48]. The AB neuron is electrically coupled to the PD neurons, and this electrical coupling causes the AB and PD neurons to burst synchronously although the PD neurons do not generate rapid bursts in the absence of the AB neuron. Together the AB and PD neurons inhibit the LP and PY neurons. The LP neuron recovers from inhibition before the PY neurons [21, 22], and the reciprocal inhibition between the LP and PY neurons produces their alternation.

The reader will note that this circuit depends critically on synaptic inhibition and electrical synapses. The pacemaker kernel consists of the electrically coupled PD and AB neurons, and the timing of the firing of the LP and PY neurons is set by the synaptic and intrinsic processes that govern the dynamics of the postinhibitory rebound bursts [10, 19, 22].

4 THE PROBLEM POSED BY MODULATION OF THE ADULT NERVOUS SYSTEM

Neuromodulators are substances that can dramatically alter the strength of synapses and/or the intrinsic properties of the individual neurons in a circuit [17]. A great deal of work over the years has shown that the STG is modulated by at least twenty different substances found both in identified input neurons or released as circulating hormones [35, 43, 44]. These include amines such as dopamine [30], serotonin [3], octopamine [2], and histamine [6, 42]. Additionally, a very large number of neuropeptides are found in fibers that project to the STG [35], and other small molecule neurotransmitters and gases also are likely modulators of the STG [47, 53].

What do neuromodulators do to the network dynamics of the STG? The changes in network dynamics produced by either the application of these substances in the bath, or by stimulating the neurons that contain them have been extensively characterized [35, 36, 43, 44]. These experiments have shown unambiguously that each neuromodulator and neuromodulatory neuron can evoke characteristic and different changes in the pyloric rhythm [35, 37, 40, 43, 44]. These involve changes in rhythm frequency, in the number of action potentials fired by each neuron in each burst, and in the relative timing or phase relationships of the constituent neurons within the motor pattern.

How do modulators produce changes in network dynamics? Neuromodulators can alter the strength of synapses [26] and/or modify the intrinsic excitability of individual neurons [11, 25]. Years of work on the effects of neuromodulatory substances on the STG have shown: (a) every neuron in the pyloric rhythm is subject to modulation by multiple substances [20, 52, 53], (b) every synapse in the pyloric network is subject to modulation [28], and (c) each modulator alters a different subset of parameters in the circuit.

It is beyond the scope of this contribution to review in detail these modulatory actions, but the general principle is illustrated in the cartoon schematic shown in figure 3. Each modulator acts on a different subset of possible targets for modulation, and consequently produces a different altered circuit output. This organization poses a deep design problem: with so many circuit components subject to modulation, in so many different combinations, how is it possible that the circuit that produces the pyloric rhythm, and other like circuits are stable? Why does this rich pattern of modulation not result in "overmodulation" or circuit "crashes"? What protections are there so that functional circuit outputs are maintained together with all of this potential for flexibility?

The answers to the above questions are not obvious. Nonetheless, some beginning insights can be drawn from what we know about the cellular mechanisms underlying neuromodulation. Many neuromodulators act through second messengers to indirectly open or close ion channels [24, 32]. This sets the stage for several neuromodulators to converge onto the same signal transduction pathway

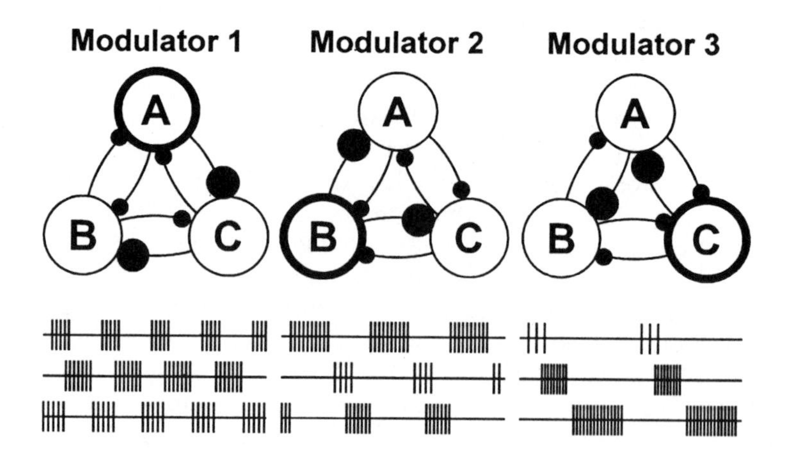

FIGURE 3 Modulation of neuronal networks. Each modulator acts on a different subset of possible targets for modulation, and consequently produces a different altered circuit output.

or on the same ion channel. The consequence of such convergence is that modulators can saturate and occlude each other's actions. This is seen in the crab STG, where six different modulators converge on the same voltage-dependent inward current [53], called the proctolin current, because it was first described in studies of the action of the neuropeptide proctolin [13]. The LP neuron is a direct target for all of these substances. Nonetheless, if one of the modulators has strongly activated the LP neuron, a second or third modulator will have little or no effect. Thus, the convergent action of neuromodulators at the cellular level provides a "ceiling effect" that prevents overmodulation. This is one of the cellular correlates of "state-dependent neuromodulation," modulation that depends on the initial state of the network when the modulator is applied. These modulatory peptides are found in different combinations in specific projection neurons [4], and because these substances act on different network targets that display receptors for a subset of them, projection neurons that release different combinations of neuropeptides still evoke different motor patterns.

The voltage dependence of the proctolin current provides a second cellular mechanism that protects against overmodulation [38, 50]. When an oscillatory neuron is depolarized by constant current or the application of a conventional excitatory neurotransmitter, the baseline membrane potential and the membrane potential at the peak of the slow wave are both depolarized (fig. 4(a)). As the amplitude of the depolarization is increased, eventually the oscillator may be blocked in the "up" state. In contrast, the proctolin current's voltage dependence

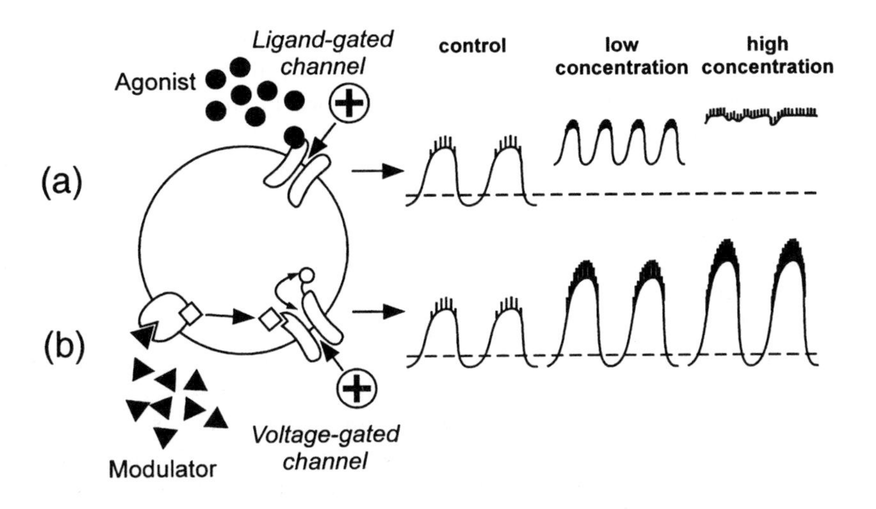

FIGURE 4 Different effects of modulation. (a) Applying an agonist to a ligand-gated channel delivers a tonic enhancement of the current through this channel. The baseline membrane potential is shifted and higher concentrations of the agonist can "crash" the rhythmic activity. (b) Modulating a voltage-gated channel via second messenger systems will only take effect in the phase of the cycle when the channel is activated. Therefore, only the amplitude of activity in a specific phase is altered and the baseline membrane potential is not affected.

means that at the trough of the oscillation there is no depolarization, but at the peak there is a large effect. Therefore, the baseline of the oscillator is not depolarized, but the amplitude of the oscillation is much enhanced (fig. 4(b)). Again, this allows the frequency and the amplitude of the pacemaker to be modulated, but prevents the pacemaker kernel from losing its ability to generate bursts.

There are undoubtedly numerous other cellular and circuit mechanisms that also protect against overmodulation. One of the challenges of the future will be to understand how these mechanisms function together to ensure that modulators can alter circuit behavior so effectively and safely.

5 THE PROBLEM POSED BY GROWTH

In most animals many of the circuits that underlie behavior must function continuously during the animal's lifetime. In lobsters, the STG is already active by midway through embryonic life [5, 46], and even small juvenile lobsters produce adult motor patterns. Figure 5(a) shows the relative sizes of a small juvenile lobster and a fully mature adult lobster. Figure 5(b) shows an intracellular dye fill of a PD neuron in the STGs of a juvenile and an adult animal shown on the same scale. First, it can be seen that the difference in the size of the STG is smaller than the difference in the size of the animals. Nonetheless, figure 5(b) shows that the growth in the animal is accompanied by significant growth in the size of the STG and in the size of the individual neurons that produce the STG's motor patterns. Despite these dramatic changes in cell and ganglion size, figure 5(c) shows that the motor patterns produced by the small and large animals are virtually indistinguishable.

It is easy to appreciate that it is important for animals to be able to produce constant motor patterns despite their growth. That said, it is not intuitively obvious how this is possible. Neurons are not isopotential. Rather, electrical signals propagate through the processes of a complex structure as a function of the number and distribution of channels [7, 28]. In principle, it is possible for a single neuron to grow in such a manner as to maintain its passive electrophysiological properties [23, 45]. However, for this to translate into constant activity, not only would the structures need to expand at the appropriate size ratios, but the number and placement of the voltage-dependent conductances would have to be properly controlled. In at least one system, changes in synaptic efficacy accompany neuronal growth [9], as synaptic sites change their relative positions in the neuronal arbor.

Among unanswered questions posed by the data shown in figure 5 are: (a) Are the motor patterns in young and old animals produced by the same underlying mechanism? It is possible that although the motor patterns are the same, different combinations of synaptic and intrinsic properties could give rise to these. (b) What are the control principles used to maintain constant physiological processes despite the large changes in the structure of the neurons that produce the circuit output?

6 THE PROBLEM POSED BY CHANNEL TURNOVER

Even in the adult animal undergoing no further growth, protein turnover is ongoing. Therefore, all of the channels that are important for excitability are constantly being replaced. Thus, it is not sufficient for neurons to establish their intrinsic properties once, but rather neurons are synthesizing, inserting, removing, and degrading ion channels and receptors continuously. What controls this constant flux of membrane proteins? If this flux were not properly controlled, then the characteristic intrinsic membrane properties of network neurons would be subject to fluctuations that might be deleterious to network function. In a series of theoretical studies, we argued that neurons can use intracellular concentrations of Ca^{2+} or other signal transduction molecules as activity sensors [1, 14, 15, 31, 33, 51]. In these models, the intracellular activity sensors

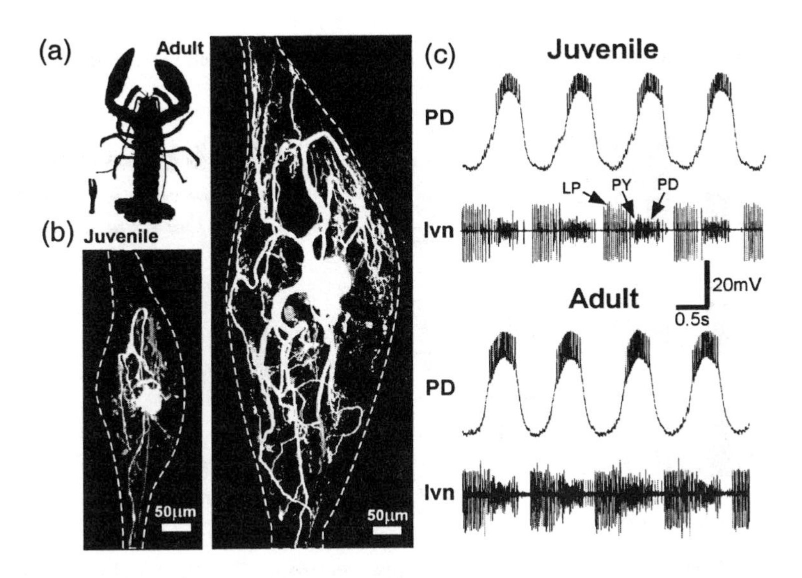

FIGURE 5 Growth of individual neurons. Despite substantial increase in size and geometry of the STG neurons, the network output is constant from small juvenile to large adult animals. (a) Size range of lobsters after metamorphosis. (b) PD neuron in a juvenile and an adult. (c) Recordings of pyloric rhythms in a juvenile and an adult.

were then used to control the densities of each kind of ion channel in the membrane, using simple negative feedback rules to achieve homeostasis [39]. These models had several essential features that made them extremely attractive: (a) model neurons with dynamically regulated current densities could self-assemble from random initial conditions [33], and (b) when perturbed by an extrinsic input, model neurons with dynamically regulating conductances could recover their initial pattern of activity, even with an ongoing perturbation [31, 33]. Since these models were first proposed, a number of experimental studies have been conducted that are consistent with the spirit of these models.

Figure 6(a) shows the results of experiments that suggest that isolated STG neurons can cell-autonomously acquire the balance of conductances necessary to produce an activity state similar to the one shown in the intact ganglion. In the intact network, all STG neurons fire rhythmically, although most of these neurons do so because of the rhythmic synaptic drive that they receive. When single STG neurons are removed from the ganglion and placed in dissociated cell culture they are initially relatively inactive [57, 58]. The top panel of figure 6(a) shows that on day 1 of culture they typically fire small, rapidly inactivating action potentials. On day 2 of culture, they tonically fire full action potentials (middle panel). On day 3 of culture, they typically fire in bursts, the pattern of activity most similar

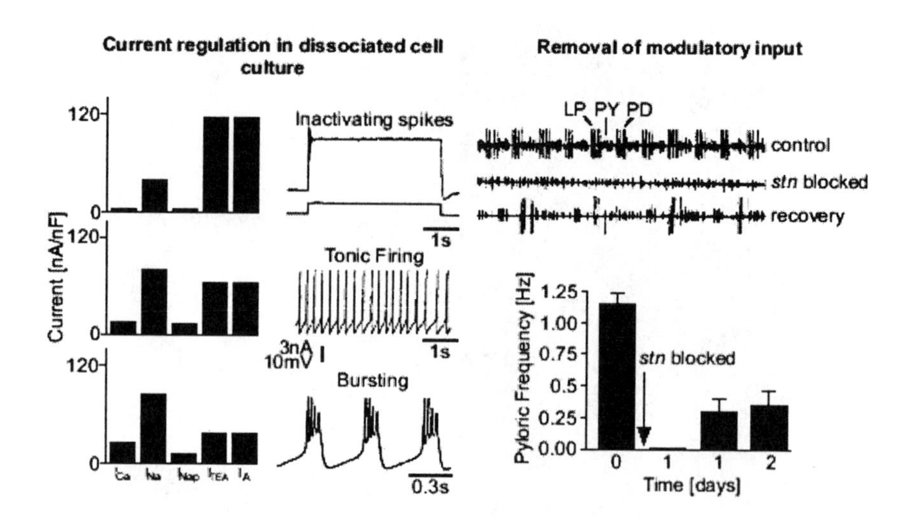

FIGURE 6 Plastic changes in intrinsic membrane properties. (a) STG neurons in dissociated cell culture are actively regulating their current densities to rebuild their rhythmic activity state. (b) Rhythmic activity of the STG resumes some time after modulatory inputs have been removed.

to that which they show in the intact STG. The histograms in figure 6(a) show the results of voltage clamp measurements of the currents in these neurons on these days. The neurons are actively regulating their current densities to rebuild their rhythmic activity state. Note that as the current densities of the inward currents are increasing, those of the outward currents are decreasing. Such changes will tend to make the neurons more excitable and more oscillatory [58].

The premise of the theoretical models described previously was that neurons use their own activity to regulate conductance densities. Consistent with this interpretation are a number of experiments. First, when STG neurons are rhythmically stimulated, their K^+ current densities are altered [14, 15]. Second, when vertebrate cortical neurons are placed in TTX to block all action potentials, they upregulate their Na^+ conductances and downregulate their K^+ conductances [8], as predicted by the earlier models.

Thus far, the models and the data we have discussed refer to regulation of the properties of single neurons. However, what matters for behavior is how whole networks of neurons are controlled. Figure 6(b) shows the results of experiments that demonstrate that network stability may be controlled as well [14, 15, 41, 55, 56]. In this experiment the top set of recordings are extracellular recordings of the pyloric rhythm in an experiment in which the anterior ganglia containing modulatory inputs were left intact. Then these inputs were removed, and the rhythmic activity was lost. However, approximately 1 day later, rhythmic activity resumed. In this case, the rhythmic activity was independent of the modulatory inputs. A model of this recovery process [14, 15] showed that cell-autonomous activity sensors were sufficient to produce network stability. These data strongly suggest that sensors of either activity and/or the release of neuromodulatory inputs may be important for the regulation of channel density necessary for stable network function. Moreover, these data argue that similar network function can be produced by different underlying mechanisms: in the control case the rhythm strongly depends on currents that are up-regulated by modulators whereas after recovery there have been changes in the network that allow it to function in the absence of neuromodulatory inputs.

7 CONCLUSIONS

Robustness of neuronal structure and function depends on all of the complexity of biological structures. Every cell in a biological structure contains thousands of control pathways that allow differentiation and appropriate responses to the environment while maintaining cellular homeostasis. The nervous system uses all of this miraculous biochemical and molecular control to implement complex networks that themselves maintain precise organization while at the same time all of the constituent molecules are in a constant state of flux. It is hard to reconcile the vast molecular and cellular complexity displayed by neurons and nervous systems with the fact that they work so well.

REFERENCES

[1] Abbott, L. F., and G. LeMasson. "Analysis of Neuron Models with Dynamically Regulated Conductances in Model Neurons." *Neural Comp.* **5** (1993): 823–842.

[2] Barker, D. L., P. D. Kushner, and N. K. Hooper. "Synthesis of Dopamine and Octopamine in the Crustacean Stomatogastric Nervous System." *Brain Res.* **161** (1979): 99–113.

[3] Beltz, B., J. S. Eisen, R. Flamm, R. M. Harris-Warrick, S. Hooper, and E. Marder. "Serotonergic Innervation and Modulation of the Stomatogastric Ganglion of Three Decapod Crustaceans (*Panulirus interruptus, Homarus americanus*, and *Cancer irroratus*)." *J. Exp. Biol.* **109** (1984): 35–54.

[4] Blitz, D. M., A. E. Christie, M. J. Coleman, B. J. Norris, E. Marder, and M. P. Nusbaum. "Different Proctolin Neurons Elicit Distinct Motor Patterns from a Multifunctional Neuronal Network." *J. Neurosci.* **19** (1999): 5449–5463.

[5] Casasnovas, B., and P. Meyrand. "Functional Differentiation of Adult Neural Circuits from a Single Embryonic Network." *J. Neurosci.* **15** (1995): 5703–5718.

[6] Claiborne, B., and A. Selverston. "Histamine as a Neurotransmitter in the Stomatogastric Nervous System of the Spiny Lobster." *J. Neurosci.* **4** (1984): 708–721.

[7] Dayan, P., and L. F. Abbott. *Theoretical Neuroscience*, 460. Cambridge, MA: MIT Press, 2001.

[8] Desai, N. S., L. C. Rutherford, and G. G. Turrigiano. "Plasticity in the Intrinsic Excitability of Cortical Pyramidal Neurons." *Nature Neurosci.* **2** (1999): 515–520.

[9] Edwards, D. H., S. R. Yeh, L. D. Barnett, and P. R. Nagappan. "Changes in Synaptic Integration During the Growth of the Lateral Giant Neuron of Crayfish." *J. Neurophysiol.* **72** (1994): 899–908.

[10] Eisen, J. S., and E. Marder. "A Mechanism for Production of Phase Shifts in a Pattern Generator." *J. Neurophysiol.* **51** (1984): 1375–1393.

[11] Elson, R. C., and A. I. Selverston. " Mechanisms of Gastric Rhythm Generation in Isolated Stomatogastric Ganglion of Spiny Lobsters: Bursting Pacemaker Potentials, Synaptic Interactions and Muscarinic Modulation." *J. Neurophysiol.* **68** (1992): 890–907.

[12] Goldman, M. S., J. Golowasch, E. Marder, and L. F. Abbott. "Global Structure, Robustness, and Modulation of Neuronal Models." *J. Neurosci.* **21** (2001): 5229–5238.

[13] Golowasch, J., and E. Marder. "Proctolin Activates an Inward Current whose Voltage Dependence is Modified by Extracellular Ca^{2+}." *J. Neurosci.* **12** (1992): 810–817.

[14] Golowasch, J., L. F. Abbott, and E. Marder. "Activity-Dependent Regulation of Potassium Currents in an Identified Neuron of the Stomatogastric Ganglion of the Crab *Cancer borealis*." *J. Neurosci.* **19** (1999): RC33.

[15] Golowasch, J., M. Casey, L. F. Abbott, and E. Marder. "Network Stability from Activity-Dependent Regulation of Neuronal Conductances." *Neural Comput.* **11** (1999): 1079–1096.

[16] Golowasch, J., M. S. Goldman, L. F. Abbott, and E. Marder. "Failure of Averaging in the Construction of a Conductance-Based Neuron Model." *J. Neurophysiol.* **87** (2002): 1129–1131.

[17] Harris-Warrick, R. M., and E. Marder. "Modulation of Neural Networks for Behavior." *Ann. Rev. Neurosci.* **14** (1991): 39–57.

[18] Harris-Warrick, R. M., E. Marder, A. I. Selverston, and M. Moulins. *Dynamic Biological Networks. The Stomatogastric Nervous System*, 328. Cambridge, MA: MIT Press, 1992.

[19] Harris-Warrick, R. M., L. M. Coniglio, R. M. Levini, S. Gueron, and J. Guckenheimer. "Dopamine Modulation of Two Subthreshold Currents Produces Phase Shifts in Activity of an Identified Motoneuron." *J. Neurophysiol.* **74** (1995): 1404–1420.

[20] Harris-Warrick, R. M., B. R. Johnson, J. H. Peck, P. Kloppenburg, A. Ayali, and J. Skarbinski. "Distributed Effects of Dopamine Modulation in the Crustacean Pyloric Network." *Ann. NY Acad. Sci.* **860** (1998): 155–167.

[21] Hartline, D. K. "Pattern Generation in the Lobster (*Panulirus*) Stomatogastric Ganglion. II. Pyloric Network Simulation." *Biol. Cybern.* **33** (1979): 223–236.
[22] Hartline, D. K., and D. V. Gassie, Jr. "Pattern Generation in the Lobster (*Panulirus*) Stomatogastric Ganglion. I. Pyloric Neuron Kinetics and Synaptic Interactions." *Biol. Cybern.* **33** (1979): 209–222.
[23] Hill, A. A., D. H. Edwards, and R. K. Murphey. "The Effect of Neuronal Growth on Synaptic Integration." *J. Comp. Neurosci.* **1** (1994): 239–254.
[24] Hille, B. *Ion Channels of Excitable Membranes*, 814. Sunderland, MA: Sinauer, 2001.
[25] Hooper, S. L., and E. Marder. "Modulation of the Lobster Pyloric Rhythm by the Peptide Proctolin." *J. Neurosci.* **7** (1987): 2097–2112.
[26] Johnson, B. R., and R. M. Harris-Warrick. "Aminergic Modulation of Graded Synaptic Transmission in the Lobster Stomatogastric Ganglion." *J. Neurosci.* **10** (1990): 2066–2076.
[27] Johnson, B. R., J. H. Peck, and R. M. Harris-Warrick. "Distributed Amine Modulation of Graded Chemical Transmission in the Pyloric Network of the Lobster Stomatogastric Ganglion." *J. Neurophysiol.* **174** (1995): 437–452.
[28] Johnston, D., and S. Wu. *Foundations of Cellular Neurophysiology*, 676. Cambridge, MA: MIT Press, 1995.
[29] Kiehn, O., and T. Eken. "Functional Role of Plateau Potentials in Vertebrate Motor Neurons." *Curr. Opin. Neurobiol.* **8** (1998): 746–752.
[30] Kushner, P., and D. L. Barker. "A Neurochemical Description of the Dopaminergic Innervation of the Stomatogastric Ganglion of the Spiny Lobster." *J. Neurobiol.* **14** (1983): 17–28.
[31] LeMasson, G., E. Marder, and L. F. Abbott. "Activity-Dependent Regulation of Conductances in Model Neurons." *Science* **259** (1993): 1915–1917.
[32] Levitan, I. B. "Modulation of Ion Channels by Protein Phosphorylation and Dephosphorylation." *Ann. Rev. Physiol.* **56** (1994): 193–212.
[33] Liu, Z., J. Golowasch, E. Marder, and L. F. Abbott. "A Model Neuron with Activity-Dependent Conductances Regulated by Multiple Calcium Sensors." *J. Neurosci.* **18** (1998): 2309–2320.
[34] Marder, E. "From Biophysics to Models of Network Function." *Annu. Rev. Neurosci.* **21** (1998): 25–45.
[35] Marder, E., and D. Bucher. "Central Pattern Generators and the Control of Rhythmic Movements." *Curr. Biol.* **11** (2001): R986–R996.
[36] Marder, E., and R. L. Calabrese. "Principles of Rhythmic Motor Pattern Generation." *Physiol. Rev.* **76** (1996): 687–717.
[37] Marder, E., and S. L. Hooper. "Neurotransmitter Modulation of the Stomatogastric Ganglion of Decapod Crustaceans." In *Model Neural Networks and Behavior*, edited by A. I. Selverston, 319–337. New York: Plenum Press, 1985

[38] Marder, E., and P. Meyrand. "Chemical Modulation of Oscillatory Neural Circuits." In *Cellular and Neuronal Oscillators*, edited by J. Jacklet, 317–338. New York: Marcel Dekker, Inc., 1989.

[39] Marder, E., and A. A. Prinz. "Modeling Stability in Neuron and Network Function: The Role of Activity in Homeostasis." *BioEssays* **24** (2002): 1145–1154.

[40] Marder, E., and J. M. Weimann. "Modulatory Control of Multiple Task Processing in the Stomatogastric Nervous System." In *Neurobiology of Motor Progamme Selection*, edited by J. Kien, C. McCrohan, B. Winlow, 3–19. New York: Pergamon Press, 1992.

[41] Mizrahi, A., P. S. Dickinson, P. Kloppenburg, V. Fenelon, D. J. Baro, R. M. Harris-Warrick, P. Meyrand, and J. Simmers. "Long-Term Maintenance of Channel Distribution in a Central Pattern Generator Neuron by Neuromodulatory Inputs Revealed by Decentralization in Organ Culture." *J. Neurosci.* **21** (2001): 7331–7339.

[42] Mulloney, B., and W. M. Hall. "Neurons with Histaminelike Immunoreactivity in the Segmental and Stomatogastric Nervous Systems of the Crayfish *Pacifastacus leniusculus* and the Lobster *Homarus americanus*." *Cell Tissue Resh.* **266** (1991): 197–207.

[43] Nusbaum, M. P., and M. P. Beenhakker. "A Small-Systems Approach to Motor Pattern Generation." *Nature* **417** (2002): 343–350.

[44] Nusbaum, M. P., D. M. Blitz, A. M. Swensen, D. Wood, and E. Marder. "The Roles of Cotransmission in Neural Network Modulation." *Trends Neurosci.* **24** (2001): 146–154.

[45] Olsen, O., F. Nadim, A. A. Hill, and D. H. Edwards. "Uniform Growth and Neuronal Integration." *J. Neurophysiol.* **76** (1996): 1850–1857.

[46] Richards, K. S., W. L. Miller, and E. Marder. "Maturation of the Rhythmic Activity Produced by the Stomatogastric Ganglion of the Lobster, *Homarus americanus*." *J. Neurophysiol.* **82** (1999): 2006–2009.

[47] Scholz, N. L., J. de Vente, J. W. Truman, and K. Graubard. "Neural Network Partitioning by NO and cGMP." *J. Neurosci.* **21** (2001): 1610–1618.

[48] Selverston, A. I., and J. P. Miller. "Mechanisms Underlying Pattern Generation in the Lobster Stomatogastric Ganglion as Determined by Selective Inactivation of Identified Neurons. I. Pyloric Neurons." *J. Neurophysiol.* **44** (1980): 1102–1121.

[49] Selverston, A. I., and M. Moulins. *The Crustacean Stomatogastric System.* Berlin: Springer-Verlag, 1987.

[50] Sharp, A. A., M. B. O'Neil, L. F. Abbott, and E. Marder. "The Dynamic Clamp: Artificial Conductances in Biological Neurons." *Trends Neurosci.* **16** (1993): 389–394.

[51] Siegel, M., E. Marder, and L. F. Abbott. "Activity-Dependent Current Distributions in Model Neurons." *Proc. Natl. Acad. Sci. USA* **91** (1994): 11308–11312.

[52] Swensen, A. M., and E. Marder. "Modulators with Convergent Cellular Actions Elicit Distinct Circuit Outputs." *J. Neurosci.* **21** (2001): 4050–4058.

[53] Swensen, A. M., and E. Marder. "Multiple Peptides Converge to Activate the Same Voltage-Dependent Current in a Central Pattern-Generating Circuit." *J. Neurosci.* **20** (2000): 6752–6759.

[54] Swensen, A. M., J. Golowasch, A. E. Christie, M. J. Coleman, M. P. Nusbaum, and E. Marder. "GABA and Responses to GABA in the Stomatogastric Ganglion of the Crab *Cancer borealis*." *J. Exp. Biol.* **203** (2000): 2075–2092.

[55] Thoby-Brisson, M., and J. Simmers. "Neuromodulatory Inputs Maintain Expression of a Lobster Motor Pattern-Generating Network in a Modulation-Dependent State: Evidence from Long-Term Decentralization *In Vitro*." *J. Neurosci.* **18** (1998): 212–225.

[56] Thoby-Brisson, M., and J. Simmers. "Transition to Endogenous Bursting after Long-Term Decentralization Requires de novo Transcription in a Critical Time Window." *J. Neurophysiol.* **84** (2000): 596–599.

[57] Turrigiano, G., L. F. Abbott, and E. Marder. "Activity-Dependent Changes in the Intrinsic Properties of Cultured Neurons." *Science* **264** (1994): 974–977.

[58] Turrigiano, G. G., G. LeMasson, and E. Marder. "Selective Regulation of Current Densities Underlies Spontaneous Changes in the Activity of Cultured Neurons." *J. Neurosci.* **15** (1995): 3640–3652.

Cross-System Perspectives on the Ecology and Evolution of Resilience

Colleen T. Webb
Simon A. Levin

Ecological systems provide a wide spectrum of benefits to humans including the very life-support systems that sustain humanity. Thus, there is a compelling need to understand how robust these systems are in the face of a changing environment, and what the mechanisms are that underlie robustness. The notion of robustness or resilience of ecological systems for us denotes the ability of the system to maintain its macroscopic functional features, such as species diversity or nutrient cycling, rather than the narrower and unattainable possibility of constancy. Indeed, it is the lack of constancy at lower levels of organization that can convey robustness to ecosystems as a whole. The robustness of ecosystem processes, such as nutrient cycling, varies across different types of ecosystems and is related to structural properties such as the strength of interactions among species. This points to the importance of examining structure-function relationships and their evolution. The robustness of the whole ecosystem may fundamentally lie at the level of species themselves, in terms of the robustness of modules of closely interacting species, or as an emergent property of the whole, complex ecosystem. Evidence from diverse ecosystems suggests that biodiversity is correlated with system robustness, but causation is complicated to untangle. In part, biodiversity itself, through functional redundancy, can enhance robustness; but conversely, increased robustness can promote biodiversity over ecological and evolutionary time. Local disturbance may promote global predictability and higher biodiversity by selecting for suites of species with traits that provide the mechanisms for ecosystem recovery following larger scale or novel perturbations. We illustrate these issues via comparisons of the response of tropical and temperate forests to disturbance by wind storms. We discuss the role of biodiversity and other mechanisms in sustaining resilience, and the importance of evolutionary history.

1 INTRODUCTION

Ecosystems are constantly in flux. Species come and go, and their population sizes continually fluctuate. Classic work [50] has emphasized this point, making clear that, in the narrow sense, the notion of a stable equilibrium does not apply, at least at the level of species abundances. Still, regularities emerge at the macroscopic level in the statistical features of species distributions, as well as biogeochemical cycles. There is no paradox here. May [51] showed that increasing numbers of species and increasing complexity reduce stability in terms of the maintenance of the constancy of species abundances; however, the very fluidity at the component level is what allows change in complex adaptive systems, and can lead to stability and robustness at higher levels of organization (see also Levin [45], Edelman and Gally [23], and Krakauer and Plotkin [42]).

Thus, even if constancy in species composition or population size does not occur, functional redundancy (termed degeneracy by Edelman and Gally [23]) can sustain basic ecosystem processes (see Steneck and Dethier [82]). Root [74, 73] introduced the concept of a guild to describe groups of species playing similar roles, and his original concept has been extended beyond foraging guilds to include, for example, the processing of nutrients by mycorrhizae [12], primary production by autotrophs [78], and herbivory [18]. Consequently, functional robustness can be maintained despite the extinction of some species from an ecosystem, because replacement by other functionally similar species maintains the same ecosystem properties. This creates a dynamic equilibrium in which species composition changes, but the roles of species within the ecosystem are preserved [57, 87]. Due to redundancy, functions within the ecosystem are preserved, and other ecosystem properties, such as nutrient cycling, are also sustained. Robustness, therefore, must be evaluated differently at different scales (see also O'Neil [59]), with the absence of robustness at lower levels potentially providing robustness at higher levels. Overall [45], robustness is determined in large part by the interplay

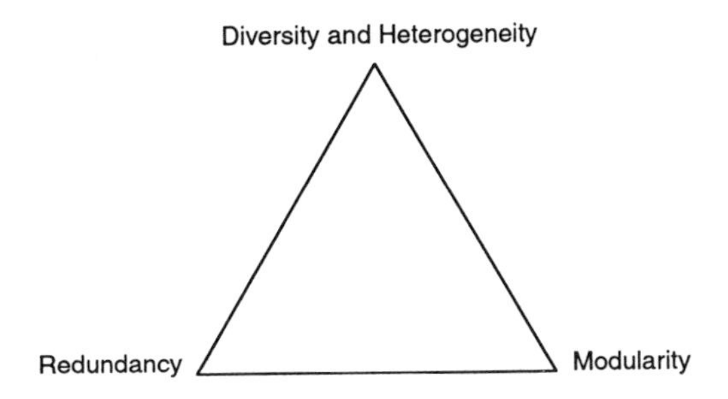

FIGURE 1 Tradeoffs exist among several factors important for ecosystem resilience.

among redundancy of components, diversity that allows system level change, and modularity (fig. 1).

Notions of stability that are tied to the exact species composition are overly restrictive in representing robustness. In ecosystems, robustness, often termed resilience, deals more broadly with the width of a basin of attraction, or the amount that an ecosystem can be perturbed without switching to a different state or type [37]. Inherent in this notion is the idea of multiple basins of attraction, and the potential for the ecosystem to flip from one type of ecosystem to another (fig. 2) [37]. Examples include the alternation of outbreak and collapse seen in pest systems, loss of rangeland under grazing pressure, and eutrophication under phosphorous overload. In the case of forest defoliating spruce budworm, a cycle emerges in which budworm density increases, leading to the destruction of vegetation, followed by budworm population crash and then forest recovery [48]. Such a cycle is analogous to what is seen in epidemic systems more generally, in which pathogens attack susceptible individuals, reducing the pool of susceptibles eventually to low enough levels that the pathogen population crashes, thereby initiating recovery of the susceptible host population through new births (or loss of immunity). In the case of rangelands, grazing pressure, coupled with lowered fire frequency and higher rainfall, can facilitate replacement of perennial grasses with annual plants and shrubs [95]. As for lakes, agricultural runoff or other inputs can trigger algal blooms, shifting a lake from clear water to turbid conditions [16, 75]. In the assessment of resilience, the level of our perspective is crucial, ranging from individual species to emergent functional properties, and we are, in particular, interested in this chapter in elucidating the relationship between change in species composition and macroscopic ecosystem properties.

Just as resilience will vary with the functional scale, it also will vary with the spatial or temporal scale. Local fluctuations may lead to global resilience, and short-term variability may confer longer-term resilience. Thus, ecosystems that are relatively protected from disturbances over long periods of time may have a reduced capacity to respond to novel perturbations. This reduced capac-

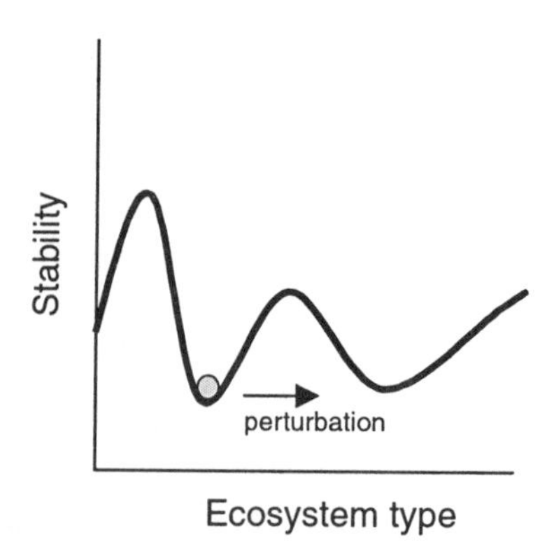

FIGURE 2 Valleys of the stability landscape correspond to stable ecosystem types. Perturbation can drive the system to flip to a different valley or ecosystem type. The width of the valley or basin of attraction (amount of perturbation necessary to switch types) and the steepness of the walls of the valley (return time) are important aspects of resilience, as is how similar the two stable states are in their species composition (after Holling [37]).

ity underscores the difficulties in extrapolating from observed levels of constancy to infer robustness, and the need for observations of the response of systems to novel disturbances, whether controlled or accidental. The relation between the ecological and evolutionary time scales is developed in more detail in Levin [45], who asks: To what extent do evolutionary forces increase the resilience of ecosystems? A fundamental issue is to understand how mechanisms of resilience will be different in complex adaptive systems, such as ecosystems and the biosphere [44], as compared to designed systems, or those systems (such as organisms) where selection pressures relate to the functioning of the whole system. It also suggests great value in a comparative approach among systems of different ecological and evolutionary regimes, and hence the need for measures that can be used across ecosystems.

Finding measures for comparing resilience among ecosystems is not straightforward, given the wide range of species compositions and the variability in potential mechanisms of resilience. However, many of the features of interest will be reflected in patterns of ecosystem processes like nutrient cycles or changes in biomass accumulation. Species composition will greatly influence these ecosystem processes, but those like nutrient cycling and biomass accumulation also average over the functions of various species and are comparable across ecosystems. Major changes in composition and function should, therefore, be reflected in changes in these ecosystem processes.

In this chapter, we address the question of how to use a comparative approach to understand the phenomenon of resilience. We begin by discussing possible metrics, and then discuss the importance of both ecological and evolutionary mechanisms underlying resilience. Generally, this includes how resilience is affected by the topology and geometry of species interactions on an ecological time scale, and how natural selection at the individual level acting over evolutionary time scales influences the ecological resilience of communities and ecosystems. Specifically, we analyze recovery following wind disturbance in order to explore the relative importance of these mechanisms. Temperate and tropical forests differ in their levels of diversity and the frequency with which wind storms occur, so this comparison allows us to look at resilience in ecosystems with different ecological regimes and evolutionary histories. Finally, we discuss the types of empirical data that will be necessary to test ideas of ecosystem resilience.

2 MEASURING RESILIENCE

To avoid circularity, we seek comparative measures of resilience that are responsive to changes in the ecosystem state (the x-axis of fig. 2), but do not rely explicitly on species composition and abundances. Typical among these measures would be indicators of total biomass or productivity, and of the cycling of nutrients.

Evaluating and comparing resilience in plant communities is often confounded by the long time scales over which succession drives ecosystems. In terms of figure 2, this implies that a basin of attraction may appear artificially large or to have unclear boundaries. One approach for measuring resilience in successional plant systems relies on theories of succession and nutrient cycling in ecosystems developed in classic studies by Likens, Vitousek, and others. Key among the ecosystem processes of interest are the storage and processing of nutrients and biomass. Pools and fluxes of nutrients, illustrated in figure 3, are affected by plant composition because rates of nutrient uptake, litterfall, and decomposition of plant material can vary among plant species [1]. These different nutrient dynamics can be assessed by means of recently developed techniques involving stable isotopes and other tracers [55, 65, 81]. When perturbation occurs, species composition and plant biomass may change, potentially leading to changes in the rate of internal recycling of nutrients, as well as the nutrient pool sizes and rates of loss from the ecosystem [47].

Stable isotope tracing, which depends on tracking the unique isotopic signatures from prey to predator, is a sensitive indicator of shifts in the strength of modularity and in the lengths of food chains. Carbon, nitrogen, and sulfur all have stable isotopes that have been used effectively in trophic studies of aquatic systems [30, 66, 69] and terrestrial systems [41].

Equilibrium theory predicts that, over successional time, the ratio of the nutrient outputs of an ecosystem to nutrient inputs should approach unity [93], and empirical studies suggest that such equilibrium is achieved at least for temperate forests [36]. Following disturbance, the length of time for this ratio to

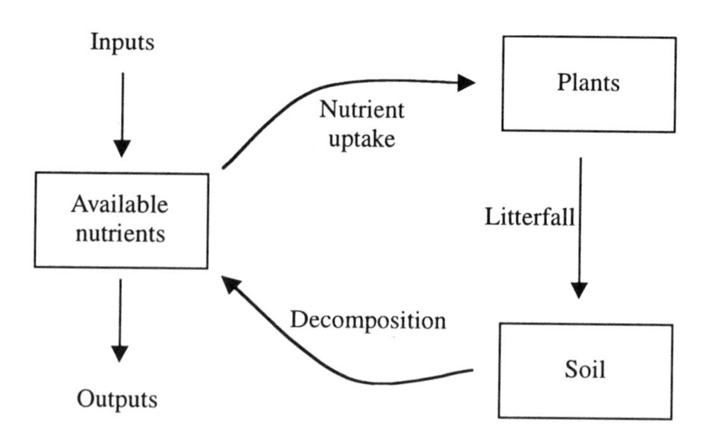

FIGURE 3 A generic nutrient cycle in an ecosystem. Plants affect nutrient cycles through nutrient uptake and litterfall rates, while decomposition in soils affects the amount of stored nutrients in soil organic matter (after Vitousek et al. [91]).

return to its initial value might be used as a type of return time and potential measure of resilience. This is not a very representative measure, however, because the new equilibrium state could be much different than the original. More generally, the return of nutrient outputs to pre-disturbance levels leads to more sensitive measures, although this too can be problematic. If inputs are constant, an entirely new equilibrium might be achieved while sustaining outputs at pre-disturbance levels. Limiting nutrients, usually nitrogen in temperate terrestrial ecosystems and phosphorous in aquatic and tropical terrestrial ecosystems, have shorter residence times in the available pool than nutrients that are essential, but not limiting, to autotrophs. Following many types of disturbance, an excess of both limiting and essential nutrients is typical; however, the excess of the limiting nutrient should be reduced faster, causing it to return to its pre-disturbance condition before essential nutrients will. Return times of limiting nutrients have indeed generally been observed to be faster than essential nutrients, although there are some exceptions [47, 93].

Similar conclusions hold for total biomass: as biomass grows, limiting and essential nutrients are incorporated into plant tissues. Over successional time, the rate of biomass accumulation initially increases and then drops off to zero as the ecosystem matures and reaches steady-state [10, 93].

One can dissect recovery following disturbance into three characteristic phases. This sequence and its implications for biomass accumulation and nutrient losses are illustrated in figure 4. Many disturbances significantly reduce the rate of biomass addition, also called net ecosystem production (NEP). Following disturbance, there is a relatively short-lived period when nutrients are lost from

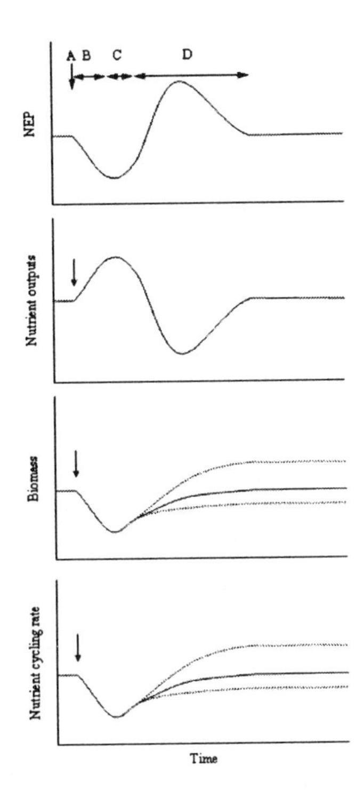

FIGURE 4 NEP and nutrient outputs can be used to measure return time, but biomass and nutrient cycling rates are needed to determine if a flip in ecosystem state has occurred. Solid lines indicate recovery to the same state and dotted lines indicate a different state. Vertical arrows indicate the time of disturbance, A. Horizontal arrows indicate length of the three phases of recovery, non-biotic losses: B, initial increase in NEP, C, and succession, D.

the ecosystem because plant growth has not yet begun, and NEP may decline farther due to continued decomposition in the absence of new plant production. This is followed by another relatively fast period when nutrient outputs begin to decrease as a result of increased NEP and corresponding increases in nutrient uptake [47]. The final, slower phase is marked by a return to the pre-disturbance state via succession [93].

Quantitatively, resilience could then be measured as the amount of time it takes to complete this sequence. Changes in biomass, NEP, nutrient outputs, and nutrient cycling all show temporal dynamics that allow quantification of return time [47]. NEP and nutrient outputs may not be the most representative measures of resilience, however, because very different types of ecosystems will experience NEP approaching zero and outputs approaching inputs with successional time [93], potentially making it difficult to determine if the ecosystem has transitioned to a different state after disturbance. However, static measures of biomass coupled with dynamic measures of the rates of nutrient cycling, as depicted in figure 4, may together be more sensitive indicators. Both total biomass and rates of internal cycling (i.e., the rate of nutrient resupply between plants and soil in fig. 3) respond to the effects of changing species composition, thereby providing more complete information.

Theory also makes a number of predictions about which types of ecosystems recover faster and are likely to be more resilient. More mature ecosystems often have greater accumulation of soil organic matter, which can be a source of nutrients during recovery [93], and, therefore, speeds return time during the successional phase. However, more mature forests have slower return times [93] than less mature forests [47], because they must undergo a longer successional phase. Thus, ecosystems in which succession is arrested by continual disturbance, such as grasslands, may recover faster. Ecosystems with smaller biomass at the mature state should also recover faster through the successional phase because maximum NEP is related to the supply rate of nutrients into the system [93], so for the same amount of input, a smaller biomass is reached sooner than a larger biomass. In a similar way, systems with high nutrient inputs also move through the successional phase more quickly [91].

3 ECOLOGICAL MECHANISMS AFFECTING RESILIENCE: STRUCTURE AND FUNCTIONING

Ecological mechanisms of resilience are fundamentally related to the structure of the ecosystem and to how this structure maintains functioning within the system. Resilience rests upon processes operating at multiple levels within the ecosystem. On the one hand, individual species within the ecosystem may be resilient and thereby confer resilience to the ecosystem [31, 33] (fig. 5(a)). In contrast, resilience may emerge from the very lack of resilience of components [45]; that is, while individual species themselves may not be resilient, resilience emerges as a property of the system as a whole [24, 86, 94] (fig. 5(b)), through processes characteristic of complex adaptive systems. This dichotomy mirrors the tradeoffs between redundancy and anti-redundancy developed by Krakauer and Plotkin [42]. Krakauer and Plotkin's use of redundancy is similar to the notion of degeneracy of Edelman and Gally [23], who distinguish redundancy (repetition of identical units) from degeneracy (functional redundancy of distinct types). There is a third aspect as well: the influence of modular structure on resilience [45] (see also Watts [98]) (fig 5c), namely the importance of the decomposition of the system into modular structures that restrain the spread of disturbance [77].

In many systems, resilience rests upon particular keystone species that are linchpins for system structure (fig. 5(a)). Paine [62] originally introduced the notion of keystone species (specifically, predators) for the intertidal communities of rocky coasts, in which starfish are keystone predators; but the notion has been broadened and proven to be very influential in a wide variety of other systems [70]. Sea otters are keystone predators in near-shore marine systems [27], disease organisms may serve keystone roles in other systems [9], or competitors [14, 56] may fill this niche in yet other situations: in many ecosystems, keystone species may not exist, but keystone functional groups, such as nitrogen fixers [97], may play analogous roles.

Despite the usefulness of the concept, detection of keystone species is not an easy task, and resilience of non-keystone species may not scale up to provide ecosystem-level resilience. The *Castanea dentate* or American chestnut exhibited over- and understory dominance [7], thereby exhibiting some measure of constancy at the species level. However, novel events occurred: eventually the chestnut was driven virtually extinct by an invasive fungus, with very few system-level consequences [92]. This example also illustrates the importance of considering the trophic status of a candidate keystone species. Keystone species are often of high trophic levels, and their removal can lead to cascading system consequences [70]. The chestnut, in contrast, as a dominant competitor but not a keystone, was simply replaced by subdominants performing similar functions.

Although conventional stability may decline with system complexity [51], resilience at the system level may increase (fig. 5(b)). Various studies exhibit positive correlations between diversity and key system measures (fig. 6). Much of this derives from experimental work [53, 86] where the constructed ecosystems are depauperate versions of real ecosystems. Consequently, they have been criticized as representing inadequate reflections of reality; yet the results of these experiments cannot be ignored. The within-ecosystem observation that resilience increases with diversity has also been suggested to hold across ecosystems [24, 94], although such comparison introduces difficulties of interpretation; Grime [33], for example, argues that species-poor ecosystems can also be extremely resilient.

Biological diversity can be key for maintaining the resilience of ecological systems for a variety of reasons. First of all, more diverse systems are more likely to enjoy the functional redundancy that can buffer systems against loss of species as well as a diversity of functional roles that may also be important for robustness. Secondly, just as diversity within a species is essential for natural selection to operate, so, too, does diversity among species increase the potential that species will emerge to fill key functional roles in a changing environment. In particular, diverse systems may be less susceptible to alien invasion because they admit fewer unexploited niches [54]. However, this last point is debatable, especially for mainland communities, where even species-poor ecosystems may be saturated. That is, low diversity systems may be ones that simply admit less diversification, rather than unsaturated ones. Thus, the relationship between high species diversity and invasibility depends strongly on the mechanisms maintaining diversity, and how they relate to saturation [52]. In fact, some mechanisms that maintain diversity also promote invasion.

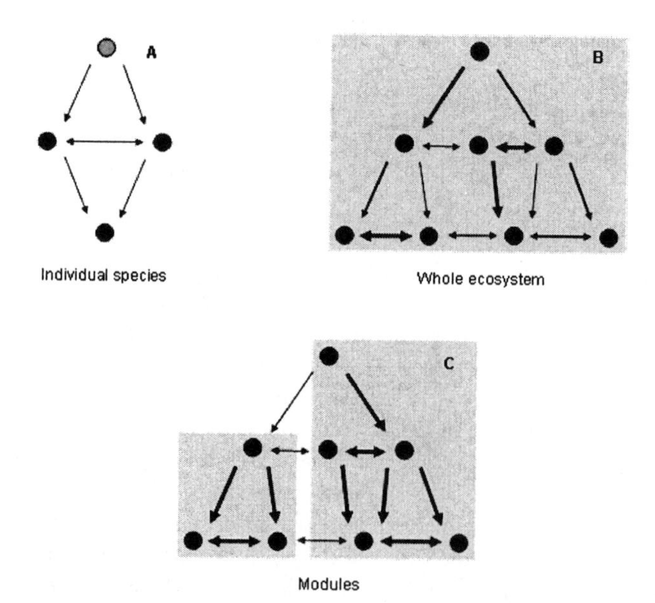

FIGURE 5 Circles represent species within an ecosystem and arrows represent their interactions. The weight of the arrow corresponds to the strength of the interaction. Grey indicates the level at which resilience lies. (a) Resilience due to a resilient, keystone species. (b) Resilience emergent at the level of the whole ecosystem. (c) Resilience due to modular structure of interactions.

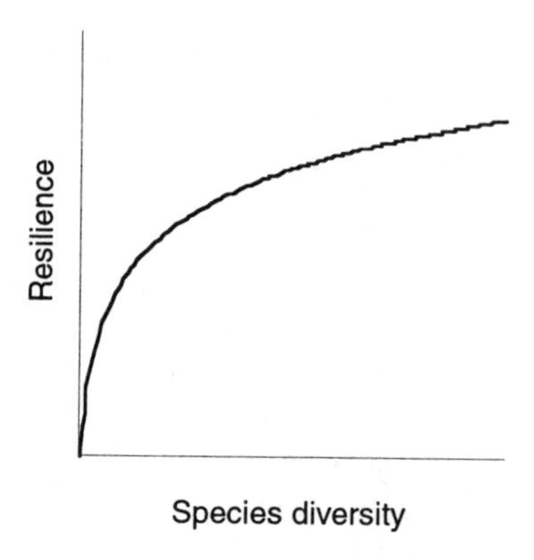

FIGURE 6 Resilience is an increasing function of species diversity in some experimental systems (after Peterson et al. [67]).

Tilman's landmark work on diversity and resilience has been criticized on the grounds that the observed relationships reflect sampling effects: that is, more diverse ecosystems are also more likely to contain those species best suited to the particular conditions [39]. Tilman argues that this is just part of the story [34, 85], and that over-yielding results in his experiments are evidence that more diverse systems exhibit higher yields in part due to niche partitioning. But even if sampling effects were responsible, at least to some extent, this is an important property. In changing environments, the sampling effect increases the likelihood that the system will continue functioning [57]. Sampling introduces a portfolio effect [88], so that an assemblage of species will show much less variation in biomass than will individual species.

Modularity (fig. 5(c)) can also lead to resilience because perturbations that affect some species within the ecosystem may be bounded within one module, insulating the rest of the ecosystem and critical functions from perturbation [45]. Examples include not only the spread of diseases and forest fires, but also the cascades of extinctions that arise in non-modular systems [80, 83]. Although modular systems, made up of smaller individual units loosely connected into metasystems [63], necessarily exhibit less redundancy within units, they can confer greater redundancy, and hence resilience, at higher levels of organization. Indeed, the relationship between modularity and resilience is at the core of debates in conservation biology about the optimal design of nature reserves. Modularity arises naturally in self-organizing systems [98], and it is a challenge to understand the evolution and consequences of modularity in ecological systems.

McCann et al. [49] also suggest that modularity in trophic structure enhances resilience in most systems. Weak interactions stabilize a community by dampening the effects of strong interactions, effectively by creating a modular structure. McCann et al. [49] argue that the limited food web data available show that weak interactions are common, conveying resilience to many systems [8, 28, 32, 61, 71, 101]. Additionally, more recent analyses of food web data find evidence of modular structure in some food webs [22, 43].

In many ecosystems, the players (key species) stay the same over time, buffered against at least familiar patterns of disturbance, but perhaps susceptible to novel disturbances. If those species are lost, ecosystem properties may or may not be altered, depending on the level of functional redundancy in the system. Resilience thus depends in part upon such redundancy, especially when the environment remains relatively constant. In the face of climate change, however, or other large-scale changes in the environment, such redundancy is of limited value. More important in such cases is variability or heterogeneity and the ability of the biota to adapt to a changing environment. To some extent of course, redundancy and heterogeneity are in opposition, creating tradeoffs in the assessment of resilience. Modularity also plays a role by containing the expansion of disturbances, but is also traded off to some degree against both redundancy and heterogeneity [45, fig. 1].

4 THE EVOLUTION OF RESILIENCE

Given that natural selection operates at the level of individual genomes, it is natural to ask what such evolution implies for resilience at the level of whole systems. A priori, there is no reason that selection at lower levels should increase the resilience of whole systems; indeed, theoretical models demonstrate that myopic intraspecific evolution can lead to the demise of the parent species, as well as to cascades of extinctions [99]. Thus, one of the fundamental open theoretical questions involves comparing designed systems, or those such as organisms that have been selected for robustness, with complex adaptive systems, in which resilience at the ecosystem level is a byproduct of selection at lower levels of organization.

Selection at lower levels of organization may or may not enhance system resilience. Simplistically, selection in a constant and homogeneous environment would be expected to reduce diversity, weaning out all but those most suited to the particular conditions. However, relatively constant tropical environments can support high diversity because of high solar radiation, larger land area, and the role of chance colonization events (founder effects) coupled with the potential for microhabitat differentiation. Thus, tropical systems may be quite resilient to a variety of disturbances, especially those that are localized and uncorrelated in nature. In temporally variable systems, selection may favor generalists; but more typically coexistence of types is facilitated [40], which is likely to enhance resilience in temporally changing environments, even when fluctuations are spatially correlated over broad scales. If localized disturbances are sufficiently frequent over a broad enough spatial area, species will emerge that can exploit those disturbances [45]. Regular disturbance leads to a diversity of species that can exploit the various stages of a community's successional development, from the opportunists that come in just after disturbance to the competitive dominants that do best under conditions of crowding, or shading. There is not, in general, selection to maintain such functional diversity, because the ecosystem is not an evolutionary unit. Still, if the environment is changing sufficiently frequently over a broad enough spatial regime, evolution can operate to select for those species that are most able to exploit that variation and perhaps respond to novel variation as well. Thus, the evolutionary history of an area dictates its capacity to respond to perturbations, both familiar and novel. This provides the foundation for making comparisons across systems, based, for example, on how frequent disturbances have been, or more generally, on how variable the environment is.

Previous disturbance history shapes the action of natural selection, and has implications for how the ecosystem will respond to changes [6]. Natural selection acting through smaller-scale disturbance selects for species that are adapted to those disturbances, and that are perhaps better able to respond to larger scale or fundamentally different types of disturbances. This may impart a form of resilience at the ecosystem level. This resilience, if it emerges, does so entirely from selection at lower levels of organization. Indeed, natural selection can operate above the level of individual genomes, selecting for cooperative or altruistic behavior through kin selection [35], or for mutualisms among species [29] or reduced virulence in parasite-host systems [13]. As one moves above these levels, however, coevolution becomes more diffuse [25, 84], and the units less well-defined. Natural selection at the individual level then swamps the effects of selection at the higher levels [100] and overwhelms the potential of higher level selection to impart resilience. How then do regularities observed at the ecosystem level, such as the constant element ratios that characterize marine systems [72], or the stability of nutrient cycles themselves, emerge? These represent some of the greatest puzzles in evolutionary theory today, at the nexus between population biology and ecosystem science, and are the subjects of intensive study.

Disturbance is one of the most fundamental forces shaping the resilience of ecosystems in the face of change, but not all disturbances are the same. Physical disturbances such as hurricanes and fires tend to be large-scale, abrupt, but heterogeneous in their damage, leaving refuges from which species may recolonize. Biological invaders, including competitors and exploiters alike, have their own characteristic patterns, influenced by features such as the dispersal characteristics of species, and coevolved relationships among them. These, too, may be patchy, as they are for plant pathogens. The history of disturbance is, therefore, captured not only in the size and frequency of disturbances, but also in other more subtle features such as intensity and genetic specificity.

Different disturbance types may interact and facilitate one another. Just as one disease may compromise an individual's immune response to others, so, too, some disturbances can affect a system's ability to respond to others. For example, the invasion of Hawaiian forests by bunchgrass, *Schigachyrium condensatum*, has increased the vulnerability of the forest to fire by increasing the fuel load [19]. On small Caribbean islands, the invasion of a large lizard predator reduced the population size of smaller prey lizards, which increased their probability of extinction following a hurricane [76].

There may also be tradeoffs in the resilience of individual species to different disturbance types. For example, species that are competitive dominants, such as the chestnut, can cover large areas of the ecosystem, both spatially and temporally, in the absence of disturbance. Because of this dominance and the heterogeneous nature of physical disturbances, such species may have refuges from which to colonize, thus giving rise to resilience in the face of physical disturbance. These same properties, however, may make them highly susceptible to pathogens because the species is highly connected in space and time; there are few refuges to escape the pathogen. Increasing the number of infected individuals within the neighborhood of a susceptible individual, which would be a consequence of increased density, increases pathogen persistence [2, 38]. Empirical evidence suggests that the spread of pathogens increases with decreasing distance among individuals [5], and modeling results suggest this is the case even when disease spread is via a vector [15]. Decreasing distance among populations also increases the chance of spreading disease in a metapopulation context [3, 64].

5 WIND STORMS AS AN EXAMPLE

Many of the basic principles discussed so far are well illustrated in a comparison of ecosystem responses to wind storms. Tree falls within forests provide an example of the importance of prior disturbance experience that selects for traits that scale up to provide resilience to more severe perturbations. Tree falls in broadleaf forests provide opportunities for early colonists, and hence select for resprouting ability and pioneer strategies of quick growth [20, 60, 90]. Resprouting is also an extremely important characteristic in recovery after wind damage, because it allows for the canopy to close quickly, keeping out vines [89] and other species that could change fundamental ecosystem properties. Pioneer strategies also generally play an important role in recolonization [102]; but when damage is extremely severe, pioneers are diluted and then out-competed, playing little role in recolonization [90]. Selection at the individual level for resprouting following tree fall does appear to scale up to provide resilience to the ecosystem; while pioneer strategies may or may not do the same, depending on the strength of the disturbance.

By comparing the response of different forest types to wind, we can evaluate the role of historical contingency in resilience. In tropical and temperate zones of the Americas, the response of broadleaf forests differs greatly from the response of conifer forests to wind damage. Broadleaf forests in general have a much higher rate of resprouting than conifer forests, and so recover more quickly [11] (and references therein). In mixed deciduous and conifer forests, dominance may pass from conifers to deciduous species following wind damage [21], with potential ecosystem-scale consequences for biomass and nutrient cycling. Broadleaf forests are dominated mostly by angiosperms, while conifer forests are dominated by gymnosperms. The high resprouting rates of angiosperms may actually suggest a strong role for historical contingency, coupled with previous disturbance experience. While conifers recover slowly following wind damage, their ecosystems often require regular fire regimes, again illustrating that tradeoffs exist in the ability of ecosystems to withstand different types of disturbance [4].

Tropical forests are more diverse than temperate forests [68], and that diversity could play an important role in their resilience. However, the patterns of recovery following wind damage in both broadleaf and conifer-dominated forests appear to be similar across temperate and tropical zones, at least in terms of return time [11], as measured by canopy closure and nitrogen losses from the system [17, 96]. This suggests that diversity per se is not as important in resilience to wind storms, but that specific traits shaped by evolution, like resprouting, are.

6 WHAT KINDS OF EMPIRICAL STUDIES ARE NEEDED?

Much of biogeochemical theory that suggests that biomass and nutrient cycling can be used to measure resilience is based on large data sets from forests in the Northeastern United States [46]. Empirical evidence suggests that these theories extend to other ecosystems as well, but the extent to which they apply is not certain. More complete measures of changes in biomass and nutrient cycles following disturbance are needed for a larger variety of ecosystems.

Several different types of comparisons of ecosystem response to disturbance would also be informative. These include the responses of systems with different inherent levels of disturbance to large-scale and novel perturbations, as well as the responses of systems with different topological organization.

7 CONCLUSIONS

We define resilience or robustness in ecological systems to mean the maintenance of macroscopic features of the ecosystem, such as productivity or basic biogeochemical processes. A wide variety of factors play roles in determining the robustness of an ecosystem including environmental variables such as nutrient inputs, as well as the structure and disturbance history of the ecosystem.

There are keys to robustness to be found in the structure of ecosystems and in the linkages among species. Functional redundancy, expressed both in terms of the repetition of individual units and in degeneracy (similar functions performed by diverse species), enhances robustness by providing backups or alternatives to replace damaged components. Yet, some forms of redundancy may compromise other structural features, in particular diversity (which allows adaptation) and modularity (which limits disturbances). Unfortunately, the signature of robustness is written in the tradeoffs among all of these characteristics (fig. 1).

Understanding robustness is facilitated by an examination of the evolutionary mechanisms that shape system-level responses. To what extent do the insights of Adam Smith [79] apply to ecosystems? That is, will selection at lower levels of organization enhance system robustness as an emergent property? How important is evolution above the level of individual species?

Although many factors clearly play a role, our understanding of how they interact to enhance robustness for a particular ecosystem remains poor. No ecosystem is robust to all types of disturbance, making robustness context-dependent in terms of the characteristics of both the ecosystem and the disturbance. This means a comparative approach is required among ecosystems with different ecological regimes and evolutionary histories and in the face of different disturbance regimes. Anthropogenic influences, from land use to species introduction, from exploitation to climate change, are altering environments dramatically, exposing species to conditions quite different from those in which they evolved. As a result, species are being lost at rates greater than they are being replaced, lowering biodiversity. This simplification of systems will fundamentally affect robustness, but to an extent we do not yet fully understand. Hence the study of the mechanisms underlying ecosystem robustness is an intellectual challenge with ramifications for the maintenance of our very way of life.

ACKNOWLEDGMENTS

We wish to thank L. O. Hedin, B. Z. Houlton, and J. C. von Fischer for their helpful discussion of biogeochemical processes. We are also pleased to acknowledge the support of the David and Lucile Packard Foundation through awards to the Santa Fe Institute and the Institute for Advanced Study, Princeton University, through grants 2000–01690 and IAS 7426–2208, and the National Science Foundation, through grant DEB–0083566 to S. A. Levin.

REFERENCES

[1] Aber, J. D., and J. M. Melillo. *Terrestrial Ecosystems.* 2d. ed. San Diego, CA: Harcourt Academic Press, 2001.

[2] Anderson, R. M., and R. M. May. "The Invasion, Persistence and Spread of Infectious Diseases within Animal and Plant Communities." *Phil. Trans. Roy. Soc. Lond. Ser. B. Biol. Sci.* **314** (1986): 533–570.

[3] Antonovics, J. "Ecological Genetics of Metapopulations: The Silene-Ustilago Plant-Pathogen System." In *Ecological Genetics*, edited by L. Real. Princeton, NJ: Princeton University Press, 1994.

[4] Attiwill, P. M. "The Disturbance of Forest Ecosystems—The Ecological Basis for Conservative Management." *Forest Ecol. & Mgmt.* **63** (1994): 247–300.

[5] Bailey, D. J., W. Otten, and C. A. Gilligan. "Saprotrophic Invasion by the Soil-Borne Fungal Plant Pathogen *Rhizoctonia solani* and Percolation Thresholds." *New Phytologist* **146** (2000): 535–544.

[6] Balmford, A. "Extinction Filters and Current Resilience: The Significance of Past Selection Pressures for Conservation Biology." *Trends Ecol. & Evol.* **11** (1996): 193–196.

[7] Barnes, B. V., D. R. Zak, S. R. Denton, and S. H. Spurr. *Forest Ecology.* 4th ed. New York: John Wiley & Sons, Inc., 1998.

[8] Berlow, E. L., S. A. Navarette, C. J. Briggs, M. E. Power, and B. A. Menge. "Quantifying Variation in the Strengths of Species Interactions." *Ecology* **80** (1999): 2206–2224.

[9] Bond, W. J. "Keystone Species." In *Biodiversity and Ecosystem Function*, edited by E. D. Schulze and H. A. Mooney, 237–253, vol. 99. New York: Springer-Verlag, 1993.

[10] Bormann, F. H., and G. E. Llikens. *Pattern and Process in a Forested Ecosystem.* New York: Springer-Verlage, 1981.

[11] Boucher, D. H., J. H. Vandermeer, K. Yih, and N. Zamora. "Contrasting Hurricane Damage in Tropical Rain Forest and Pine Forest." *Ecology* **71** (1990): 2022–2024.

[12] Bruns, T. D. "Thoughts On the Processes that Maintain Local Species-Diversity of Ectomycorrhizal Fungi." *Plant and Soil* **170** (1995): 63–73.

[13] Bull, J. J. "Perspective—Virulence." *Evolution* **48** (1994): 1423–1437.

[14] Callaghan, T. V., M. C. Press, J. A. Lee, D. L. Robinson, and C. W. Anderson. "Spatial and Temporal Variability in the Responses of Arctic Terrestrial Ecosystems to Environmental Change." *Polar Resh.* **18** (1999): 191–197.

[15] Caraco, T., M. C. Duryea, S. Glavanakov, W. Maniatty, and B. K. Szymanski. "Host Spatia Heterogeneity and the Spread of Vector-Borne Infection." *Theor. Pop. Biol.* **59** (2001): 185–206.

[16] Carpenter, S. R., D. Ludwig, and W. A. Brock. "Management of Eutrophication for Lakes Subject to Potentially Irreversible Change." *Ecol. App.* **9** (1999): 751–771.

[17] Cooper-Ellis, S., D. R. Foster, G. Carlton, and A. Lezberg. "Forest Response to Catastrophic Wind: Results from an Experimental Hurricane." *Ecology* **80** (1999): 2683–2696.

[18] Cornell, H. V., and D. M. Kahn. "Guild Structure in the British Arboreal Arthropods—Is It Stable and Predictable." *J. Animal Ecol.* **58** (1989): 1003–1020.

[19] D'Antonio, C. M., R. F. Hughes, and P. M. Vitousek. "Factors Influencing Dynamics of Two Invasive C-4 Grasses in Seasonally Dry Hawaiian Woodlands." *Ecology* **82** (2001): 89–104.

[20] Denslow, J. S. "Tropical Rain Forest Gaps and Tree Species Diversity." *Ann. Rev. Ecol. & Syst.* **18** (1987): 431–451.

[21] Dunn, C. P., G. R. Guntenspergen, and J. R. Dorney. "Catastrophic Wind Disturbance in an Old-Growth Hemlock Hardwood Forest, Wisconsin." *Can. J. Bot.—Revue Canadienne de Botanique* **61** (1983): 211–217.

[22] Dunne, J. A., R. J. Williams, and N. D Martinez. "Food Web Structure and Network Theory: The Role of Convectance and Size." *PNAS* **99** (2002):12917–12922.

[23] Edelman, G. M., and J. A. Gally. "Degeneracy and Complexity in Biological Systems." *Proc. Natl. Acad. Sci.* **98** (2001): 13763–13768.

[24] Ehrlich, P., and A. Ehrlich. *Extinction: The Causes and Consequences of the Disappearance of Species.* New York: Random House, 1981.

[25] Ehrlich, P. R., and P. H. Raven. "Butterflies and Plants: A Study in Coevolution." *Evolution* **18** (1965): 586–608.

[26] Eigen, M. "On the Nature of Virus Quasispecies." *Trends Microbiol.* **4** (1996): 216–218.

[27] Estes, J. A., and J. F. Palmisan. "Sea Otters—Their Role in Structuring Nearshore Communities." *Science* **185** (1974): 1058–1060.

[28] Fagan, W. F., and L. E. Hurd. "Hatch Density Variation of a Generalist Arthropod Predator—Population Consequences and Community Impact." *Ecology* **75** (1994): 2022–2032.

[29] Frank, S. A. "Genetics of Mutualism—The Evolution of Altruism Between Species." *J. Theor. Biol.* **170** (1994): 393–400.

[30] Fry, B. "Using Stable Isotopes to Monitor Watershed Influences on Aquatic Trophodynamics." *Can. J. Fisheries & Aquatic Sci.* **56** (1999): 2167–2171.

[31] Givnish, T. J. "Does Diversity Beget Stability." *Nature* **371** (1994): 113–114.

[32] Goldwasser, L., and J. Roughgarden. "Construction and Analysis of a Large Caribbean Food Web." *Ecology* **74** (1993): 1216–1233.

[33] Grime, J. P. "Dominant and Subordinate Components of Plant Communities: Implications for Succession, Stability, and Diversity." In *Colonization, Succession and Stability*, edited by A. J. Gray, M. J. Crawley and P. J. Edwards, 413–428. Oxford: Blackwell, 1987.

[34] Haddad, N. M., D. Tilman, J. Haarstad, M. Ritchie, and J. M. H. Knops. "Contrasting Effects of Plant Richness and Composition on Insect Communities: A Field Experiment." *Am. Natural.* **158** (2001): 17–35.

[35] Hamilton, W. D. "Evolution of Altruistic Behavior." *Am. Natural.* **97** (1964): 354–356.

[36] Hedin, L. O., J. J. Armesto, and A. H. Johnson. "Patterns of Nutrient Loss from Unpolluted, Old-Growth Temperate Forests—Evaluation of Biogeochemical Theory." *Ecology* **76** (1995): 493–509.

[37] Holling, C. S. "Resilience and Stability of Ecological Systems." *Ann. Rev. Ecol. & Syst.* **4** (1973): 1–23.

[38] Holmes, E. E. "Basic Epidemiological Concepts in a Spatial Context." In *Spatial Ecology: The Role of Space in Population Dynamics and Interspecific Interactions*, edited by D. Tilman and P. Kareiva. Princeton, NJ: Princeton University Press, 1997.

[39] Huston, M. A. "Hidden Treatments in Ecological Experiments: Reevaluating the Ecosystem Function of Biodiversity." *Oecologia* **110** (1997): 449–460.

[40] Hutchinson, G. E. "Population Studies—Animal Ecology and Demography—Concluding Remarks." *Cold Spring Harbor Symp. Quant. Biol.* **22** (1957): 415–427.

[41] Kelly, J. F. "Stable Isotopes of Carbon and Nitrogen in the Study of Avian and Mammalian Trophic Ecology." *Can. J. Zool.—Revue Canadienne de Zoologie* **78** (2000): 1–27.

[42] Krakauer, D. C., and J. B. Plotkin. "Redundancy, Anti-redundancy, and the Robustness of Genomes." *Proc. Natl. Acad. Sci. USA* **99** (2002): 1405–1409.

[43] Krause, A. E., K. A. Frank, D. M. Mason, R. E. Ulanowicz, and W. W. Taylor. "Compartments Revealed in Food Web Structure." *Nature* **426** (2003): 282–285.

[44] Levin, S. A. "Ecosystems and the Biosphere as Complex Adaptive Systems." *Ecosystems* **1** (1998): 431–436.

[45] Levin, S. A. *Fragile Dominion.* Cambridge, MA: Perseus Publishing, 1999.

[46] Likens, G. E., and F. H. Bormann. *Biogeochemistry of a Forested Ecosystem.* 2d. New York: Springer-Verlag, 1995.

[47] Likens, G. E., F. H. Bormann, R. S. Pierce, and W. A. Reiners. "Recovery of a Deforested Ecosystem." *Science* **199** (1978): 492–496.

[48] Ludwig, D., D. D. Jones, and C. S. Holling. "Qualitative Analysis of Insect Outbreak Systems: The Spruce Budworm and Forest." *J. Animal Ecol.* **47** (1978): 315–332.

[49] McCann et al. 1998.

[50] MacArthur, R. H., and E. O. Wilson. "An Equilibrium Theory of Insular Zoogeography." *Evolution* **17** (1963): 373–387.

[51] May, R. M. *Stability and Complexity in Model Ecosystems.* Princeton, NJ: Princeton University Press, 1973.

[52] Moore, J. L., N. Mouquet, J. H. Lawton, and M. Loreau. "Coexistence, Saturation and Invasion Resistance in Simulated Plant Assemblages." *Oikos* **94** (2001): 303–314.

[53] Naeem, S., L. J. Thompson, S. P. Lawler, J. H. Lawton, and R. M. Woodfin. "Declining Biodiversity Can Alter the Performance of Ecosystems." *Nature* **368** (1994): 734–737.

[54] Naeem, S., J. M. H. Knops, D. Tilman, K. M. Howe, T. Kennedy, and S. Gale. "Plant Diversity Increases Resistance to Invasion in the Absence of Covarying Extrinsic Factors." *Oikos* **91** (2000): 97–108.

[55] Nasholm, T., A. Ekblad, A. Nordin, R. Giesler, M. Hogberg, and P. Hogberg. "Boreal Forest Plants Take Up Organic Nitrogen." *Nature* **392** (1998): 914–916.

[56] Newsham, K. K., A. R. Watkinson, H. M. West, and A. H. Fitter. "Symbiotic Fungi Determine Plant Community Structure—Changes in a Lichen-Rich Community Induced by Fungicide Application." *Functional Ecol.* **9** (1995): 442–447.

[57] Norberg, J., D. P. Swaney, J. Dushoff, J. Lin, R. Casagrandi, and S. A. Levin. "Phenotypic Diversity and Ecosystem Functioning in Changing Environments: A Theoretical Framework." *Proc. Natl. Acad. Sci. USA* **98** (2001): 11376–11381.

[58] Nowak, M. A. "What Is a Quasi-Species." *Trends Ecol.& Evol.* **7** (1992): 118–121.

[59] O'Neil, R. V. "Is It Time to Bury the Ecosystem Concept? (With full military honors, of course!)." *Ecology* **82** (2001): 3275–3284.

[60] Paciorek, C. J., R. Condit, S. P. Hubbell, and R. B. Foster. "The Demographics of Resprouting in Tree and Shrub Species of a Moist Tropical Forest." *J. Ecol.* **88** (2000): 765–777.

[61] Paine, R. T. "Food-Web Analysis through Field Measurement of Per-Capita Interaction Strength." *Nature* **355** (1992): 73–75.

[62] Paine, R. T. "Food Web Complexity and Species Diversity." *Am. Natural.* **100** (1966): 65–75.

[63] Paine, R. T. "Food Webs—Linkage, Interaction Strength and Community Infrastructure—The 3rd Tansley Lecture." *J. Animal Ecol.* **49** (1980): 667–685.

[64] Park, A. W., S. Gubbins, and C. A. Gilligan. "Invasion and Persistence of Plant Parasites in a Spatially Structured Host Population." *Oikos* **94** (2001): 162–174.

[65] Perakis, S. S., and L. O. Hedin. "Fluxes and Fates of Nitrogen in Soil of an Unpolluted Old-Growth Temperate Forest, Southern Chile." *Ecology* **82** (2001): 2245–2260.

[66] Peterson, B. J., and R. W. Howarth. "Sulfur, Carbon, and Nitrogen Isotopes Used to Trace Organic Matter Flow in the Salt Marsh Estuaries of Sapelo Island, Georgia." *Limnology & Oceanography* **32** (1987): 1195–1213.

[67] Peterson, G., C. R. Allen, and C. S. Holling. "Ecological Resilience, Biodiversity, and Scale." *Ecosystems* **1** (1998): 6–18.

[68] Pitman, N. C. A., J. W. Terborgh, M. R. Silman, P. V. Nunez, D. A. Neill, C. E. Ceron, W. A. Palacios, and M. Aulestia. "Dominance and Distribution of Tree Species in Upper Amazonian Terra Firma Forests." *Ecology* **82** (2001): 2101–2117.

[69] Post, D. M., M. L. Pace, and N. G. Hairston. "Ecosystem Size Determines Food-Chain Length in Lakes." *Nature* **405** (2000): 1047–1049.

[70] Power, M. E., D. Tilman, J. A. Estes, B. A. Menge, W. J. Bond, L. S. Mills, G. Daily, J. C. Castilla, J. Lubchenco, and R. T. Paine. "Challenges in the Quest for Keystones." *Bioscience* **46** (1996): 609–620.

[71] Raffaelli, D. G., and S. J. Hall. "Assessing the Importance of Trophic Links in Food Webs." In *Food Webs: Integration of Patterns and Dynamics*, edited by G. A. Polis and K. O. Winemiller, 185–191. New York: Chapman & Hall, 1995.

[72] Redfield, A. C. "On the Proportions of Organic Derivatives in Sea Water and Their Relation to the Composition of Plankton." In *James Johnstone Memorial Volume*, edited by R. J. Daniel, 177–192. Liverpool: University Press of Liverpool, 1934.

[73] Root, R. B. "Guilds." In *Are There Expendable Species?*, edited by P. Kareiva and S. Levin, 295–301. Princeton, NJ: Princeton University Press, 2001.

[74] Root, R. B. "The Niche Exploitation Pattern of The Blue-Grey Gnatcatcher." *Ecol. Monographs* **37** (1967): 317–350.

[75] Scheffer, M., S. H. Hosper, M. L. Meijer, and B. Moss. "Alternative Equilibria in Shallow Lakes." *Trends Ecol. & Evol.* **8** (1993): 275–279.
[76] Schoener, T. W., D. A. Spiller, and J. B. Losos. "Predators Increase the Risk of Catastrophic Extinction of Prey Populations." *Nature* **412** (2001): 183–186.
[77] Simon, H. A. "The Architecture of Complexity." *Proc. Am. Phil. Soc.* **106** (1962): 467–482.
[78] Small, P. F., and C. E. Malone. "Seasonal Variation in the Structure of Diatom Communities in Two Pennsylvania Streams." *J. Freshwater Ecol.* **13** (1998): 15–20.
[79] Smith, A. *The Wealth of Nations.* New York: Knopf, 1776. Reprinted 1991.
[80] Solé, R. V., S. C. Manrubia, M. Benton, and P. Bak. "Self-Similarity of Extinction Statistics in the Fossil Record." *Nature* **388** (1997): 764–767.
[81] Stark, J. M., and S. C. Hart. "High Rates of Nitrification and Nitrate Turnover in Undisturbed Coniferous Forests." *Nature* **385** (1997): 61–64.
[82] Steneck, R. S., and M. N. Dethier. "A Functional-Group Approach to the Structure of Algal-Dominated Communities." *Oikos* **69** (1994): 476–498.
[83] Terborgh, J., L. Lopez, P. Nunez, M. Rao, G. Shahabuddin, G. Orihuela, M. Riveros, R. Ascanio, G. H. Adler, T. D. Lambert, and L. Balbas. "Ecological Meltdown in Predator-Free Forest Fragments." *Science* **294** (2001): 1923–1926.
[84] Thompson, J. N. "Concepts of Coevolution." *Trends Ecol. & Evol.* **4** (1989): 179–183.
[85] Tilman, D. "The Ecological Consequences of Changes in Biodiversity: A Search for General Principles." *Ecology* **80** (1999): 1455–1474.
[86] Tilman, D., and J. A. Downing. "Biodiversity and Stability in Grasslands." *Nature* **367** (1994): 363–365.
[87] Tilman, D., J. Knops, D. Wedin, P. Reich, M. Ritchie, and E. Siemann. "The Influence of Functional Diversity and Composition on Ecosystem Processes." *Science* **277** (1997): 1300–1302.
[88] Tilman, D., C. L. Lehman, and C. E. Bristow. "Diversity-Stability Relationships: Statistical Inevitability or Ecological Consequence? " *Am. Natural.* **151** (1998): 277–282.
[89] Vandermeer, J., I. Granzow de la Cerda, and D. Boucher. "Contrasting Growth Rate Patterns in Eighteen Tree Species from a Post-Hurricane Forest in Nicaragua." *Biotropica* **29** (1997): 151–161.
[90] Vandermeer, J., I. Granzow de la Cerda, D. Boucher, I. Perfecto, and J. Ruiz. "Hurricane Disturbance and Tropical Tree Species Diversity." *Science* **290** (2000): 788–791.
[91] Vitousek, P. M., L. O. Hedin, P. A. Matson, J. H. Fownes, and J. Neff. "Within-System Element Cycles, Input-Output, Budgets, and Nutrient Limitation." In *Successes, Limitations and Frontiers in Ecosystem Science*, edited by M. L. Pace and P. M. Groffman, 432–451. New York: Springer, 1998.

[92] Vitousek, P. "Biological Invasions and Ecosystem Properties." In *Ecology of Biological Invasions of North America and Hawaii*, edited by H. A. Mooney and J. A. Drake, 163–178. New York: Springer-Verlag, 1986.

[93] Vitousek, P. M., and W. A. Reiners. "Ecosystem Succession and Nutrient Retention Hypothesis." *Bioscience* **25** (1975): 376–381.

[94] Walker, B. H. "Biodiversity and Ecological Redundancy." *Conservation Biol.* **6** (1992): 18–23.

[95] Walker, B. H. "Rangeland Ecology—Understanding and Managing Change." *Ambio* **22** (1993): 80–87.

[96] Walker, L. R., J. K. Zimmerman, D. J. Lodge, and S. GuzmanGrajales. "An Altitudinal Comparison of Growth and Species Composition in Hurricane-Damaged Forests in Puerto Rico." *J. Ecol.* **84** (1996): 877–889.

[97] Ward, D., and C. Rohner. "Anthropogenic Causes of High Mortality and Low Recruitment in Three Acacia Tree Taxa in the Negev Desert, Israel." *Biodiversity & Conservation* **6** (1997): 877–893.

[98] Watts, D. J., and S. H. Strogatz. "Collective Dynamics of 'Small-World' Networks." *Nature* **393** (1998): 440–442.

[99] Wedd, C. T. "A Complete Classification of Darwinian Extinction in Ecological Interactions." *Am. Natural.* **161** (2003): 181–205.

[100] Williams, G. C. *Adaptation and Natural Selection.* Princeton, NJ: Princeton University Press, 1966.

[101] Wootton, J. T. "Estimates and Tests of Per Capita Interaction Strength: Diet, Abundance, and Impact of Intertidally Foraging Birds." *Ecol. Monographs* **67** (1997): 45–64.

[102] Zimmerman, J. K., E. M. Everham, R. B. Waide, D. J. Lodge, C. M. Taylor, and N. V. L. Brokaw. "Responses of Tree Species to Hurricane Winds in Subtropical Wet Forest in Puerto Rico—Implications for Tropical Tree Life Histories." *J. Ecol.* **82** (1994): 911–922.

Robustness in Ecosystems

Brian Walker
Garry Peterson
John M. Anderies
Ann Kinzig
Steve Carpenter

1 INTRODUCTION

In exploring the difference between "stable" and "robust," Jen [16] considers robustness in systems to be a measure of feature persistence under perturbations. It is a property that emerges from the system's architecture and evolutionary history, and involves mechanisms for learning, problem solving, and creativity. The feature that persists is typically seen as the central feature for the system and Jen [16] stresses that it makes no sense to speak of a system being either stable or robust without specifying both the feature and the perturbation of interest.

This view of robustness is very similar to how ecologists define resilience, as illustrated in the account by Carpenter et al. [4] stressing resilience "of what, to what?" Learning and creativity (at least) are attributes that apply to ecosystems as well as to systems involving humans, as we'll discuss later. However, our focus in this chapter is biophysical. Though we are generally concerned with the dynamics of linked social-ecological systems, as fully integrated systems, our task here is to consider robustness in ecosystems.

In the most recent overview of resilience, Gunderson and Holling [9] define it as the capacity of a system to undergo disturbance and still maintain its functions and controls, in line with the original definition of ecological resilience by Holling [11]. It is measured by the magnitude of disturbance the system can experience and still functionally persist. Thus, a system may be resilient with respect to some attributes and not others. They contrast this with what they call "engineering resilience," attributed to Pimm [28], where the appropriate measure is the ability of the system to resist disturbance, and the rate at which it returns to equilibrium following disturbance (see also Tilman and Downing [36]). One reason for the appeal of the Pimm definition is that it lends itself more easily to being estimated. If the system concerned can be described as an interaction matrix (predator/prey, competition), and values can be assigned to the elements of this matrix (i.e., the effects of the species on each other), then the speed of return to equilibrium following a small perturbation can be measured by the size of the dominant eigenvalue of the matrix. The emphasis on "small" perturbation is important because the eigenvalue can't be used when the system is far from equilibrium. Trying to measure the Holling version of resilience (the width of the basin of attraction) is conceptually not as simple, and we return to this in the final section. From an ecologist's point of view, though it is of some interest to know how strongly a system will return to its former (equilibrium) state after a small disturbance, many of the most important questions have to do with the performance of the system when it is far from equilibrium, and especially when it is in the neighborhood of a threshold.

Ecological and engineering resilience reflect different properties of systems. Ecological resilience concentrates on the ability of a set of mutually reinforcing structures and processes to persist. It allows ecologists or managers to focus upon transitions between different system organizations, which are maintained by different sets of organizing processes and structures. Engineering resilience, on the other hand, concentrates on conditions near a steady state where transient measurements of rate of return are made following small disturbances. Engineering resilience focuses upon conditions near the present state of a system, while ecological resilience focuses on plausible alternative system configurations. Consequently, ecological resilience is more useful than engineering resilience for assessing either the response of a system to large perturbations, or when gradual changes in a system's properties may cause the system to move from one organization to another.

Ecological resilience focuses attention on the dynamics of systems at multiple scales. In complex systems it appears to be produced by a hierarchically organized network of interactions. These interactions enhance resilience by producing and maintaining the persistence of functions at different scales, as well as functions within a scale, and replicating them [27].

The replication of ecological function at different scales contributes to resilience, because ecological disruption usually occurs across a limited range of scales, allowing ecological functions that operate at other, undisturbed scales to persist. Ecological functions that are replicated across a range of scales can withstand a variety of disturbances. For example, Fragoso [6] has shown that in Brazil's Maracá Island Ecological Reserve, palm seeds are dispersed across a range of scales by a variety of species. The seed dispersers range in size from small rodents, which typically disperse seeds within 5 m of parent trees, to tapirs (*Tayassu tajacu*), which disperse seeds up to 2 km. Dispersal at multiple scales allows the palm population to persist despite a variety of disturbances occurring at different scales. Small-scale dispersal maintains palm patches, while larger-scale dispersal bypasses barriers at small scales to create new palm patches.

Functional diversity within a scale also contributes to resilience. In the previous example, a variety of species, including deer, peccaries, primates, and rodents, disperse palm seeds at the same scale (short distances) [6]. Changes in one of these species do not have a large effect on short distance seed dispersal because of compensatory dispersal by others. An increase in seed availability allows other species to increase their seed dispersal. This compensating complementarity [7] among seed dispersers operating at similar scales reduces the impact of population fluctuations on ecosystem function, increasing its resilience. Similarly, Walker et al. [37] demonstrated that when grazing eliminated the dominant species of grasses in an Australian rangeland they were replaced by functionally similar species that were less abundant on the previously lightly grazed site. Variation in environmental sensitivities among species that perform similar functions allowed functional compensation to occur.

Cross-scale effects extend beyond compensation. In some systems disturbance at small scales may be required to sustain states at larger scales. An example is provided by the recovery of South Florida's mangrove forests following Hurricane Andrew. The hurricane killed many large trees, but young flexible trees survived. These young trees occurred in gaps in the forest produced by frequent, small-scale fires produced by lightning strikes. Recovery of the forest from large-scale disturbance was facilitated by the diversity produced by past small-scale disturbances [33].

These types of dynamics are reflected in other literatures as well—Tainter [34, 35] in archaeology, and Schumpeter [31] in economics. It makes operationalizing resilience more difficult, because in addition to deciding what key functions we want to sustain, we also have to decide at what scale we wish to sustain them. We then must acknowledge that maintaining them at the target scale may mean not only paying attention to other scales but allowing change at those other scales. Cross-scale behavior of systems is a key feature emerging from resilience analyses [9] and we return to it later.

While we think the Hollingesque definition of resilience better captures many of the key dynamic ecosystem features of interest to ecologists and managers, we recognize the additional importance of the concept of resistance contained in the "engineering" definition. Resistance concerns the amount of external pressure needed to bring about a given amount of disturbance in the system. Some ecosystems are particularly persistent because they are intrinsically resistant—they absorb high levels of external pressure without changing. Ecosystems on deep, sandy soils are an example. Whatever grazing or fire pressure is applied to the above-ground parts, the majority of the plant biomass remains underground. Such systems can be subjected to much greater herbivore numbers without showing the same degree of change as occurs on shallow soils. Because resilience is measured in terms of displacement in state space (a distance on an n-dimensional graph), and managers are faced with stresses (herbivore numbers, low rainfall), we need a relationship between physical stress and perturbation size. Resistance is a measure of this relationship. We, therefore, need to add resistance as the complementary attribute of resilience in order to assess long-term persistence. We are concerned with the magnitude of disturbance that can be experienced before a system moves into a different region of state space and different set of controls. The twin effects of resilience and resistance are illustrated in figure 1 (from Carpenter et al. [4]).

Based on the above interpretation, and in line with the definition adopted by a group of ecologists and social scientists working on resilience in regional social-ecological systems (www.resalliance.org), resilience has been described as having three defining characteristics:

1. the amount of change the system can undergo (and implicitly, therefore, the amount of extrinsic force the system can sustain, or resist) and still remain within the same domain of attraction (i.e., retain the same controls on structure and function);
2. the degree to which the system is capable of self-organization; and
3. the degree to which the system can build capacity to learn and adapt.

Adaptive capacity is a component of resilience that reflects the learning aspect of system behavior in response to disturbance [9]. Resilience, as such, can be desirable or undesirable. For example, system states that decrease social welfare, such as polluted water supplies or (in social systems) dictatorships, can be highly resilient. Sustainability, in contrast, is an overarching goal that includes assumptions or preferences about which system states are desirable.

2 EXAMPLES OF RESILIENCE IN ECOSYSTEMS

In systems with multiple stable states, gradually changing conditions may have little effect on the state of the ecosystem, but nevertheless reduce the size of the "basin of attraction"—the amount of state space in which target functional attributes can persist. This loss of resilience makes a particular configuration

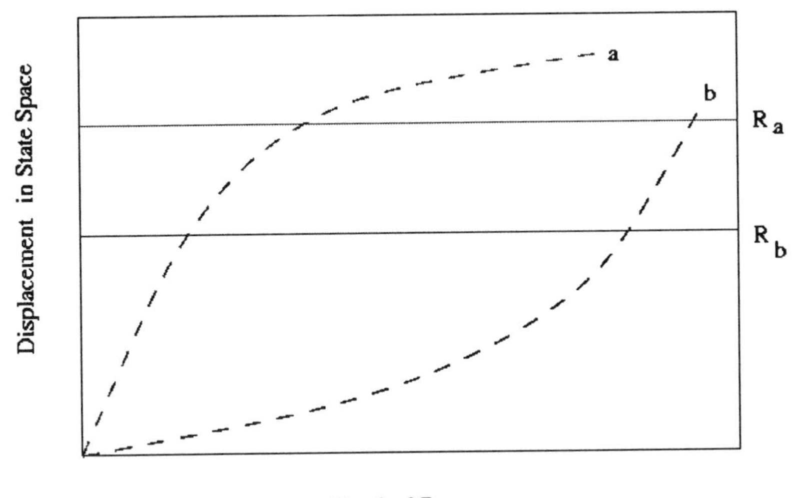

FIGURE 1 The two components of resilience—size of the basin of attraction and displacement versus exogenous force. The curves depict, for two hypothetical systems, *a* and *b*, the relationship between force and the distance in state space the system moves from equilibrium. System *b* is more resistant to displacement than system *a*. The horizontal lines indicate the resilience of the systems. System *a* is more resilient than system *b*. The intersection of a curve and line for a given system indicates its persistence—the level of exogenous forcing the system can tolerate before moving to a new domain of attraction. After Carpenter et al. [4]

more fragile in the sense that it can be easily tipped into a different configuration by stochastic events.

Scheffer et al. [30] describe examples of loss of resilience, leading to state shifts, in a variety of different ecosystem types—lakes, coral reefs, woodlands, deserts, and oceans—at various spatial and temporal scales. Taking the first of these, a well-known example is the sudden loss of transparency and vegetation observed in shallow lakes subject to human-induced eutrophication. The clear water and rich submerged vegetation state is changed through nutrient loading. Water clarity seems to be hardly affected by increased nutrient concentrations until a critical threshold is passed, at which the lake shifts abruptly from clear to turbid. With this increase in turbidity, submerged plants largely disappear and there is an associated loss of animal diversity. The restoration of clear water requires substantially lower nutrient levels than those at which the collapse of the vegetation occurred. Experimental work suggests that plants increase water clarity, thereby enhancing their own growing conditions. This causes the clear state to be a self-stabilizing alternative to the turbid situation. The reduction of phytoplankton biomass and turbidity by vegetation involves a suite of mechanisms including reduction of nutrients in the water column, protection of phytoplankton grazers such as *Daphnia* against fish predation, and prevention of sediment resuspension. In contrast, fish play a key role in maintaining the turbid state as they control *Daphnia* in the absence of plants. Also, in search for benthic food they re-suspend sediments, adding to turbidity. Whole lake experiments show that a temporary strong reduction of fish biomass as "shock therapy" can bring such lakes back into a permanent clear state if the nutrient level is not too high.

Semi-arid rangelands provide an analogous example of resilience in a terrestrial ecosystem. The same basic pattern has occurred in the rangelands of Africa, Australia, and North and South America. Rangelands are mixtures of grasses and woody plants (small trees and shrubs). The dynamics of the wood:grass ratio are driven by fire and grazing pressure under variable rainfall [40]. The long-run stable attractor (because with fire the system does not approach an equilibrium—it approaches a periodic orbit) of pre-human rangelands, set primarily by the soil moisture regime (dependent on rainfall and soil type), was likely a fairly open layer of shrubs and trees with a well-developed grass layer allowing for periodic fires. The fires would have created a spatially dynamic mosaic of age classes and vegetation structures. When burning by Aboriginal peoples (especially in Australia and Africa) increased fire frequencies above the pre-human, lightning-induced regimes, more open, grassy states were established. Subsequent exclusion of fire under pastoral management caused an increase in woody biomass at the expense of grass. A threshold effect in shrub density occurs at the point where the competitive effect of shrubs on grass prevents sufficient grass growth for the fuel needed to carry a fire. Anderies et al. [1] have modeled these dynamics, and show how the amount of change in grass shoots and roots (and, therefore, the flexibility in grazing levels) decreases as shrubs increase.

Figure 2 illustrates the impact shrub invasion has on the resilience of a rangeland system when fire is suppressed. The dotted lines are the nullclines for shoot biomass (the above-ground portion of the grass plant) for two different levels of shrub density and the solid line is the nullcline for crown biomass (the roots and growing points of the grass plant). The dotted line to the lower right corresponds to a case when there are few shrubs in the system while that in the upper left corresponds to a case after 25 years of unchecked shrub growth. The nullclines show when the change in state variables is zero. Thus, points where they intersect are equilibrium points for the system.

There are three equilibria for each of the two cases: two stable equilibria (solid circles) and one unstable equilibrium (open circle). The dashed line that emerges from each of the unstable equilibria is the separatrix, dividing the state space into two regions. Any trajectory beginning above and to the right of the separatrix will converge to the upper equilibrium with high grass biomass; beginning below and to the left of the separatrix, to the lower equilibrium with no grass. This distance between the stable equilibrium and the separatrix is a

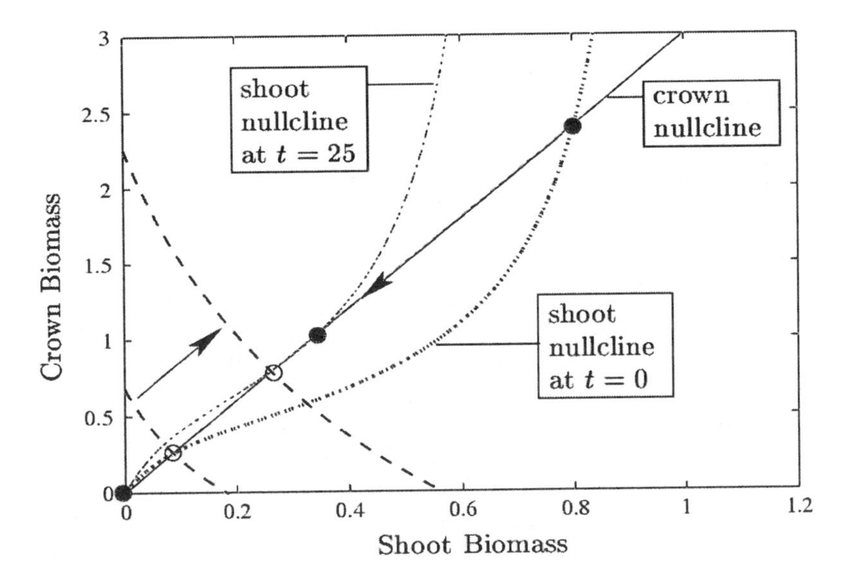

FIGURE 2 The influence of woody weed invasion on the shoot/crown system. As woody shrubs invade the system over time, the shoot nullcline is forced upward and to the left. The two isoclines shown in the figure show the position of the shoot nullcline for two different cases. The solid and open circles indicate the locations of stable and unstable equilibria, respectively. The lower right curve corresponds to a low shrub density at time zero. After 25 years have passed, the invasion of shrubs has pushed the isocline to the new position shown in the upper left curve. After Anderies et al. [1]

measure of the resilience of the configuration with grass to perturbations such as sudden increases in grazing pressure or drought.

The effect of shrub invasion is to move the shrub nullcline up and to the left. This has two effects: it moves the separatrix up and to the right, and moves the upper equilibrium point down and to the left as indicated by the arrows in figure 2. This moves the upper equilibrium closer to the unstable equilibrium and closer to the separatrix, i.e., the boundary of the basin of attraction. This shrinking of the basin of attraction reduces the resilience of the desirable state suitable for grazing.

The only mechanism to stop the progressive loss of resilience of the desirable state is fire. In an ungrazed situation (i.e., pre-European settlement), natural periodic fires would help prevent woody vegetation from dominating the system [22]. As grazing pressure increased, fire regimes changed and may have resulted in encroachment of native shrubs into formerly productive grassland areas. Thus, the removal of a natural driver in the system, fire, can dramatically reduce the resilience (robustness) of the system.

The above example assumes a spatially homogeneous system. Systems with spatially separated subpopulations, of plants or of animals, can exhibit more complicated dynamics. Such a meta-population of a species (a connected set of populations) can exist at either a high-density connected state or a low-density fragmented state. In a landscape composed of potential habitats, the population of a particular habitat depends on its neighboring sites. If the population at a site becomes extinct, the probability of recolonization increases with the aggregate size of the surrounding populations. This effect produces a positive feedback between the density of a region's population and the likelihood that this region's population can maintain itself. Consequently, a regional population can rapidly decline if its population begins to fail to recolonize potential sites, because this further reduces the probability of recolonizing sites [10].

Ocean reefs are a good example of systems in which resilience is influenced by meta-population dynamics. In general, they can be dominated by corals, surface algae, or macro-algae. The three factors that can control switches between states are: (1) changes in the extent of predation on algae by fish and sea urchins, (2) changes in nutrient concentrations and (3) the presence of new areas to grow [18]. Consequently, disturbance events such as storms influence shifts between stable states by providing new areas for recruitment, an influx of nutrients leading to eutrophication, and subsequent variation in the density of algae eaters [25]. Fishing and variation in recruitment can strongly influence fish populations, while the interaction of density-dependent recruitment and circulation patterns allows sea urchins to exist at self-maintaining high- or low-density states [20]. These interactions suggest that reefs can exist in three self-maintaining states, coral-fish, turf algae-urchins, and macro-algae [5, 18].

As a final example, in North Florida sandhill vegetation communities can be dominated by either oaks (*Quercus* spp.) or pines (*Pinus* spp). Fire mediates the competitive relationships between these two states [25]. Longleaf pine (*Pinus palustris*) is a particularly fire tolerant species. Mature longleaf pines shed needles that provide good fuel for ground fires, and young longleaf pines can survive such ground fires. Young hardwoods, however, are intolerant of fire, and mature hardwoods shed leaves that suppress the build-up of fuel for ground fires. This lack of fuel tends to suppress fire in hardwood stands, encouraging the growth of more hardwoods, while fuel accumulation in stands of pine tends to encourage fire, suppressing hardwoods and encouraging its own growth [8, 29]. Similar dynamics can be seen in seasonal tropical rainforests, where opening the forest and introducing fire can allow the forest to flip to a self-maintaining shrub or savanna state [21, 32]. In the northern Florida forests, the aggregation of longleaf and hardwood dominated sites into patches amplifies the forces driving particular sites towards either longleaf or hardwood dominance. Historically, wildfire maintained relatively homogenous longleaf pine savannas. However, restoring the historical fire regime will not restore the landscape due to the reduced effect of fire on the fire-inhibiting alternative stable state (hardwood dominated forest) and the reduced ability of fire to spread across the landscape. These changes in the forest reduce the average size of fires and decrease the ability of fire to burn many areas. Consequently, restoring the historic fire regime does not expose sites in the landscape to the same frequency of fire they experienced historically.

3 DETERMINANTS OF RESILIENCE

The lakes and rangelands examples illustrate the significance of slow (mud phosphate, woody plants) and fast (water phosphate, grass biomass) variables in managed ecosystems. The fast variables are those that confer the benefits we seek, in terms of the supply of ecosystem goods and services, and it is, therefore, not surprising that it is these fast variables on which managers focus. But it is the slow variables that determine how much the fast variables can be changed before the system changes to an alternate state. The slow variables determine the resilience of the systems. The relationship between changes in fast and slow variables, and the positions of thresholds in the slow variables in terms of major changes in the fast variables (i.e., the resilience of the system) are influenced by a number of attributes, or ecosystem features.

Chief amongst the attributes that confer resilience are such features as:

- species and genetic diversity;
- spatial patterns and community diversity;
- modularity—the degree and nature of internal connectedness in the system, including trophic web patterns and connectivity between spatial sub-units [19];
- ecological "memory" (for example, seed banks, refugia for mobile species);
- species mobility and migration; and
- evolutionary and ecological history (if systems have been "evolutionarily tested," or have experienced several invasion waves, they may have a greater genetic capacity than a comparable system without such a history, and this translates to a greater creative capacity following major perturbations).

Resilience is also determined by the physical environment, such as the nature of the physical substrate, including such properties as soil texture and cation exchange capacity. The full set of such features determines how much creativity and adaptation can take place during periods of major change.

In managed ecosystems, attempting to achieve some particular product or outcome can lead unwittingly to reduced resilience, as in the case of rangelands, where prevention of occasional heavy grazing gradually leads to loss of grazing-resistant species and a lowered resilience to any subsequent bouts of heavy grazing pressure.

In the case of coral reefs, resilience there has been reduced by fishing and nutrient pollution. Fishing removes fish that eat algae, while nutrient pollution stimulates algal growth. Both of these changes decrease the resilience of coral reefs, but increase the resilience of algae dominated reefs [14]. In forests, fragmentation often reduces resilience. In the North Florida example, fragmentation due to road building and logging decreased the ability of fire to spread across the landscape. This reduced the resilience of the fire-maintained longleaf pine forest [25]. In Amazon forests, the converse has occurred. Fragmentation due to road building and the use of fire for agricultural clearing has opened up mesic forests allowing fire to spread further into the forest, decreasing the resilience of mesic forests and increasing the resilience of scrublands [21].

4 CHANGES IN RESILIENCE THROUGH TIME

Much of the foregoing emphasizes that maintaining resilience, or robustness, in ecosystems requires change. It is only through the ecosystem probing the boundaries of its domain of attraction that the attributes of the system that confer resilience can be maintained. Rangelands that are never heavily grazed or burned lose the species that are able to withstand defoliation and fire and the amount of grazing and fire the system can withstand accordingly declines. A metaphor for change that usefully captures the dynamics of ecosystems in relation to changes in resilience is that of the adaptive cycle [13]. The main feature of this cycle is the progression of the system through four characteristic phases—rapid growth and exploitation(r), accumulation and conservation (K), breakdown and release (Ω), and reorganization (α). The cycle is usually presented as a figure 8 in terms of two axes, capital accumulation (or available capital) and connectivity. However, variants can be envisaged where it might be a simple loop (e.g., Walker and Abel [39]). The essential feature of the adaptive cycle, however, is the contrast between the "forward" (r to K) and "back" (Ω to α) loops. The forward part is slow, the backloop very fast. Ecosystems spend perhaps 95% of their time in the $r - K$ phases and hence it is not surprising that it is this part of the adaptive cycle that has attracted nearly all the attention of ecosystem managers. During these phases the system is relatively stable and predictable in terms of future conditions. It is relatively easy to manage. The backloop is generally fast, chaotic in terms of behavior and yet crucial in terms of determining the next $r - K$ trajectory the system will follow. Managers have little idea of how to manage during these short periods, and there are few analytical techniques to help them.

From experience so far [9] what we do know is that the state of the system during the K phase, as it enters into the phase of rapid change following a disturbance, strongly influences the outcome of the reorganization phase. It is here where things like ecological memory play an important role. The lesson for managers is that, while it is appropriate to manage an ecosystem for yield or other attributes during the $r - K$ phase of dynamics (an agro-ecosystem, managed forest, lake, or ocean fishery, etc.), the effects of their management actions also influence the resilience of the system and its capacity to recover following some external perturbation during the Ω to α phase. The dynamics of the system during Ω to α are unpredictable and because of this uncertainty in outcome, the most appropriate actions are to enhance the attributes of the system that confer resilience on the desired states of the system, or to reduce those attributes that enhance the resilience of undesired states. Put in another way, managing the backloop of the adaptive cycle amounts to attempting to prevent the system from emerging from the reorganization phase along undesirable trajectories. Trying to make the system stay in, or move to, some particular state is likely to fail.

While guiding the system during the backloop (and being ready to do it very quickly) becomes a new focus for ecosystem managers, an important implication of the adaptive cycle is that one also has to manage for the backloop during other phases of the cycle. Managers need to act before the backloop to build resilience of desired configurations, so that during the fast backloop the system is not constrained to emerge on to some undesired trajectory. An important, related question for managers is if and when they should initiate a Ω phase. Creating small to moderate disturbances reduces the magnitude of Ω, and initiating change before the K phase has reached a moribund, accident-waiting-to-happen state, which can increase the options available during reorganization.

5 RESILIENCE IN COMPLEX SYSTEMS

Using the metaphor of the adaptive cycle to think about resilience in a system again raises the issue of scale. Referring back to Jen's [16] observations on the need to specify both the feature and the perturbation of interest, both depend on scale—temporal and spatial. There are many examples of scale and cross-scale effects in ecosystems, and an early one will suffice as an illustration. The dynamics of the spruce budworm—spruce/fir forest system in the NE USA and Canada, described by Holling and others [12, 26]—is strongly influenced by predation on the budworms, largely by birds. The birds are migratory and what happens to them while in their southern habitats affects the resilience of the forest system to the budworms. However, of more importance in the long-term persistence of ecosystems are cross-scale effects between the ecological and socio-economic systems. Enhancing resilience at one scale may reduce it at another. For example, drought subsidies to farmers might increase the short-term resilience of an individual farming operation but make the whole region (or farming system) more vulnerable, especially where (as has happened in Australia and the USA) the subsidy allows for the purchase of fodder to keep livestock alive, leading to further degradation of the ecological system. A subsidy restricted to enabling farmers to transport livestock to market for sale would have a different outcome. Carpenter and others [2, 3] have described and modeled the cross-scale dynamics of lake ecosystems and the social and ecological dynamics of the agricultural regions in which the lakes are embedded.

Cross-scale effects become particularly interesting in the dynamics of nested adaptive cycles. Like ecosystems, social-ecological systems operate in a nested hierarchy of scales, described by Gunderson and Holling [13] as a panarchy in which each scale operates according to its own spatial and temporal dynamics, but with the dynamics being moderated by those of the scales above and below. Those below, being faster, provide opportunities for innovation and change within the scale in question, and those above provide a protecting mechanism, enhancing resilience through the sorts of mechanisms described earlier.

6 REMAINING CONCEPTUAL QUESTIONS

Of the many areas requiring further attention, the following four are likely to most challenge ecologists during the next phase of resilience research.

6.1 SPATIAL PATTERN AND DYNAMICS

Managing for resilience is limited by the tools available for assessing it, and this is especially a problem in systems that are spatially heterogeneous. In landscapes composed of a mosaic of subsystems in various states, connections between sites contribute significantly to any one site's resilience. This is illustrated by the work described earlier on the spatial dimension of resilience in coral reefs, metapopulations, and forest dynamics. A sharper focus is needed on the ways in which spatial pattern and dynamics influence resilience, as illustrated by the following.

In coral reefs Nystrom and Folke [23] have defined spatial resilience as "the dynamic capacity of a reef matrix to reorganize and maintain ecosystem function following disturbance." They describe how links between reefs can help reefs regenerate following disturbance. An important aspect of spatial resilience in coral reefs is that reefs in different locations are affected by disturbances in different ways. A hurricane may severely damage reefs in shallow waters while leaving reefs in deeper waters intact. If distributed reefs are connected to intact reefs their regeneration can be accelerated.

Peterson [24, 25] has shown that when alternative stable states are distributed across a landscape and states themselves mediate the connections among sites (by alternatively enhancing or attenuating connections) the resilience of the states will be enhanced compared to their existence in isolated situations. Consequently, when sites are distributed across a landscape, the local density of the different states will determine the resilience of local sites and the resilience of states will vary across the landscape.

Landscapes composed of a mosaic of alternate stable states will frequently exhibit key areas that act to fragment or connect areas in the same state. Such areas are potentially key points for management intervention. What defines these areas will depend upon the ecosystem being analyzed. However, in general, their existence will make the management of landscapes that exhibit alternate states more complex than simpler landscapes. While this complexity is a barrier to some types of change, it will also offer opportunities for innovative management practices that use landscape pattern and the internal dynamics of alternate states to amplify management activities. The challenge for managing such landscapes is discovering and applying such policies.

Even in the absence of alternate vegetation states, spatial pattern can exhibit resilience if the pattern entrains the processes that produce that pattern. Landscape pattern emerges at a spatial and temporal scale that is self-maintaining, while resilient to small modifications in spatial pattern or variation in ecological processes. Larger changes in either cause the pattern to abruptly collapse [24].

The further development and testing of simple methods that can help understand and assess spatial resilience are likely to be an important part of future ecological research into resilient, or robust, landscapes.

6.2 THRESHOLDS

Thresholds are fundamental aspects of multistable state systems, and they are a common feature of ecological systems. Examples of thresholds exist in animal behavior, in ecological processes, community assemblage, population dynamics, meta-population dynamics, and landscape pattern. However, thresholds are difficult to detect and study. Two questions, in particular, need to be addressed:

1. Can a threshold be detected before it is reached—especially if it is one that has not been crossed previously? Different regions in state space are divided up by stable manifolds, called invariant sets. They allow us to decompose phase space. Unfortunately, the definition of invariant sets is not something that allows easy quantification of whether a system is near one or not. This then becomes a key conceptual question. There have been suggestions that the second (or nth) derivative of the key variable may be useful, but it is an area that has not been explored, at least in ecology.
2. Can we develop a typology of thresholds, for various kinds of systems, that will help managers to know whether to suspect that a threshold might exist?

6.3 ROBUSTNESS VS. RESILIENCE

The notion of robustness has an engineering connotation. In robust control, the idea is to make some function or process that needs to be controlled robust to uncertainty in model parameters. For example, the wing of a jet must be designed with the knowledge that we don't exactly know how air flows over the wing—but we have a fairly good idea. A design to optimize some property of the wing based on the idea of how things work might cause severe decreases in performance if the idea is too far wrong. The design needs to be robust to errors in the parameter estimates. In this sense, robustness is not an intrinsic property of a system but rather is highly contrived. That is, we want a very specific solution to be robust to a very well known set of uncertainties for a fairly well understood system. The examples of resilience in ecosystems are of internal mechanisms. Can the engineering-based concepts of robustness help in making the understanding of resilience in ecosystems more useful to managers and planners?

6.4 MANAGING THE ADAPTIVE CYCLE

Given the nature and consequences of the adaptive cycle, a tentative question is: When should a policy maker or manager initiate an Ω phase? Managed systems do not have natural time scales in their adaptive cycle dynamics. They are driven by managers' decisions. The tendency in social-ecological systems is for people to attempt to prolong the K phase for as long as possible. One example is the earlier management of Yellowstone National Park, where fires were controlled to maintain mature forest stands, leading to severe insect problems and devastating fires when they did occur. This, however, could be attributed to lack of understanding at the time. A more difficult situation is where the K phase is actively prolonged through short-term self interest despite knowledge of a looming Ω. This occurred, for example, in some Australian rangelands (the Cobar region of New South Wales was one such area) used for wool production over the past couple of decades. Declining terms of trade for wool, coupled with declining range condition in the form of increasing shrub densities, would likely have resulted in a series of small scale Ω's and a consequent regional scale reorganization. The presence of a floor price for wool, however, acted as a subsidy to individuals, allowing them to persist as wool growers beyond the point where they constituted economically viable systems. The wool industry lobbied for the maintenance of the floor price. When it was eventually withdrawn in 1991, due to an expanding wool mountain, many more farmers went bankrupt and a greater area of rangeland had changed into a woody thicket state than would have occurred had the floor price not been in place. The scale of the Ω was greater. While wool prices were declining the price of goat meat had been increasing, and in the ensuing reorganization phase many of the surviving farmers, especially those with extensive shrub areas, switched from sheep production for wool to goat production for meat (sheep graze, goats browse and graze).

Given that initiating change is under the control of either managers (making voluntary changes) or policy makers (e.g., withdrawing the floor price), when is the best time to do it? Is there an optimal time for intervention that generates sufficient change to ensure novelty and reorganization, and still allows sufficient stability and productivity gain? Or is it preferable to try to find the set of "rules" that will ensure the system self-organizes (ecologically, economically, and socially) along desirable trajectories that involve Ω's and α's of varying kinds and intensities? The outcome of the agent-based model of Janssen et al. [15] supports this view, but raises the issue of how to design such rule sets.

REFERENCES

[1] Anderies, J. M., M. A. Janssen, and B. H. Walker. "Grazing Management, Resilience and the Dynamics of a Fire Driven Rangeland System." *Ecosystems* **5** (2002): 23–44.

[2] Carpenter, S., W. Brock, and P. Hanson. "Ecological and Social Dynamics in Simple Models of Ecosystem Management." *Conservation Ecol.* **3** (1999): 4.

[3] Carpenter, S. R., D. Ludwig, and W. A. Brock. "Management of Eutrophication for Lakes Subject to Potentially Irreversible Change." *Ecol. Applications* **9** (1999): 751–771.

[4] Carpenter, S. C., B. H. Walker, M. Anderies, and N. Abel. "From Metaphor to Measurement: Resilience of What to What?." *Ecosystems* **4** (2001): 465–481.

[5] Done, T. J. "Phase Shifts in Coral Reef Communities and Their Ecological Significance." *Hydrobiologia* **247** (1992): 121–132.

[6] Fragoso, J. M. V. "Tapir-Generated Seed Shadows: Scale-Dependent Patchiness in the Amazon Rain Forest." *J. Ecol.* **85** (1997): 519–529.

[7] Frost, T. M., S. R. Carpenter, A. R. Ives, and T. K. Kratz. "Species Compensation and Complementarity in Ecosystem Function." In *Linking Species and Ecosystems*, edited by C. G. Jones and J. H. Lawton, 224–239. New York: Chapman & Hall, 1994.

[8] Glitzenstein, J. S., W. J. Platt, and D. R. Streng. "Effects of Fire Regime and Habitat on Tree Dynamics in North Florida Longleaf Pine Savannas." *Ecol. Monographs* **65** (1995): 441–476.

[9] Gunderson, L. H., and C. S. Holling, eds. *Panarchy: Understanding Transformations in Human and Natural Systems.* Washington, DC: Island Press, 2001.

[10] Hanski, I., J. Poyry, T. Pakkala, and M. Kuussaari. "Multiple Equilibria in Metapopulation Dynamics." *Nature* **377** (1995): 618–621.

[11] Holling, C. S. "Resilience and Stability of Ecological Systems." *Ann. Rev. Ecol. Syst.* **4** (1973): 1–23.

[12] Holling, C. S. "Temperate Forest Insect Outbreaks, Tropical Deforestation and Migratory Birds." *Mem. Ent. Soc. Can.* **146** (1988): 21–32.

[13] Holling, C. S. "Understanding the Complexity of Economic, Ecological, and Social Systems." *Ecosystems* **5** (2001): 390–405.

[14] Hughes, T. P. "Catastrophes, Phase Shifts, and Large-Scale Degradation of a Caribbean Coral Reef." *Science* **265** (1994): 1547–1551.

[15] Janssen, M., B. H. Walker, J. Langridge, and N. Abel. "An Adaptive Agent Model for Analyzing Coevolution of Management and Policies in a Complex Rangeland System." *Ecol. Model.* **131** (2000): 249–268.

[16] Jen, Erica. "Stable or Robust? What's the Difference?" This volume.

[17] Jones, C. G., and J. H. Lawton, eds. *Linking Species and Ecosystems.* New York: Chapman & Hall, 1994.

[18] Knowlton, N. "Thresholds and Multiple Stable States in Coral-Reef Community Dynamics." *Am. Zool.* **32** (1992): 674–682.
[19] Levin, S. A. *Fragile Dominion: Complexity and the Commons.* Cambridge. MA: Perseus Publishing, 1999.
[20] McClanahan, T. R., A. T. Kamukuru, N. A. Muthiga, M. G. Yebio, and D. Obura. "Effect of Sea Urchin Reductions on Algae, Coral, and Fish Populations." *Con. Bio.* **10** (1996): 136–154.
[21] Nepstad, D. C., A. Verissimo, A. Alencar, C. Nobre, E. Lima, P. Lefebvre, P. Schlesinger, C. Potter, P. Moutinho, E. Mendoza, M. Cochrane, and V. Brooks. "Large-Scale Impoverishment of Amazonian Forests by Logging and Fire." *Nature* **398** (1999): 505–508.
[22] Noble, J., and A. Grice. "Fire Regimes in Semi-Arid and Tropical Pastoral Lands: Managing Biological Diversity and Ecosystem Function." In *Flammable Australia*, edited by Ross A. Bradstock, Jann E. Williams, and Malcolm A. Gill. Cambridge, MA: Cambridge University Press, 2001.
[23] Nystrom, M., and C. Folke. "Spatial Resilience of Coral Reefs." *Ecosystems* **4** (2001): 406–417.
[24] Peterson, G. D. "Contagious Disturbance, Ecological Memory, and the Emergence of Landscape Pattern." *Ecosystems* **5** (2002): 329–338.
[25] Peterson, G. D. "Managing a Landscape that can Exhibit Alternate Atable States." In *Resilience and the Behavior of Large Scale Ecosystems*, edited by L. H. Gunderson, C. S. Holling, and C. Folke. Washington, DC: Island Press, in press.
[26] Peterson, G. D. "Scaling Ecological Dynamics: Self-Organization, Hierarchical Structure and Ecological Resilience." *Climatic Change* **44** (2000): 291–309.
[27] Peterson, G. D., C. R. Allen, and C. S. Holling. "Ecological Resilience, Biodiversity and Scale." *Ecosystems* **1** (1998): 6–18.
[28] Pimm, S. L. "The Complexity and Stability of Ecosystems." *Nature* **307** (1984): 321–326.
[29] Rebertus, A. J., G. B. Williamson, and E. B. Moser. "Longleaf Pine Pyrogenicity and Turkey Oak Mortality in Florida Xeric Sandhills." *Ecology* **70** (1989): 60–70.
[30] Scheffer, M., S. C. Carpenter, J. Foley, C. Folke, and B. H. Walker. "Catastrophic Shifts in Ecosystems." *Nature* **413** (2001): 591–596
[31] Schumpeter, J. A. *Business Cycles; A Theoretical, Historical, and Statistical Analysis of the Capitalist Process.* New York, NY: McGraw-Hill, 1964.
[32] Siegert, F., G. Ruecker, A. Hinrichs, and A. A. Hoffmann. "Increased Damage from Fires in Logged Forests During Droughts Caused by El Niño." *Nature* **414** (2001): 437–440.
[33] Smith, T., III, M. B. Robblee, H. R. Wanless, and T. W. Doyle. "Mangroves, Hurricanes and Lightning Strikes." *BioScience* **44** (1994): 256–262.
[34] Tainter, J. A. . *The Collapse of Complex Societies.* Cambridge, MA: Cambridge University Press, 1988.

[35] Tainter, J. A. "Sustainability of Complex Societies." *Futures* **27** (1995): 397–407.
[36] Tilman, D., and J. A. Downing. "Biodiversity and Stability in Grasslands." *Nature* **367** (1994): 363–365.
[37] Walker, B., A. Kinzig, and J. Langridge. "Plant Attribute Diversity, Resilience, and Ecosystem Function: The Nature and Significance of Dominant and Minor Species." *Ecosystems* **2** (1999): 95–113.
[38] Walker, B., S. Carpenter, J. M. Anderies, N. Abel, G. Cumming, M. Janssen, L. Lebel, J. Norberg, G. Peterson, and R. Pritchard. "Analyzing Resilience in a Social-Ecological System: An Evolving Framework." *Conservation Ecol.* **6(1)** (2002): 14.
[39] Walker, B. H., and N. A. Abel. "Resilient Rangelands: Adaptation in Complex Systems." In *Panarchy: Understanding Transformations in Human and Natural System*, edited by L. Gunderson and C. S. Holling, ch. 11, 293–313. Washington, DC: Island Press, 2001.
[40] Walker, Brian H. "Rangeland Ecology: Understanding and Managing Change." *Ambio* **22** (1993): 2–3.

Robustness in the History of Life?

Douglas H. Erwin

1 INTRODUCTION

The definitions of robustness are many and varied, but most focus on the ability of a system to retain function in the face of an external perturbation. Assessing function in the fossil record is a difficult but not intractable task and such a definition poses considerable difficulty for considering robustness in the fossil record. Paleontologists have identified patterns of stability in the fossil record of diverse marine and terrestrial plants and animals through the past 600 million years in taxonomic diversity, in the stability of major morphological architectures, or body plans, and in the structure of ecological relationships. Inferring the extent to which the documented resilience (return to a previous state after a perturbation) of these patterns implies a robustness of function must always be indirect. The ability to infer ecological or developmental function in deep time is limited by the fact that the fossil record is primarily composed of durably skeletonized marine organisms, bones of various vertebrates, insects, and the often disarticulated remains of plants, but, of course, missing the soft anatomy. In almost all cases the many soft-bodied organisms, those without durable skeletons, are missing from the record. We are left with changing patterns in the number of taxa (generally families or genera, less commonly species) within clades, patterns of dominance within assemblages of organisms at a particular locality (assemblages that have often been accumulated over a span of time that can extend to hundreds or even thousands of years), and the conservation of morphological architectures. For many of these systems we may be able to make plausible arguments that stability reflects an underlying robustness, but identifying the functional attributes of this robustness can generally only be done by comparison to living organisms and extant ecosystems. This is an increasingly difficult task the farther back in time one goes.

The pervasiveness of such stability is surprising when one considers the extent of perturbations which affect life. Human history is a poor guide to conditions in the past. Sea-level changes, many of well over 100 meters, have been frequent and abrupt. Climatic fluctuations range from widespread continental glaciations to pervasive global warming producing temperate to subtropical conditions near the poles. Wide swings in atmospheric carbon dioxide levels are just one of many major fluctuations in ocean and atmospheric chemistry. Massive volcanism has rapidly buried continental-sized regions on land and created vast undersea deposits. And, of course, there have been numerous impacts of extraterrestrial objects. Some of these disturbances, like the meteorite that triggered the extinction of dinosaurs 65 million years ago, have led to pervasive mass extinctions, reorganizations of marine and terrestrial ecosystems, and the rapid spread of novel morphologies. Yet other disturbances have had little apparent effect. Over the past several million years, sea level has fluctuated by 100 meters or more, for example, yet there are few apparent cases of either extinction or speciation of marine animals. The spatial scale of environmental perturbations also varies widely, from purely local events to global catastrophes, and the temporal scale is similarly broad.

In the face of such ubiquitous perturbations (or perhaps because of them), life has proven remarkably resilient. Apparent robustness in deep time is reflected in patterns of taxonomic diversity (properly correcting for preservational and sampling biases), the coherence and stability of ecological associations, and within the characteristic architectural body plans produced by development. The last is particularly interesting, for despite the power of adaptive evolution to mold morphology and the developmental processes which generate them, the groundplans of most clades of marine animals have persisted for over half a billion years. Patterns of robustness are also apparent at a variety of scales, including apparent punctuated patterns of species evolution, coordinated changes in community composition, and the resilience of community structure to environmental disturbances at a variety of scales, including mass extinctions.

These apparent patterns of robustness in the fossil record raise several critical issues: Are they real, or artifacts of preservation or analysis? Does robustness exist as a process, or are these patterns an epiphenomenon of some other process?

Why do some disturbances have little or no ecological or evolutionary impact while others, seemingly of the same magnitude, trigger wide-ranging changes? Here I discuss the patterns of apparent robustness seen in the fossil record at different scales and across different aspects of biotic diversity, before addressing the issue of whether robustness is itself a process or simply an apparent outcome.

2 ROBUSTNESS ACROSS MULTIPLE SCALES

Estimates of the number of species alive today range from 5 million to as many as 30 million species, with the diversity of microbial life greatly exceeding what was even suspected only a decade ago. In the 3.5 billion years since the origin of life, hundreds of millions of species have made their brief appearances here, but most will never be known. One of the more startling discoveries of the past decade is the penetration of microbial ecosystems thousands of feet into the crust of the earth, encompassing organisms that live on hydrogen, others that live off methane, and many which favor temperatures of over 100°C.

There are two primary sources of evidence on patterns of diversity through the history of life. The fossil record has been the principal source of such information, particularly for the past 600 million years since the origin of animals. Marine organisms with skeletons that are durable enough to be preserved provide the bulk of the fossils; fossils from land can be abundant, but, in general, the record is more spotty. Comparison of molecular sequences (DNA and RNA) from living species is the other essential tool for reconstructing relationships between groups and thus the history of life. While it is difficult to establish from molecular data how many species have gone extinct, when the branching order is calibrated against fossils it can be a critical source of information on the evolutionary history of groups with poor fossil records.

The general pattern of marine animal diversity through the past 600 million years is shown in figure 1, with the rapid diversification of animal body plans at the base of the Paleozoic ("the Cambrian radiation"), with a continuing diversification of classes, orders, families, and genera into the Ordovician. This diversification establishes the major clades dominating marine ecosystems for the remainder of the Paleozoic. Following the great mass extinction at the close of the Permian, a new pulse of diversification begins in the Triassic, and appears to be continuing today. This general pattern is interrupted by five great mass extinctions, at the close of the Ordovician (ca. 439 Ma), during the late Devonian (ca. 372 Ma), at the end of the Permian (251 Ma), Triassic (199 Ma), and Cretaceous (65 Ma). Many paleontologists also suspect that a mass extinction of early animals during the latest Proterozoic may have triggered the Cambrian explosion of bilaterian animals. Equally remarkable is the stability that persists between these episodes of change. What produced the long period of apparent stability from the Ordovician through the Permian, punctuated only by a few

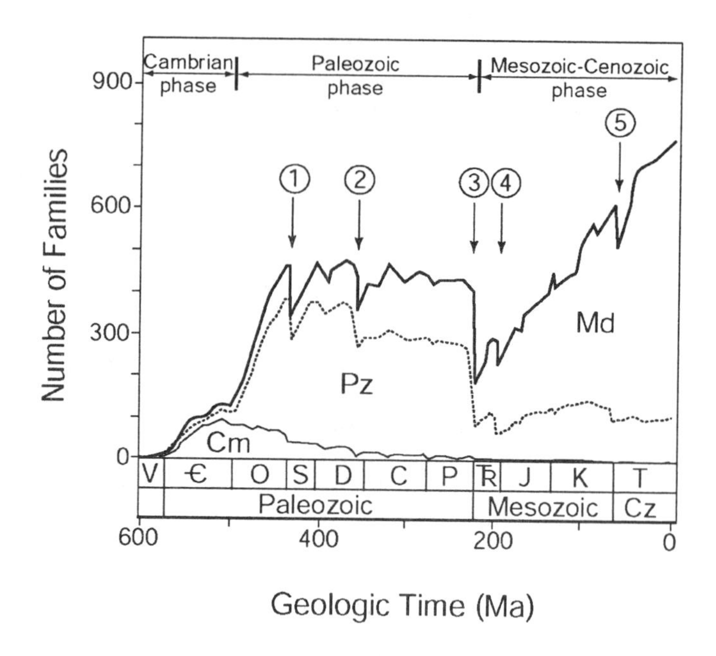

FIGURE 1 Diversity of marine families through the Phanerozoic, showing the major mass extinctions. 1—End-Ordovician mass extinction; 2—Late Devonian mass extinctions; 3—Late Permian mass extinctions; 4—End-Triassic mass extinction; and 5—End-Cretaceous mass extinction.·After Sepkoski [34].

mass extinctions? Why the unrestrained growth in marine diversity since the end-Permian mass extinction?

Before spending too much time seeking the answers to such questions, it is important to have some degree of confidence in the data. Several recent analyses, have raised questions about how well we actually know the history of even well-skeletonized groups in the fossil record. The data in figure 1 was largely collected by the late Jack Sepkoski of the University of Chicago and represents the first and last occurrences of hundreds of marine families and thousands of genera. A number of artifacts can affect such patterns, as Sepkoski was well aware. One important source of bias is variation in how wellv different intervals have been sampled: intensively studied times will be better known than less intensively sampled intervals. Alroy et al. [1] collected a massive data set to explore the impact of correcting for unequal sampling. Although the data remains incomplete, and disagreement persists over the proper techniques for sample standardization, the results of Alroy et al. suggest that the dramatic post-Permian increase in diversity in figure 1, for example, may be largely an artifact. Jackson and Johnson [26] noted that published and museum data may, in principle, be inadequate to correctly describe past diversity patterns and that an entirely new collecting effort may be required. In essence, they argue that targeted, standardized collecting will be required to correct for the biases in the published paleontological record. A second form of bias is imposed by variations in the amount of available rock area, generally imposed by changes in sea level for marine rock. Several studies have suggested an important role for both sea level and rock area in altering apparent diversity patterns [32, 33, 37, 38]. These studies clearly show that short-term (million-year) variability in diversity patterns is affected by the biases; less clear is whether larger-scale diversity patterns are similarly affected.

These preservational and sampling biases constrain the nature of questions that can be asked of the fossil record, but the record is sufficiently robust to address a number of important issues. Particularly striking is the appearance of the major architectures of animals, from sea anemones to fish, of the major groups of plants, of dinosaurs, and of modern mammals in relatively brief bursts, at least from a geological perspective. Why are these major evolutionary innovations underlying the diversity of life so often clustered into relatively narrow intervals of time? Many of these creative bursts of evolution follow mass extinctions and other biotic crises, suggesting a link between environmental disturbances and innovation. In this section I will begin at the finest scale, considering patterns of robustness in species patterns, followed by community composition and then across biodiversity crises.

Traditional population genetic models of the appearance of new species lead to the expectation of gradual adaptation to changing conditions during the lifespan of a species, with speciation occurring as different populations adapted to different conditions in distinct geographic regions (e.g., Mayr [31]). Eldredge and Gould [12]s challenged this orthodoxy with their model of punctuated equilibrium, pointing out that the fossil record of most marine invertebrates (which generally have a far better fossil record than terrestrial organisms) displayed a pattern of prolonged, dynamic stability in morphology, interrupted by rapid morphologic shifts during speciation events. This pattern suggests that morphological change, and arguably speciation as well, is concentrated in discrete events, rather than being a lengthy process, from which it follows that much of the adaptive modulation of species characteristics may be relatively irrelevant during speciation [17, 19, 20]. The response to this proposal was lengthy and occasionally tendentious (see Charlesworth et al. [8], Erwin and Anstey [14], and Maynard Smith [30] for a review) but much research tended to support the claims of Eldredge and Gould (e.g., Cheetham and Jackson [10], Erwin and Anstey [14], and Jackson and Cheetham [25]). Population geneticists soon found that they could incorporate rapid change within their models (e.g., Lande [24] and Hendry and Kinneson [21]), suggesting there was less difficulty accommodating evidence from the fossil record than some commentators had argued. The pervasiveness of cryptic speciation (e.g., Knowlton et al. [27, 28]), where species are morphologically similar but do not interbreed and evidently have independent evolutionary trajectories, poses an additional problem for punctuated equilibrium that is difficult to solve. Although the debate over the tempo and mode of speciation persists, the lack of sufficiently rigorous studies on the part of paleontologists means that the number of valid studies remains fairly low.

The relative frequency of stasis is the most compelling issue for a consideration of robustness, and if stasis does occur, what are the forces that produce it? Jackson and Cheetham [25] noted that stasis and abrupt morphological change can occur within a single lineage, with no branching or formation of new species, a pattern they describe as punctuated anagenesis. Bryozoans, a colonial, filter-feeding organism, have been the basis for some of the most detailed studies of speciation patterns. For *Metrarabdotos*, random genetic drift with average mutation rates is sufficient to generate the observed changes in morphology. Although many observers have invoked environmental tracking as a driving force, this appears to be unnecessary. The difficulty arises in explaining stasis, since stabilizing selection appears to be required, yet is difficult to envisage on time scales of millions of years. Cheetham [9] notes that depletion of sufficient genetic variation to generate new species could be countered if phenotypic plasticity (the ability to produce different morphologies under varying environmental conditions) was heritable. Although *Metrarabdotos* is but a single example, what is particularly striking about the stability documented by Cheetham is that it persists through periods of intense environmental change in the Caribbean (where *Metrarabdotos* lived) suggesting a degree of robustness. Regrettably, the processes producing apparent robustness in morphology and species characteristics remain obscure.

Ecosystems are adapted to disturbances ranging from fire to severe storms and outbreaks of disease. Such minor disturbances generally increase the number of species in an area, and are even required in many habitats. Recovery is generally rapid. More severe biotic disturbances can destroy entire ecosystems, with the rate of recovery dependent upon the immigration of species into the affected area, and how quickly the ecological fabric can be rewoven. Modern ecology provides an explanation for disturbances at this level, but ongoing, human-induced disturbances are on a far greater scale, more similar to biotic crises and great mass extinctions documented by the fossil record. The fossil record suggests a hierarchical structuring of stable community assemblages with restructuring due to physical disturbances of varying magnitude.

In 1970, Walker and Laporte [40] compared shallow marine carbonate rocks from the Ordovician and Devonian of New York. Both sets of rocks recorded an increase in sea level across a carbonate platform in supratidal, intertidal, and subtidal environments. The two deposits were separated by 72 million years, the end-Ordovician mass extinction and glaciation, a precipitous decline in the abundance of trilobites, and an increase in the abundance of nautiloid cephalopods. Despite all this, between 80 and 100 percent of the taxa are ecologically equivalent between the two areas. Functionally similar taxa are occupying the same roles in the similar environments. For example, the supratidal communities are primarily composed of scavenging leperditiid ostracodes and algal mats, with rare suspension feeders. This remarkable similarity was the first hint of extensive conservation of assemblage structure over long spans of geologic time, and through extensive periods of environmental change and the appearance of new clades of animals.

The Devonian of New York also served as the setting for another study of community change by Brett and Baird [5], who compared the transition between the Onondaga and Hamilton formations. Fewer than one-third of the species found in the Hamilton came from earlier units within the Appalachian region. The changes in sea level marking the shift in formations allowed the immigration of about half the species from other areas. Once this assemblage formed, it persisted for several millions of years, a pattern that has become known to paleontologists as coordinated stasis.

This pattern of assemblage stability alternating with pulses of rapid change occurs on larger scales as well. Boucot [3] synthesized this information, identifying a series of ecological-evolutionary units (EEU's) characterized by relative stasis with clades evolving within a particular niche, but without major innovations. Change was concentrated at the boundaries between EEU's, which are often associated with mass extinctions [35, 36]. Sheehan suggested that during the post-extinction recoveries, clades developed new adaptations, allowing access to niches not previously occupied. Brett and Baird [5] described the shorter intervals they studied as ecological evolutionary subunits, on the scale of 2–6 million years of stability, followed by relatively rapid reorganization.

Despite some claims for ecological control over this process (for multiple views on this issue see Ivany and Schopf [22]), communities are not phylogenetic or geneological entities but assemblages drawn together by common environmental requirements [2]. The absence of genealogical connectivity between assemblages is what makes the persistence of similar ecological structures so perplexing. Although it has not been widely discussed by paleontologists, Knowlton [28] noted that ecologists have long been aware that multiple stable states can persist in a single environmental setting, but whether this stability on ecological time scales is relevant to the far longer time scales of the fossil record is uncertain, and perhaps doubtful. What forces enable similar patterns to persist over such long time scales? Is this apparent pattern of hierarchical stability and change a figment of the record or how paleontologists study it, or does it reflect something about the response of the biota to biotic disturbances of varying scales?

The five great mass extinctions represent the next step up in biotic disturbance. Rebuilding ecosystems after these events involves both the origination of new species and the creation of new ecological relationships, often producing significant evolutionary innovation. Studies of recoveries after the five great mass extinctions, as well as after several small biotic crises documented in the fossil record have produced a general model of the recovery process [13]. Following the end of an extinction, there is often a burst of highly opportunistic species. Perhaps the best example of this is the burst of ferns in the earliest Cenozoic immediately after the extinction that removed the dinosaurs, and many plants, 65 million years ago. A very similar process occurs today in disturbed fields, or along the side of a new road. A survival interval following mass extinction generally lasts only a few hundred thousand years, at most, and is characterized by relatively few species, but by high numbers of each species. Few new species appear during this interval. A recovery stage follows the survival interval, marked by the appearance of new species and the re-emergence of other surviving species in the fossil record. Recent studies have shown that this simplistic model of an initial survival interval of opportunists, followed by rapid generation of new species during biotic recoveries, is far too simple [13]. Opportunists often are only found in one part of the world, with no apparent opportunists in other regions. More curiously, the re-emergence of new species appears to occur remarkably rapidly, suggesting that the rapid diversification of new species during the recovery interval receives a substantial boost from positive feedback: as new species evolve, and ecosystems are rebuilt, roles are created for yet more new species.

Previous attempts to model this process have failed to incorporate positive feedback: as species evolve they create opportunities for yet more species. This creative, self-reinforcing aspect of evolutionary recoveries provides much of the impetus for evolutionary radiations, and links them to other episodes of biological innovation. The absence of models of this positive feedback process hampers our understanding and limits paleontologists' ability to design new investigations into the fossil record. Current models follow one of two approaches: (1) fixed ecospace models (akin to a chessboard, with the possible ecological spaces specified in advance) in which occupation of model niche space is driven by lineage branching and logistic growth and limited by competition; and (2) simple logistic growth models of interacting lineages. Neither of these approaches is particularly realistic, and they largely fail to reveal anything about the actual processes of recovery or innovation. Additionally, these models assume that a single model applies globally, failing to consider the importance of different patterns in different regions.

Major evolutionary innovations occur in the aftermath of mass extinctions or as the result of major adaptive breakthroughs. The latter case also involves the creation of new ecological space, so biotic recoveries and evolutionary innovations can be closely related. Many significant evolutionary innovations occur in discrete bursts which fundamentally reorganize pre-existing ecological relationships, essentially creating a new world. Examples of this include the rapid appearance of the major animal groups at the base of the Cambrian (ca. 530 Ma), the diversification of all the major architectures of plants during the Devonian (ca. 360 Ma), the formation of the major dinosaur clades during the late Triassic (ca. 220 Ma) and the extraordinary explosion of new groups of mammals after the extinction of the dinosaurs (55–65 Ma). These events share a number of similarities, including the diversity of new groups that appear, rapid evolutionary change, and dramatic shifts in ecological interactions.

At these scales of disturbance, robustness seems hard to identify. Although many major clades and some ecosystems recover following such extensive biotic crises, the available evidence suggests that the rebuilding often structures ecosystems in very different ways (although not always, as with the Ordovician and Devonian example above). If so, this suggests that the apparent robustness seen at the smaller scale of millions to tens of millions of years within Ecological-Evolutionary units breaks down during at least some mass extinctions.

3 BODY PLANS: STABILITY AND CHANGE

The final area of stability documented by the fossil record, in conjunction with modern comparative developmental studies, involves the persistence of architectural frameworks, or body plans, within the major clades of marine organisms. Each of the major durably skeletonized marine clades, or phyla, can be traced to the Cambrian explosion of animal life (except bryozoans, which first appear slightly later in the Early Ordovician). Preservation of a variety of generally unfossilizable worms and other groups in the Middle Cambrian Burgess Shale deposit as well as molecular phylogenies based largely on 18S rRNA suggest that all of the phyla trace their ancestry to an interval between 600 and 520 million years ago, spanning the late Neoproterozoic and the Early Cambrian [23, 39].

Not only were the major clades (phyla and classes) of marine organisms established during this period, but quantitative assessments of the occupation of morphological space, in other words, the variety of different architectural themes within a body plan, provide a complementary view of the explosion in diversity. Arthropods are the most diverse group today, and form the basis for one of the quantitative studies of this issue. Briggs et al. [6, 16] compared modern and Cambrian arthropods, using the soft-bodied forms of the Middle Cambrian Burgess Shale. The comparison employed landmarks outlining the morphology, determining the shape of each group, then constructing a multidimensional morphospace vs. statistical reduction of the resulting shape variables. Although these studies succeeded in their goal in disproving Stephen Jay Gould's assumption in *Wonderful Life* [18] that morphological breadth (or disparity) of Cambrian arthropods was greater than today, Briggs and his colleagues showed that disparity was just as great in the Middle Cambrian as it is today. This is a fairly surprising result, particularly since the Cambrian sample was from one locality and the modern sample surveyed the broad spectrum of marine habitats occupied by arthropods. There have been remarkable developments in arthropod morphology since the Cambrian, notably the spread of crustaceans, but the disappearance of other morphologies such as trilobites and other Paleozoic groups means that Recent and Cambrian arthropods occupy roughly similar volumes of morphospace.

Few similar comprehensive comparisons of major groups have been conducted, but the arthropod results suggest another pattern of long-term stability in deep time: of the occupation of morphospace. Why this should be so is unclear, and, consequently, it is not clear that this constitutes a pattern of robustness even by the looser definitions that need to be adopted in dealing with the fossil record. Support for the suggestion that these morphological patterns reveal a degree of robustness comes from recent studies in comparative developmental biology which have revealed a completely unexpected conservation of developmental systems among bilaterian organisms (for a recent review see Wilkins [41]).

Two examples will suffice to illustrate this phenomenon. The Hox complex is a cluster of eight to ten genes which in most organisms is normally arrayed along a single chromosome. The first gene in the sequence controls formation of structures in the anterior part of the head, and each gene in sequence basically controls regionalization of progressively more posterior parts of the developing embryo. The effect of turning on the genes in sequence is to progressively pattern a developing embryo from an unprepossessing lump to a structured organism. Many of the genes in the cluster have very similar sequences, indicating that the number of genes has expanded via gene duplication. Vertebrates illustrate this well, for the entire cluster of 12 genes in primitive cephalochordates has undergone two rounds of duplication and subsequent loss of some individual genes so that most vertebrates have four clusters with a total of 39 hox genes [11]. Another gene, *pax6*, has become relatively famous from studies showing that inducing expression of this gene is sufficient to form a complete fly eye, suggesting this is a master regulator of an entire cascade of genes active in forming an eye.

There are two alternative views of the implications of such extensive conservation of transcription factors and other signaling molecules. Carroll et al. [7] summarized the argument that sequence conservation implies a fair degree of functional conservation. From this perspective, the presence of transcription factors with similar roles in widely divergent animals implies that the last common ancestor of the two groups contained these transcription factors playing a similar role. Thus *pax6* controlled eye formation in the ancestral group of all bilaterian animals, and the Hox genes controlled pattern formation along an anterior-posterior axis. As Carroll et al. illustrated, the sum of the wide range of conserved developmental regulators produces a fairly complex animal as the last common ancestor of flies and mice (the two favorite animals of developmental biologists) and thus the ancestral bilaterian animal.

An alternative to this view has recently been proposed [15] in which the ancestral role of many of these developmental regulators was much simpler. Davidson and I proposed that many of these regulators originally controlled cell types, rather than complex networks leading to elaborate morphologies, and noted the evidence in favor of this view. We suspect that the role of some regulators, including the Hox complex was as advertised by Carroll. *Pax6*, in contrast, may have originally been involved in generating one of the variety of cell types required to construct a complex eye. The developmental regulators are seen as a developmental toolkit, equipped with vectoral patterning systems and other devices that can be employed by different clades to construct a remarkable range of morphologies. This view is favored by the recognition that complex eyes have evolved many times and there a few morphological attributes shared among them. Certainly the eyes of a fly and a mouse bear little similarity to each other. If this alternative should largely be correct, the complex ancestor favored by Carroll et al. [7] must be replaced by a far less complex organism.

Resolving these different views will be necessary to assess the robustness of developmental systems. The first alternative involves functional conservation of key developmental controls across more than 600 million years—a remarkable demonstration of developmental robustness. In the simpler alternative proposed by Erwin and Davidson [15], little robustness is present, since the morphogenetic pathways that produce complex morphologies are largely constructed independently in different sub-clades of the Bilateria.

4 DOES THE FOSSIL RECORD DOCUMENT ROBUSTNESS, OR IS IT AN EPIPHENOMENON?

Do the patterns described here truly reflect robustness in the history of life, taken as conservation of function in the face of ongoing environmental and biological change? Alternatively, are these patterns of apparent robustness simply an epiphenomenon of other processes? The latter alternative cannot be easily dismissed. Since the fossil record does not preserve unambiguous evidence of function, paleontologists must infer it from patterns of stability and resilience. I have described a series of phenomena, covering taxic diversity, ecological structure, and developmental patterns, which suggest the existence of robustness across spans of millions to hundreds of millions of years. We do not yet possess the tools to determine whether functional conservation is an appropriate inference from these patterns. Paleontologists may have been lead astray by preservational biases, by an overemphasis on abundant and easily preserved taxa and by a conservation of final morphology despite a ferment of underlying genomic and developmental change. Since robustness is not an issue upon which paleontologists have focused much attention, it remains quite possible that these patterns of apparent robustness are not reliable, or that if they do reliably preserve a record of stability and resilience, that these do not demonstrate true robustness.

A central issue to resolving process from pattern is to document the rapidity of evolutionary change, which is, in turn, critically dependent upon having a reliable time scale. (As a colleague of mine recently said: "No dates—No rates!"). A number of new techniques have provided the first real opportunity we have had for understanding how rapidly evolutionary change has occurred in deep time. Until just the past five years or so, evolutionary events older than about 80 million years were difficult to investigate with sufficient temporal resolution: we simply couldn't tell time well enough to distinguish rapid events from slow events. The recent development of new techniques in the very high-precision dating of ancient volcanic ash beds now yields dates with a precision of about 200,000 years for rocks over 500 million years old [4]. Previously, good dates for the Paleozoic were often precise to plus or minus 20 million years. Older estimated ages have now been shown to have been incorrect by as much as 70-to-80 million years. Coincidentally, paleontologists have developed new analytical techniques for examining rates of change in fossil groups. Thus, we are on the verge of being able to reliably study evolutionary rates in deep time for the first time.

The development of process-based models which can be tested with paleontological data suggest a way forward. Such a model-based approach may, in specific cases, reveal that robustness can be documented in deep time, as well as the processes underlying it. Alternatively, many of these apparent examples of robustness may fade under closer scrutiny as artifacts. Paleontologists have tended not to design research programs around testing such process-based models, but collaborative development of such a research program may prove very rewarding. Incorporation of new dates into such models and properly accounting for preservational biases may prove an enormously productive method to study robustness in deep time.

5 CONCLUSIONS

The fossil record of life illuminates to a host of non-analogue conditions, rates and processes that we would not otherwise know existed because they fall far beyond the range of variability that humans have experienced. But while the fossil record is wonderful at documenting the range of patterns, it is more difficult to extract the processes responsible, and this is the problem plaguing the issue of robustness in deep time. Do these patterns of long-term stability that we see reflect a conservation of functional relationships. Comparative developmental studies may often suggest (subject to the qualifications above) that similar processes have been conserved over hundreds of millions of years, implying a robustness to morphological architectures that was unexpected even a decade ago. Ecological and taxonomic diversity patterns are more problematic. Patterns deduced from the fossil record may imply particular patterns, particularly when the time available is well-constrained, but a range of processes may be compatible with the data. Rarely can paleontologists uniquely specify the process involved in generating an ecological or evolutionary pattern.

ACKNOWLEDGMENTS

This research was funded in part by the robustness project at the Santa Fe Institute, and by a grant to the Santa Fe Institute from the Thaw Charitable Trust.

REFERENCES

[1] Alroy, J., C. R. Marshall, R. K. Bambach, K. Bezusko, M. Foote, F. T. Fursich, T. A. Hansen, S. M. Holland, L. C. Ivany, D. Jablonski, D. K. Jacobs, D. C. Jones, M. A. Kosnik, S. Lidgard, S. Low, A. I. Miller, P. M. Novack-Gottshall, T. D. Olszewski, M. E. Patzkowsky, D. M. Raup, K. Roy, J. J. Sepkoski, Jr., M. G. Sommers, P. J. Wagner, and A. Webber. "Effects of Sampling Standardization on Estimates of Phanerozoic Marine Diversification." *PNAS* **98** (2001): 6261–6266.

[2] Bambach, R. K. "Do Communities Evolve?" In *Palaeobiology II*, edited by D. E. G. Briggs and P. R. Crowther, 437–440. Oxford: Blackwell, 2001.

[3] Boucot, A. J. "Does Evolution Take Place in an Ecological Vacuum?" *J. Paleontology* **57** (1983): 1–30.

[4] Bowring, S. A., and D. H. Erwin. "A New Look at Evolutionary Rates in Deep Time: Uniting Paleontology and High-Precision Geochronology." *GSA Today* **8(9)** (1998): 1–6.

[5] Brett, C. E., and G. C. Baird. "Coordinated Stasis and Evolutionary Ecology of Silurian to Middle Devonian Faunas in the Appalachian Basin." In *New Approaches to Speciation in the Fossil Record*, edited by D. H. Erwin and A. L. Anstey, 285–315. New York: Columbia University Press, 1995.

[6] Briggs, D. E. G., R. A. Fortey, M. A. Wills, A. M. Sola, and M. Kohler. "How Big Was the Cambrian Evolutionary Explosion? A Taxonomic and Morphologic Comparison of Cambrian and Recent Arthropods." In *Evolutionary Patterns and Processes*, edited by D. R. Lees and D. Edwards, 34–44. London: Linnean Society of London, 1992.

[7] Carroll, S. B., J. K. Grenier, and S. D. Weatherbee. *From DNA to Diversity*, 214. Malden, MA: Blackwell Science, 2001.

[8] Charlesworth, B., R. Lande, and M. Slatkin. "A Neo-Darwinian Commentary on Macroevolution." *Evolution* **36** (1982): 460–473.

[9] Cheetham, A. H. "Evolutionary Stasis vs. Change." In *Palaeobiology II*, edited by D. E. G. Briggs and P. R. Crowther, 137–142. Oxford: Blackwell, 2001.

[10] Cheetham, A. H., and J. B. C. Jackson. "Process from Pattern: Tests for Selection versus Random Change in Punctuated Bryozoan Speciation." In *New Approaches to Speciation in the Fossil Record*, edited by D. H. Erwin and R. L. Anstey, 184–207. New York: Columbia University Press, 1995.

[11] de Rosa, R., J. K. Grenier, T. Andreeva, C. E. Cook, A. Adoutte, M. Akam, S. B. Carroll, and G. Balavoine. "Hox Genes in Brachiopods and Priapulids and Protostome Evolution." *Nature* **399** (1999): 772–776.

[12] Eldredge, N., and S. J. Gould. "Punctuated Equilibria: An Alternative to Phyletic Gradualism." In *Models in Paleobiology*, edited by T. J. M. Schopf, 82–115. San Francisco, CA: Freeman, 1972.

[13] Erwin, D. H. "Lessons from the Past: Biotic Recoveries from Mass Extinctions." *PNAS* **98** (2001): 5399–5403.

[14] Erwin, D. H., and R. L. Anstey. "Speciation in the Fossil Record." In *New Approaches to Speciation in the Fossil Record*, edited by D. H. Erwin and R. L. Anstey, 11–38. New York: Columbia University Press, 1995.

[15] Erwin, D. H., and E. Davidson. "The Last Common Bilaterian Ancestor." *Development* (2003): In press.

[16] Fortey, R. A., D. E. G. Briggs, and M. A. Wills. "The Cambrian Evolutionary 'Explosion' Recalibrated." *BioEssays* **19** (1997): 429–434.

[17] Gould, S. J. "Gulliver's Further Travels: The Necessity and Difficulty of a Hierarchical Theory of Selection." *Proc. Roy. Soc. Lond. B* **353** (1998): 307–314.

[18] Gould, S. J. *Wonderful Life: The Burgess Shale and the Nature of History*, 347. New York: W. W. Norton, 1989.

[19] Gould, S. J., and N. Eldredge. "Punctuated Equilibria: The Tempo and Mode of Evolution Reconsidered." *Paleobiol.* **3** (1997): 115–151.

[20] Gould, S. J., and N. Eldredge. "Punctuated Equilibrium Comes of Age." *Nature* **366** (1993): 223–227.

[21] Hendry, A. P., and M. T. Kinnison. "Perspective: The Pace of Modern Life: Measuring Rates of Contemporary Microevolution." *Evolution* **53** (1999): 1637–1653.

[22] Ivaney, L. C., and K. M. Schopf, eds. "New Perspectives on Faunal Stability in the Fossil Record." *Palaeogeography, Palaeoclimatol. & Palaeoecol.* **127** (1996): 1–361.

[23] Knoll, A. H., and S. B. Carroll. "Early Animal Evolution: Emerging Views from Comparative Biology and Geology." *Science* **284** (1999): 2129–2137.

[24] Lande, R. "The Dynamics of Peak Shifts and the Pattern of Morphologic Evolution." *Paleobiol.* **12** (1986): 343–354.

[25] Jackson, J. B. C., and A. H. Cheetham. "Tempo and Mode of Speciation in the Sea." *Trends in Ecol. Evol.* **14** (1999): 72–77.

[26] Jackson, J. B. C., and K. G. Johnson. "Paleoecology. Measuring Past Biodiversity." *Science* **293** (2001): 2401–2404.

[27] Knowlton, N. "Sibling Species in the Sea." *Annual Rev. Ecol. Sys.* **24** (1993): 189–216.

[28] Knowlton, N. "Thresholds and Multiple Stable States in Coral Reef Community Dynamics." *Amer. Zool.* **32** (1992): 674–682.

[29] Knowlton, N., E. Weil, L. A. Weigt, and H. M. Guzman. "Sibling Species in *Monastraea annularis*, Coral Bleaching and the Coral Climate Record." *Science* **255** (1992): 330–333.

[30] Maynard Smith, J. "The Genetics of Stasis and Punctuation." *Ann. Rev. Genet.* **17** (1983): 11–25.

[31] Mayr, E. *Animal Species and Evolution.* Cambridge, MA: Harvard University Press, 1963.

[32] Peters, S. E., and M. Foote. "Biodiversity in the Phanerozoic: A Reinterpretation." *Paleobiol.* **27** (2001): 583–601.

[33] Peters, S. E., and M. Foote. "Determinants of Extinction in the Fossil Record." *Nature* **416** (2002): 420–424.
[34] Sepkoski, J. J. "A Model of Phanerozoic Taxonomic Diversity." *Paleobiology* **10** (1984): 246–267.
[35] Sheehan, P. M. "A New Look at Evolutionary Units." *Palaeoceanography, Palaeoclimatol. & Palaeoecol.* **127** (1996): 21–32.
[36] Sheehan, P. M. "History of Marine Biodiversity." *Geological J.* **36** (2001): 231–249.
[37] Smith, A. B. "Large-Scale Heterogeneity of the Fossil Record: Implications for Phanerozoic Biodiversity Studies." *Phil. Trans. Roy. Soc. Lond. Ser. B* **356** (2001): 351–367.
[38] Smith, A. B., A. S. Gale, and N. E. A. Monks. "Sea-Level Change and Rock-Record Bias in the Cretaceous: A Problem for Extinction and Biodiversity Studies." *Paleobiol.* **27** (2001): 241–253.
[39] Valentine, J. W., D. Jablonski, and D. H. Erwin. "Fossils, Molecules and Embryos: New Perspectives on the Cambrian Explosion." *Development* **126** (1999): 851–859.
[40] Walker, K. R., and L. F. Laporte. "Congruent Fossil Communities from Ordovician and Devonian Carbonates of New York." *J. Paleo.* **44** (1970): 928–944.
[41] Wilkins, A. S. *The Evolution of Developmental Pathways.* Sunderland, MA: Sinauer Associates, 2002.

Computation in the Wild

Stephanie Forrest
Justin Balthrop
Matthew Glickman
David Ackley

1 INTRODUCTION

The explosive growth of the Internet has created new opportunities and risks by increasing the number of contacts between separately administered computing resources. Widespread networking and mobility have blurred many traditional computer system distinctions, including those between operating system and application, network and computer, user and administrator, and program and data. An increasing number of executable codes, including applets, agents, viruses, e-mail attachments, and downloadable software, are escaping the confines of their original systems and spreading through communications networks. These programs coexist and coevolve with us in our world, not always to good effect. Our computers are routinely disabled by network-borne infections, our browsers crash due to unforeseen interactions between an applet and a language implementation, and applications are broken by operating system upgrades. We refer to this situation as *computation in the wild*, by which we mean to convey the fact that software is developed, distributed, stored, and executed in rich and dynamic environments populated by other programs and computers, which collectively form a *software ecosystem.* The thesis of this chapter is that networked computer systems can be better understood, controlled, and developed when viewed from the perspective of living systems.

Taking seriously the analogy between computer systems and living systems requires us to rethink several aspects of the computing infrastructure—developing design strategies from biology, constructing software that can survive in the wild, understanding the current software ecosystem, and recognizing that all nontrivial software must evolve. There are deep connections between computation and life, so much so that in some important ways, "living computer systems" are already around us, and moreover, such systems are spreading rapidly and will have major impact on our lives and society in the future.

In this chapter we outline the biological principles we believe to be most relevant to understanding and designing the computational networks of the future. Among the principles of living systems we see as most important to the development of robust software systems are: modularity, autonomy, redundancy, adaptability, distribution, diversity, and use of disposable components. These are not exhaustive, simply the ones that we have found most useful in our own research. We then describe a prototype network intrusion-detection system, known as LISYS, which illustrates many of these principles. Finally, we present experimental data on LISYS' performance in a live network environment.

2 COMPUTATION IN THE WILD

In this section we highlight current challenges facing computer science and suggest that they have arisen because existing technological approaches are inadequate. Today's computers have significant and rapidly increasing commonalities with living systems, those commonalities have predictive power, and the computational principles underlying living systems offer a solid basis for secure, robust, and trustworthy operation in digital ecosystems.

2.1 COMPLEXITY, MODULARITY, AND LINEARITY

Over the past fifty years, the manufactured computer has evolved into a highly complex machine. This complexity is managed by deterministic digital logic, which performs extraordinarily complicated operations with high reliability, using modularity to decompose large elements into smaller components. Methods for decomposing functions and data dominate system design methods, ranging from object-oriented languages to parallel computing. An effective decomposition requires that the components have only limited interactions, which is to say that overall system complexity is reduced when the components are nearly independent [24]. Such well-modularized systems are "linear" in the sense that they obey an analog of the superposition principle in physics, which states that for affine differential equations, the sum of any two solutions is itself a solution (see Forrest [10] for a detailed statement of the connection). We use the term "linear" (and "nonlinear") in this sense—linear systems are decomposable into independent modules with minimal interactions and nonlinear systems are not. In the traditional view, nearly independent components are composed to create or execute a program, to construct a model, or to solve a problem. With the emergence of large-scale networks and distributed computing, and the concomitant increase in aggregate complexity, largely the same design principles have been applied. Although most single computers are designed to avoid component failures, nontrivial computer networks must also survive them, and that difference has important consequences for design, as a much greater burden of autonomy and self reliance is placed on individual components within a system.

The Achilles heel of any modularization lies in the interactions between the elements, where the independence of the components, and thus the linearity of the overall system, typically breaks down. Such interactions range from the use of public interfaces in object-oriented systems, to symbol tables in compilers, to synchronization methods in parallel processes. Although traditional computing design practice strives to minimize such interactions, those component interactions that are not eliminated are usually assumed to be deterministic, reliable, and trustworthy: A compiler assumes its symbol table is correct; an object in a traditional object-oriented system does not ask where a message came from; a closely coupled parallel process simply waits, indefinitely, for the required response. Thus, even though a traditionally engineered system may possess a highly decoupled and beautifully linear design, at execution time its modules are critically dependent on each other and the overall computation is effectively a monolithic, highly nonlinear entity.

As a result, although software systems today are much larger, perform more functions, and execute in more complex environments compared to a decade ago, they also tend to be less reliable, less predictable, and less secure, because they critically depend on deterministic, reliable and trustworthy interactions while operating in increasingly complex, dynamic, error prone, and threatening environments.

2.2 AUTONOMY, LIVING SOFTWARE, AND REPRODUCTIVE SUCCESS

Eliminating such critical dependencies is difficult. If, for example, a component is designed to receive some numbers and add them up, what else can it possibly do but wait until they arrive? A robust software component, however, could have multiple simultaneous goals, of which "performing an externally assigned task" is only one. For example, while waiting for task input a component might perform garbage collection or other internal housekeeping activities. Designing components that routinely handle multiple, and even conflicting, goals leads to robustness.

This view represents a shift in emphasis about what constitutes a practical computation, away from traditional algorithms which have the goal of finding an answer quickly and stopping, and toward processes such as operating systems that are designed to run indefinitely. In client-server computing, for example, although client programs may start and stop, the server program is designed to run forever, handling requests on demand. And increasingly, many clients are designed for open-ended execution as well, as with web browsers that handle an indefinite stream of page display requests from a user. Peer-to-peer architectures such as the ccr system [1] or Napster [19], which eliminate or blur the client/server distinction, move even farther toward the view of computation as interaction among long-lived and relatively independent processes. Traditional algorithmic computations, such as sorting, commonly exist not as stand-alone terminating computations performed for a human user, but as tools used by higher-level nonterminating computational processes. This change, away from individual terminating computations and toward loosely coupled ecological interactions among open-ended computations, has many consequences for architecture, software design, communication protocols, error recovery, and security.

Error handling provides a clear example of the tension between these two approaches. Viewed algorithmically, once an error occurs there is no point in continuing, because any resulting output would be unreliable. From a living software perspective, however, process termination is the last response to an error, to be avoided at nearly any cost.

As in natural living systems, successful computational systems have a variety of lifestyles other than just the "run once and terminate" algorithms on the one hand and would be immortal operating systems and processes on the other. The highly successful open-source Apache web server provides an example of an alternative strategy. Following the pattern of many Unix daemons, Apache employs a form of queen and workers organization, in which a single long-lived process does nothing except spawn and manage a set of server subprocesses, each of which handles a given number of web page requests and then dies. This hybrid strategy allows Apache to amortize the time required to start each new worker process over a productive lifetime serving many page requests, while at the same time ensuring that the colony as a whole can survive unexpected fatal errors in server subprocesses. Further, by including programmed subprocess death in the architecture, analogous to cellular apoptosis (programmed cell death), Apache deals with many nonfatal diseases as well, such as memory leaks. A related idea from the software fault tolerance community is that of software rejuvenation [22], in which running software is periodically stopped and restarted again after "cleaning" of the internal state. This proactive technique is designed to counteract software aging problems arising from memory allocation, fragmentation, or accumulated state.

In systems such as the Internet, which are large, open, and spontaneously evolving, individual components must be increasingly self-reliant and autonomous. They must also be able to function without depending on a consistent global design to guarantee safe interactions. Individual components must act autonomously, protect themselves, repair themselves, and increasingly, as in the example of Apache, decide when to replicate and when to die. Natural biological systems exhibit organizational hierarchies (e.g., cells, multicellular organisms, ecosystems) similar to computers, but each individual component takes much more responsibility for its own interactions than what we hitherto have required of our computations.

The propertiesv of autonomy, robustness, and security are often grouped together under the term "survivability," the ability to keep functioning, at least to some degree, in the face of extreme systemic insult. To date, improving application robustness is largely the incremental task of growing the set of events the program is expecting—typically in response to failures as they occur. For the future of large-scale networked computation, we need to take a more basic lesson from natural living systems, that survival is more important than the conventional strict definitions of correctness. As in Apache's strategy, there may be leaks, or even outright terminal programming errors, but they won't cause the entire edifice to crash.

Moving beyond basic survival strategies, there is a longer term survivability achieved through the reproductive success of high-fitness individuals in a biological ecosystem. Successful individuals reproduce more frequently, passing on their genes to future generations. Similarly, in our software ecosystem, reproductive success is achieved when software is copied, and evolutionary dead-ends occur when a piece of software fails to be copied—as in components that are replaced (or "patched") to correct errors or extend functionality.

2.3 DISPOSABILITY, ADAPTATION, AND HOMEOSTASIS

As computers become more autonomous, what happens when a computation makes a mistake, or is attacked in a manner that it cannot handle? We advocate building autonomous systems from disposable components, analogous to cells of a body that are replicating and dying continually. This is the engineering goal of avoiding single points of failure, taken to an extreme. Of course, it is difficult to imagine that individual computers in a network represent disposable components; after all, nobody is happy when it's their computer that "dies," for whatever reason, especially in the case of a false alarm. However, if the disposability is at a fine enough grain size, an occasional inappropriate response is less likely to be lethal. Such lower-level disposability is at the heart of Apache's survival strategy, and it is much like building reliable network protocols on top of unreliable packet transfer, as in TCP/IP. We will also see an example of disposability in the following section, when we discuss the continual turnover of detectors in LISYS, known as *rolling coverage*.

Computation today takes place in highly dynamic environments. Nearly every aspect of a software system is likely to change during its normal life cycle, including: who uses the system and for what purpose, the specification of what the system is supposed to do, certain parts of its implementation, the implementors of the system, the system software and libraries on which the computation depends, and the hardware on which the system is deployed. These changes are routine and continual.

Adaptation and homeostasis are two important components of a robust computing system for dynamic environments. Adaptation involves a component modifying its internal rules (often called "learning") to better exploit the current environment. Homeostasis, by contrast, involves a component dynamically adjusting its interactions with the environment to preserve a constant internal state. As a textbook says, homeostasis is "the maintenance of a relatively stable internal physiological environment or internal equilibrium in an organism" [5, p. G–11].

Given the inherent difficulty of predicting the behavior of even static nonlinear systems, and adding the complication of continuously evolving computational environments, the once plausible notion of finding *a priori* proofs of program correctness is increasingly problematic. Adaptive methods, particularly those used in biological systems, appear to be the most promising near-term approach for modeling inherent nonlinearity and tracking environmental change. The phenomenon of "bit rot" is widespread. When a small piece of a program's context changes, all too often either the user or a system administrator is required to change pathnames, apply patches, install new libraries (or reinstall old ones), etc. Emerging software packaging and installation systems, such as RPM [15] or Debian's `apt-get` utility, provide a small start in this direction. These systems provide common formats for "sporifying" software systems to facilitate mobility and reproduction, and they provide corresponding developmental pathways for adapting the software to new environments. We believe that this sort of adaptability should take place not just at software installation but as an ongoing process of accommodation, occurring automatically and autonomously at many levels, and that programs should have the ability to evolve their functionality over time.

Turning to homeostasis, we see that current computers already have many mechanisms that can be thought of as homeostatic, for example, temperature and power regulation at the hardware level, and virtual memory management and process scheduling at the operating systems level. Although far from perfect, they are essential to the proper functioning of almost any application program. However, mechanisms such as virtual memory management and process scheduling cannot help a machine survive when it is truly stressed. This limitation stems in part from the policy that a kernel should be fair. When processor time is shared among all processes, and memory requests are immediately granted if they can be satisfied, stability cannot be maintained when resource demands become extreme. Fairness necessitates poor performance for all processes, either through "thrashing," when the virtual memory system becomes stressed, or through unresponsiveness, failed program invocations, and entire system crashes, when there are too many processes to be scheduled. One fair response to extreme resource contention, used by AIX [4], is random killing of processes. This strategy, however, is too likely to kill processes that are critical to the functioning of the system. Self-stabilizing algorithms [23] are similar in spirit, leading to systems in which an execution can begin in any arbitrary initial state and be guaranteed to converge to a "legitimate" state in a finite number of steps.

A more direct use of homeostasis in computing is Anil Somayaji's pH system (for process homeostasis) [25, 26]. pH is a Linux kernel extension which monitors every executing process on a computer and responds to anomalies in process behavior, either by delaying or aborting subsequent system calls. Process-level monitoring at the system-call level can detect a wide range of attacks aimed at individual processes, especially server processes such as sendmail and Apache [11]. Such attacks often leave a system functional, but give the intruder privileged access to the system and its protected data. In the human body, the immune system can kill thousands of cells without causing any outward sign of illness; similarly, pH uses execution delays for individual system calls as a way to restore balance. By having the delays be proportional to the number of recent anomalous system calls, a single process with gross abnormalities will effectively be killed (e.g., delayed to the point that native time-out mechanisms are triggered), without interfering with other normally behaving processes. Likewise, small deviations from normal behavior are tolerated in the sense that short delays are transient phenomena which are imperceptible at the user level. pH can successfully detect and stop many intrusions before the target system is compromised. In addition to coping with malicious attacks, pH detects a wide range of internal system-level problems, slowing down the offending program to prevent damage to the rest of the system.

2.4 REDUNDANCY, DIVERSITY, AND THE EVOLUTIONARY MESS

Biological systems use a wide variety of redundant mechanisms to improve reliability and robustness. For example, important cellular processes are governed by molecular cascades with large amounts of redundancy, such that if one pathway is disrupted an alternative one is available to preserve function. Similarly, redundant encodings occur throughout genetics, the triplet coding for amino acids at the nucleotide level providing one example. These kinds of redundancy involve more than the simple strategy of making identical replicas that we typically see in redundant computer systems. Identical replicas can protect against single component failure but not against design flaws. The variations among molecular pathways or genetic encodings thus provide additional protection through the use of diverse solutions.

This diversity is an important source of robustness in biology. As one example, a stable ecosystem contains many different species which occur in highly conserved frequency distributions. If this diversity is lost and a few species become dominant, the ecosystem becomes unstable and susceptible to perturbations such as catastrophic fires, infestations, and disease. Other examples include the variations among individual immune systems in a population and various forms of genetic diversity within a single species. Computers and other engineered artifacts, by contrast, are notable for their lack of diversity. Software and hardware components are replicated for several reasons: economic leverage, consistency of behavior, portability, simplified debugging, and cost of distribution and maintenance. All these advantages of uniformity, however, become potential weaknesses when they replicate errors or can be exploited by an attacker. Once a method is created for penetrating the security of one computer, all computers with the same configuration become similarly vulnerable [13]. Unfortunately, lack of diversity also pervades the software that is intended to defend against attacks, be it a firewall, an encryption algorithm, or a computer virus detector. The potential danger grows with the population of interconnected and homogeneous computers.

Turning to evolution, we see that the history of manufactured computers is a truly evolutionary history, and evolution does not anticipate, it reacts. To the degree that a system is large enough and distributed enough that there is no effective single point of control for the whole system, we must expect evolutionary forces—or "market forces"—to be significant. In the case of computing, this happens both at the technical level through unanticipated uses and interactions of components as technology develops and at the social level from cultural and economic pressures. Having humans in the loop of an evolutionary process, with all their marvelous cognitive and predictive abilities, with all their philosophical ability to frame intentions, does not necessarily change the nature of the evolutionary process. There is much to be gained by recognizing and accepting that our computational systems resemble naturally evolving systems much more closely than they resemble engineered artifacts such as bridges or buildings. Specifically, the strategies that we adopt to understand, control, interact with, and influence the design of computational systems will be different once we understand them as ongoing evolutionary processes.

In this section we described several biological design principles and how they are relevant to computer and software systems. These include modularity, redundancy, distribution, autonomy, adaptability, diversity, and use of disposable components. As computers have become more complex and interconnected, these principles have become more relevant, due to an increasing variety of interactions between software components and a rising number of degrees of freedom for variation in computational environments. Software components are now confronted with challenges increasingly similar to those faced by organic entities situated in a biological ecosystem. An example of how biological organisms have evolved to cope with their environments is the immune system. In the next section we explore a software framework that is specifically patterned after the immune system. In addition to exhibiting many of the general biological design principles advocated above, this framework illustrates a specific set of mechanisms derived from a particular biological system and applied to a real computational problem.

3 AN ILLUSTRATION: LISYS

In the previous section, we outlined an approach to building and designing computer systems that is quite different from that used today. In this section we illustrate how some of these ideas could be implemented using an artificial immune system framework known as ARTIS and its application to the problem of network intrusion detection in a system known as LISYS. ARTIS and LISYS were originally developed by Steven Hofmeyr in his dissertation [16, 17].

The immune system processes peptide patterns using mechanisms that in some cases correspond closely to existing algorithms for processing information (e.g., the genetic algorithm), and it is capable of exquisitely selective and well-coordinated responses, in some cases responding to fewer than ten molecules. Some of the techniques used by the immune system include learning (affinity maturation of B cells, negative selection of B and T cells, and evolved biases in the germline), memory (cross-reactivity and the secondary response), massively parallel and distributed computations with highly dynamic components (on the order of 10^8 different varieties of receptors [27] and 10^7 new lymphocytes produced each day [20]), and the use of combinatorics to address the problem of scarce genetic resources (V-region libraries). Not all of these features are included in ARTIS, although ARTIS could easily be extended to include them (affinity maturation and V-region libraries are the most notable lacunae).

ARTIS is intended to be a somewhat general artificial immune system architecture, which can be applied to multiple application domains. In the interest of brevity and concreteness, we will describe the instantiation of ARTIS in LISYS, focusing primarily on how it illustrates our ideas about computation in the wild. For more details about LISYS as a network intrusion-detection system, the reader is referred to Balthrop et al [2, 3] and Hofmeyr [17].

3.1 THE NETWORK INTRUSION PROBLEM

LISYS is situated in a local-area broadcast network (LAN) and used to protect the LAN from network-based attacks. In contrast with switched networks, broadcast LANs have the convenient property that every location (computer) sees every packet passing through the LAN.[1] In this domain, we are interested in building a model of normal behavior (known as *self* through analogy to the normally occurring peptides in living animals) and detecting deviations from normal (known as *non-self*). Self is defined to be the set of normal pairwise TCP/IP connections between computers, and non-self is the set of connections, potentially an enormous number, which are not normally observed on the LAN. Such connections may represent either illicit attacks or significant changes in

[1] There are several ways in which the architecture could be trivially modified to run in switched networks. For example, some processing could be performed on the switch or router, SYN packets could be distributed from the switch to the individual nodes, or a combination could be tried.

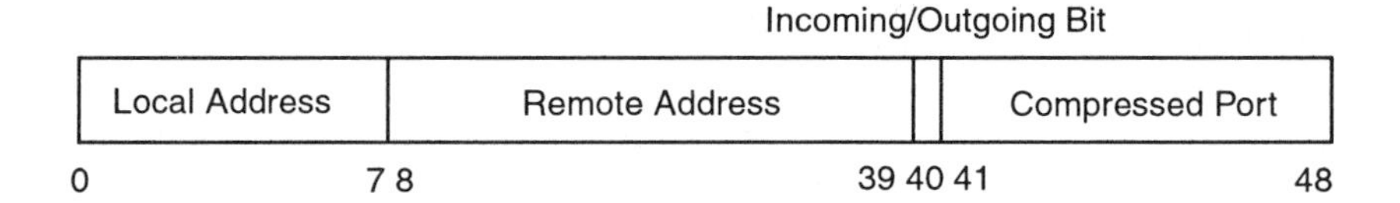

FIGURE 1 The 49-bit compression scheme used by LISYS to represent TCP SYN packets.

network configuration or use. A connection can occur between any two computers in the LAN as well as between a computer in the LAN and an external computer. It is defined in terms of its "data-path triple"—the source IP address, the destination IP address, and the port by which the computers communicate [14, 18].

LISYS examines only the headers of IP packets. Moreover, at this point we constrain LISYS to examine only a small set of the features contained in the headers of TCP SYN packets. The connection information is compressed to a single 49-bit string that unambiguously defines the connection. This string is compressed in two ways [17]. First, it is assumed that one of the IP addresses is always internal, so only the final byte of this address needs to be stored. Secondly, the port number is also compressed from 16 bits to 8 bits. This is done by re-mapping the ports into several different classes. Figure 1 shows the 49-bit representation.

3.2 LISYS ARCHITECTURE

LISYS consists of a distributed set of *detectors*, where each detector is a 49-bit string (with the above interpretation) and a small amount of local state. The detectors are distributed in the sense that they reside on multiple hosts; there exists one *detector set* for each computer in the network. Detector sets perform their function independently, with virtually no communication. Distributing detectors makes the system more robust by eliminating the single point of failure associated with centralized monitoring systems. It also makes the system more scalable as the size of the network being protected increases.

A perfect *match* between a detector and a compressed SYN packet means that at each location in the 49-bit string, the symbols are identical. However, in the immune system matching is implemented as binding affinities between molecules, where there are only stronger and weaker affinities, and the concept of perfect matching is ill defined. To capture this, LISYS uses a partial matching rule known as r-contiguous bits matching [21]. Under this rule, two strings match if they are identical in at least r contiguous locations. This means there are $l - r + 1$ windows where a match can occur, where l is the string length. The value r is a threshold which determines the specificity of detectors, in that it controls how many strings can potentially be matched by a single detector. For example, if $r = l$ (49 in the case of LISYS' compressed representation), the match is maximally specific, and a detector can match only a single string—itself. As shown in Esponda [9], the number of strings a detector matches increases exponentially as the value of r decreases.

LISYS uses *negative detection* in the sense that valid detectors are those that fail to match the normally occurring behavior patterns in the network. Detectors are generated randomly and are initially *immature.* Detectors that match connections observed in the network during the *tolerization period* are eliminated. This procedure is known as *negative selection* [12]. The tolerization period lasts for a few days, and detectors that survive this period become *mature.* For the r-contiguous bits matching rule and fixed self sets which don't change over time, the random detector generation algorithm is inefficient—the number of random strings that must be generated and tested is approximately exponential in the size of self. More efficient algorithms based on dynamic programming methods allow us to generate detectors in linear time [6, 7, 28, 29, 30]. However, when generating detectors asynchronously for a dynamic self set, such as the current setting, we have found that random generation works sufficiently well. Negative detection allows LISYS to be distributed, because detectors can make local decisions independently with no communication to other nodes. Negative selection of randomly generated detectors allows LISYS to be adaptive to the current network condition and configuration.

LISYS also uses *activation thresholds.* Each detector must match multiple connections before it is activated. Each detector records the number of times it matches (the *match count*) and raises an alarm only when the match count exceeds the activation threshold. Once a detector has raised an alarm, it returns its match count to zero. This mechanism has a time horizon: Over time the match count slowly returns to zero. Thus, only repeated occurrences of structurally similar and temporally clumped strings will trigger the detection system. The activation threshold mechanism contributes to LISYS's adaptability and autonomy, because it provides a way for LISYS to tolerate small errors that are likely to be false positives.

LISYS uses a "second signal," analogous to costimulatory mechanisms in the immune system. Once a detector is activated, it must be costimulated by a human operator or it will die after a certain time period. Detectors which are costimulated become *memory detectors.* These are long-lived detectors which are more easily activated than non-memory detectors. This secondary response improves detection of true positives, while costimulation helps control false positives.

Detectors in LISYS have a finite lifetime. Detectors can die in several ways. As mentioned before, immature detectors die when they match self, and activated detectors die if they do not receive a costimulatory signal. In addition, detectors have a fixed probability of dying randomly on each time step, with the average lifespan of a single detector being roughly one week. Memory detectors are not subject to random death and may thus survive indefinitely. However, once the

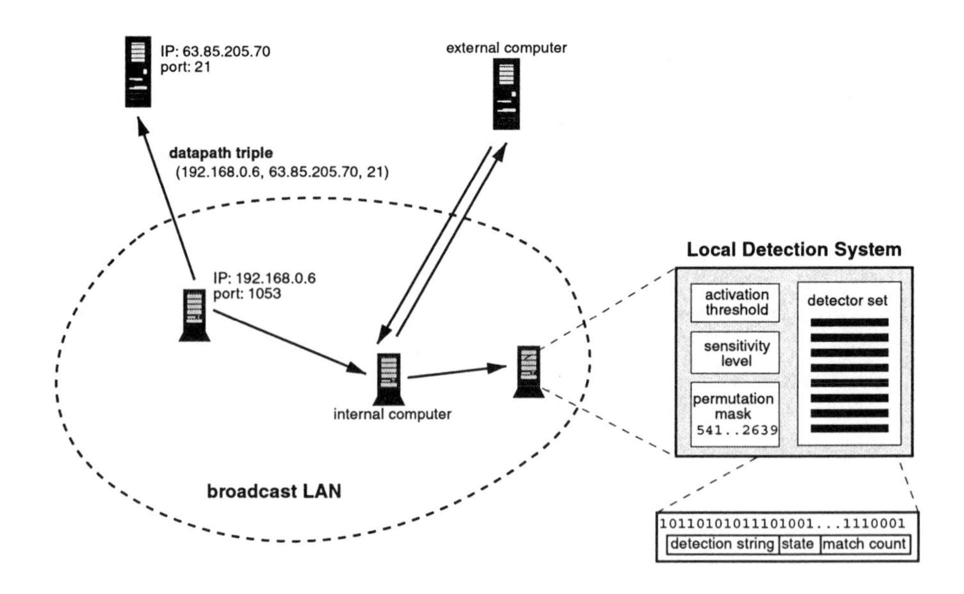

FIGURE 2 The architecture of LISYS.

number of memory detectors exceeds a specified fraction of the total detector population, some are removed to make room for new ones. The finite lifetime of detectors, when combined with detector regeneration and tolerization, results in *rolling coverage* of the self set. This rolling coverage clearly illustrates the principles of disposable components and adaptability.

Each independent detector set (one per host) has a *sensitivity level*, which modifies the local activation threshold. Whenever the match count of a detector in a given set goes from 0 to 1, the effective activation threshold for all the other detectors in the same set is reduced by one. Hence, each different detector that matches for the first time "sensitizes" the detection system, so that all detectors on that machine are more easily activated in future. This mechanism has a time horizon as well; over time, the effective activation threshold gradually returns to its default value. This mechanism corresponds very roughly the effect that inflammation, cytokines, and the other molecules have on the sensitivity of nearbyb individual immune system lymphocytes. Sensitivity levels in LYSYS are a simple adaptive mechanism intended to help detect true positives, especially distributed coordinated attacks.

LISYS uses *permutation masks* to achieve diversity of representation.[2] A permutation mask defines a permutation of the bits in the string representation of the network packets. Each detector set (network host) has a different, randomly

[2]Permutation masks are one possible means of generating *secondary representations.* A variety of alternative schemes are explored in Hofmeyr [17].

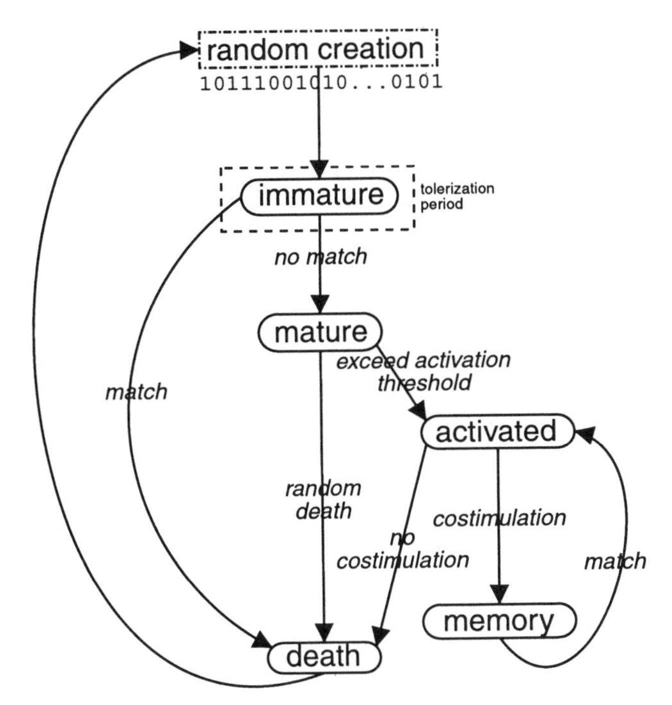

FIGURE 3 The lifecycle of a detector in LISYS.

generated, permutation mask. One feature of the negative-selection algorithm as originally implemented is that it can result in undetectable patterns called *holes*, or put more positively, *generalizations* [6, 7]. Holes can exist for any symmetric, fixed-probability matching rule, but by using permutation masks we effectively change the match rule on each host, which gives us the ability to control the form and extent of generalization in the vicinity of self [8]. Thus, the permutation mask controls how the network packet is presented to the detection system, which is analogous to the way different MHC types present different sets of peptides on the cell surface.

Although LISYS incorporates many ideas from natural immune systems, it also leaves out many features. One important omission is that LISYS has only a single kind of detector "cell," which represents an amalgamation of T cells, B cells, and antibodies. In natural immune systems, these cells and molecules have distinct functions. Other cell types missing from LISYS include effector cells, such as natural killer cells and cytotoxic T cells. The mechanisms modeled in LISYS are almost entirely based on what is known as "adaptive immunity." Also important is innate immunity. Moreover, an important aspect of adaptive immunity is clonal selection and somatic hypermutation, processes which are absent from LISYS. In the future, it will be interesting to see which of these additional features turn out to have useful computational analogs.

4 EXPERIMENTS

In this section we summarize some experiments that were performed in order to study LISYS's performance in a network intrusion-detection setting and to explore how LISYS' various mechanisms contribute to its effectiveness. For the experiments described in this section, we used a simplified form of LISYS in order to study some features more carefully. Specifically, we used a version that does not have sensitivity levels, memory detectors, or costimulation. Although we collected the data on-line in a production distributed environment, we performed our analysis off-line on a single computer. This made it possible to compare performance across many different parameter values. The programs used to generate the results reported in this chapter are available from ⟨http://www.cs.unm.edu/~immsec⟩. The programs are part of LISYS and are found in the LisysSim directory of that package.

4.1 DATA SET

Our data were collected on a small but realistic network. The normal data were collected for two weeks in November 2001 on an internal restricted network of computers in our research group at UNM. The six internal computers in this network connected to the Internet through a single Linux machine acting as a firewall, router and masquerading server. In this environment we were able understand all of the connections, and we could limit attacks. Although small, the network provides a plausible model of a corporate intranet where activity is somewhat restricted and external connections must pass through a firewall. And, it resembles the increasingly common home network that connects to the Internet through a cable or DSL modem and has a single external IP address. After collecting the normal data set, we used a package "Nessus" to generate attacks against the network.

Three groups of attacks were performed. The first attack group included a denial-of-service (DOS) attack from an internal computer to an external computer, attempted exploitations of weaknesses in the configuration of the firewall machine, an attack against FTP (file transfer protocol), probes for vulnerabilities in SSH (secure shell), and probes for services such as chargen and telnet. The second attack group consisted of two TCP port scans, including a stealth scan (difficult to detect) and a noisy scan (easy to detect). The third attack group consisted of a full nmap[3] port scan against several internal machines.

Most of these attacks are technically classified as probes because they did not actually execute an attack, simply checking to see if the attack could be

[3]nmap is a separate software package used by Nessus.

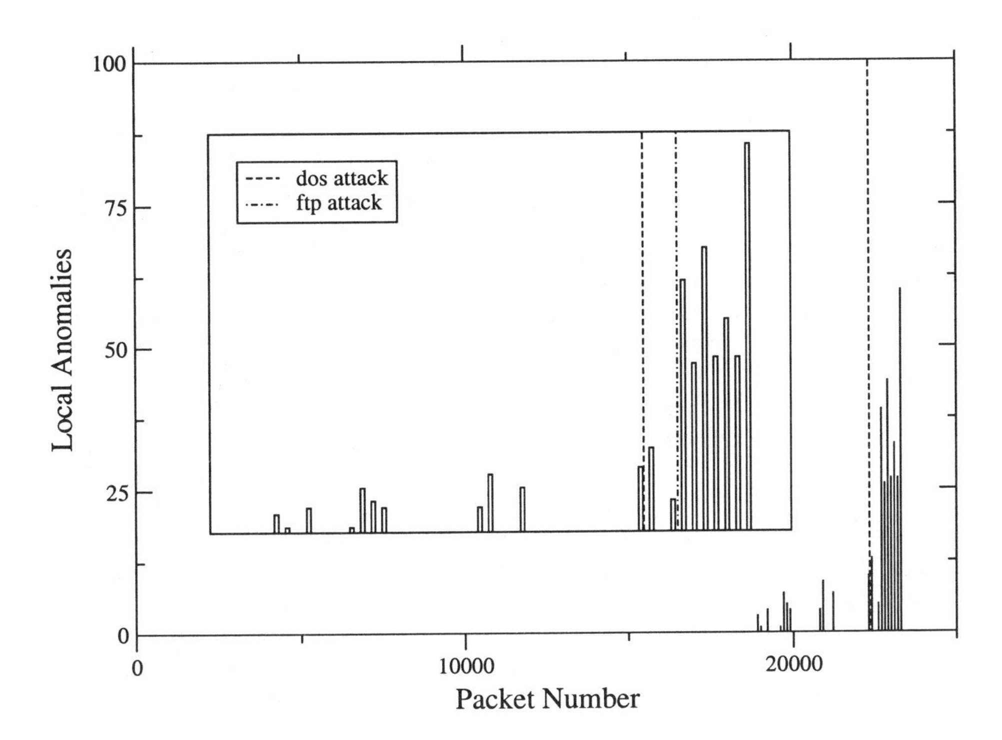

FIGURE 4 Local anomalies for attack group one. The inset is a zoomed view of the same data.

performed. However, in order to succeed, a probe would typically occur first; actually executing the attack would create additional TCP traffic, making it easier for LISYS to detect the attack. More details about this data set are given in Balthrop et al. [3].

4.2 DETECTING ABNORMAL BEHAVIOR

Here we reports how a simplified version of LISYS performs. In addition to the simplifications mentioned earlier, we also eliminated the use of permutation masks for this set of experiments and used a single large detector set running on a single host. Because we are operating in a broadcast network environment and not using permutation masks, this configuration is almost equivalent to distributing the detectors across different nodes (there could be some minor differences arising from the tolerization scheme and parameter settings for different detector sets).

We used the following parameter values [3]: $r = 10$, `number of detectors` $=$ 6,000, `tolerization period` $=$ 15,000 connections, and `activation threshold`

= 10. The experiments were performed by running LISYS with the normal network data followed by the attack data from one of the attack groups.

Figure 4 shows how LISYS performed during the first group of attacks. The x-axis shows time (in units of packets) and the y-axis shows the number of anomalies per time period. An anomaly occurs whenever the match count exceeds the activation threshold. The vertical line indicates where the normal data ended and the attack data began. Anomalies are displayed using windows of 100. This means that each bar is an indication of the number of anomalies signaled in the last 100 packets. There are a few anomalies flagged in the normal data, but there are more anomalies during the group of attacks.

The graph inset shows that LISYS was able to detect the denial-of-service and FTP attacks. The vertical lines indicate the beginning of these two attacks, and there are spikes shortly after these attacks began. The spikes for the FTP attack are significantly higher than those for the DOS attack, but both attacks have spikes that are higher than the spikes in the normal data, indicating a clear separation between true and false positives. This view of the data is interesting because the height of the spikes indicates the system's confidence that there is an anomaly occurring at that point in time.

Figure 5 shows the LISYS anomaly data for attack group two. By looking at this figure, we can see that there is something anomalous in the last half of the attack data, but LISYS was unable to detect anomalies during the first half of the attack.[4] Although the spikes are roughly the same height as the spikes in the normal data, the temporal clustering of the spikes is markedly different. Figure 6 shows the LISYS anomaly data for attack group three. The figure shows that LISYS overwhelmingly found the nmap scan to be anomalous. Not only are the majority of the spikes significantly higher than the normal data spikes, but there is a huge number of temporally clustered spikes.

These experiments support the results reported in Hofmeyr [17], suggesting that the LISYS architecture is effective at detecting certain classes of network intrusions. However, as we will see in the next section, LISYS performs much better under slightly different circumstances.

5 THE EFFECT OF DIVERSITY

The goal of the experiments in this section was to assess the effect of diversity in our artificial immune system. Recall that diversity of representation (loosely analogous to MHC diversity in the real immune system) is implemented in LISYS by permutation masks. We were interested to see how LISYS' performance would be affected by adding permutation masks. Because performance is measured in

[4]In LISYS, detectors are continually being generated, undergoing negative selection, and being added to the repertoire. Some new detectors were added during the first half of the attack, but these detectors turned out not to be crucial for the detection in the second half.

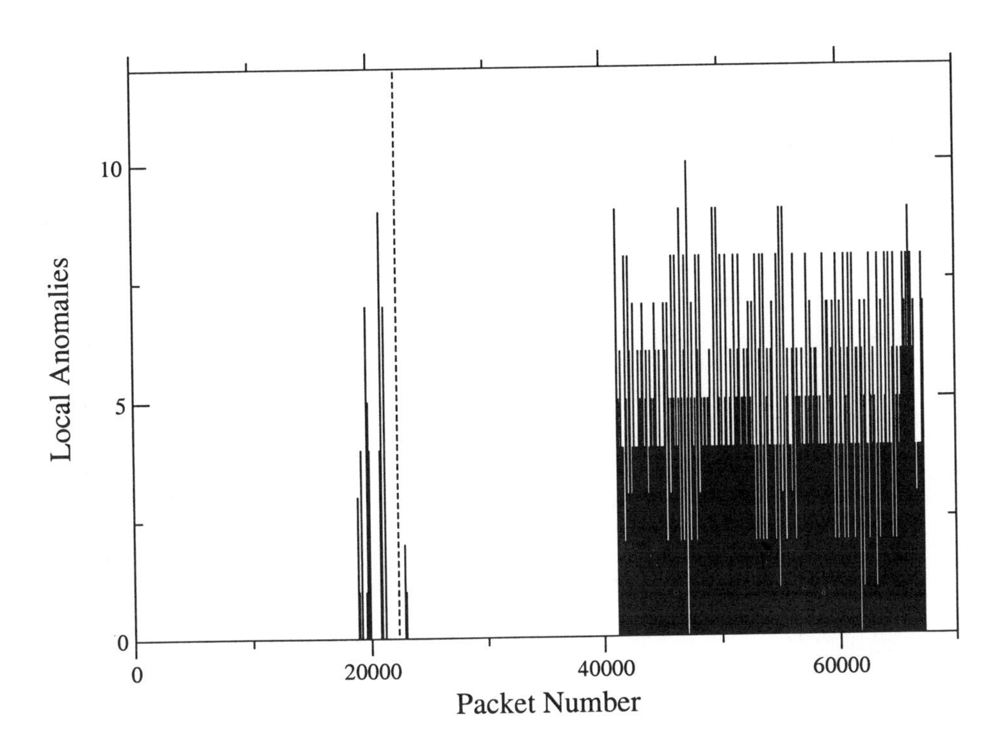

FIGURE 5 Local anomalies for attack group two.

terms of true and false positives, this experiment tested the effect of permutations on the system's ability to generalize (because low false-positive rates correspond to good generalization, and high false positives correspond to poor generalization). For this experiment, 100 sets of detectors were tolerized (generated randomly and compared against self using negative selection) using the first 15,000 packets in the data set (known as the *training set*), and each detector set was assigned a random permutation mask. Each detector set had exactly 5,000 mature detectors at the end of the tolerization period [3], and for consistency we set the random death probability to zero. Five-thousand detectors provides maximal possible coverage (i.e., adding more detectors does not improve subsequent matching) for this data set and r threshold. Each set of detectors was then run against the remaining 7,329 normal packets, as well as against the simulated attack data. In these data (the *test sets*), there are a total of 475 unique 49-bit strings. Of these 475, 53 also occur in the training set and are thus undetectable (because any detectors which would match them are eliminated during negative selection). This leaves 422 potentially detectable strings, of which 26 come from

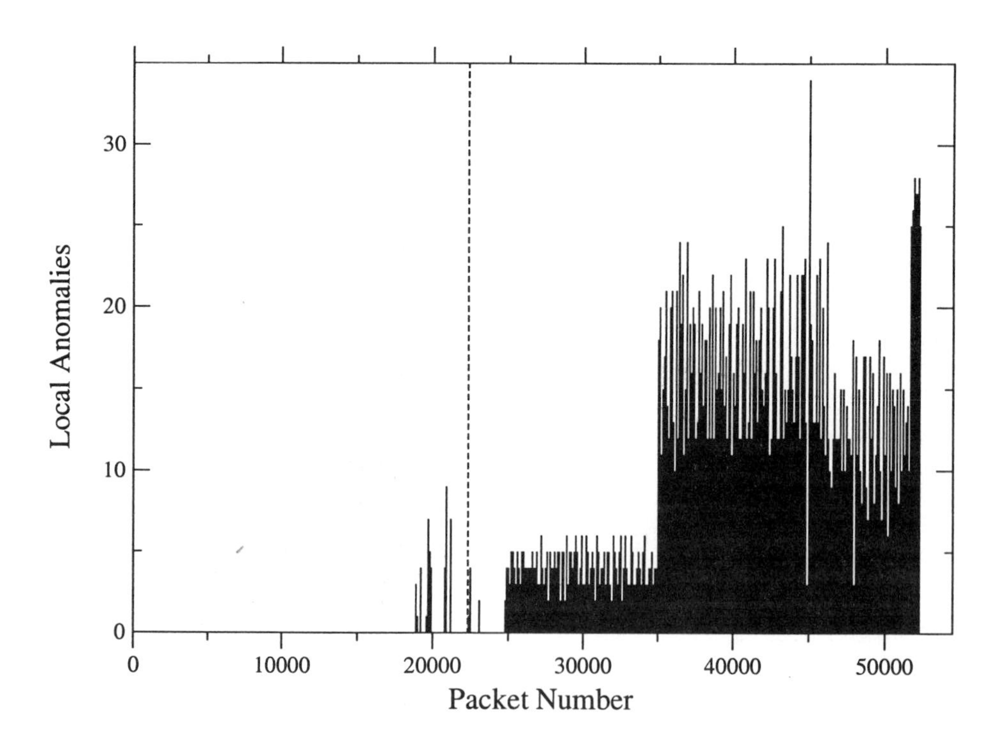

FIGURE 6 Local anomalies for attack group three.

the normal set and 396 are from the attack data, making the maximal possible coverage by a detector set 422 unique matches.

An ideal detector set would achieve zero false positives on the normal test data and a high number of true positives on the abnormal data. Because network attacks rarely produce a single anomalous SYN packet, we don't need to achieve perfect true-positive rates at the packet level in order to detect all attacks against the system. For convenience, however, we measure false positives in terms of single packets. Thus, a perfect detector set would match the 396 unique attack strings, and fail to match the 26 new normal strings in the test set.

Figure 7 shows the results of this experiment. The performance of each detector set is shown as a separate point on the graph. Each detector set has its own randomly generated permutation of the 49 bits, so each point shows the performance of a different permutation. The numbers on the x-axis correspond to the number of unique self-strings in the test set which are matched by the detector set, i.e., the number of false positives (up to a maximum of 26). The y-axis plots a similar value with respect to the attack data, i.e. the number of unique true positive matches (up to a maximum of 396). The graph shows that there is

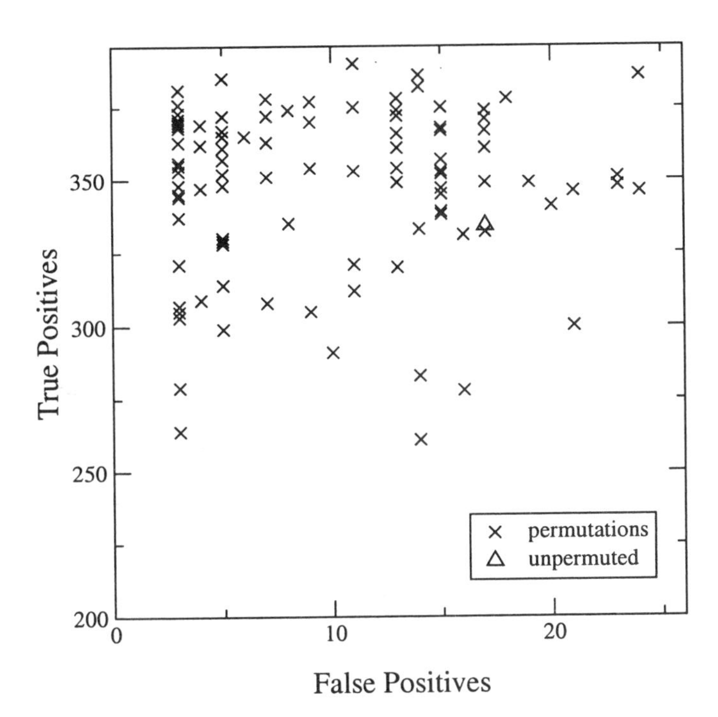

FIGURE 7 LISYS performance under different permutations. Each plotted point corresponds to a different permutation, showing false positives (x-axis) and true positives (y-axis).

a large difference in the discrimination ability of different permutations. Points in the upper left of the graph are the most desirable, because they correspond to permutations which minimize the number of false positives and maximize the number of true positives; points toward the lower right corner of the graph indicate higher false positives and/or lower true positives. Surprisingly, the performance of the original (unpermuted) mapping is among the worst we found, suggesting that the results reported in the previous section are a worst case in terms of true vs. false positives. Almost any other random permutation we tried outperformed the original mapping. Although we don't yet have a complete explanation for this behavior, we believe that it arises in the following way.

The LISYS design implicitly assumes that there are certain predictive bit patterns that exhibit regularity in self, and that these can be the basis ofs distinguishing self from non-self. As it turns out, there are also deceptive bit patterns which exhibit regularity in the training set (observed self), but the regularity does not generalize to the rest of self (the normal part of the test set). These patterns tend to cause false positives when self strings that do not fit the predicted regularity occur. We believe that the identity permutation is bad because the predictive bits are at the ends of the string, while the deceptive region is in the middle. Under such an arrangement, it is difficult to find a window that covers many predictive bit positions without also including deceptive ones. It is highly likely that a random permutation will break up the deceptive region, and bring the predictive bits closer to the middle, where they will appear in more windows.

The preceding analysis is based on the choice of a single permutation. In the original system developed by Hofmeyr, however, each host on the network used a different (randomly chosen) permutation of the 49 bits. Using a diverse set of random permutation masks reduces the overall number of undetectable patterns, thus increasing population-level robustness, but there is a risk of increasing false positives. In the experiments described here, using all permutations (with each permutation consisting of 5,000 detectors) would raise the number of true positives to 396 out of 396, and it would also raise the number of false positives to 26 out of 26 (compared with 334/396 true positives and 17/26 false positives when the original representation is used). However, if we chose only those permutations with low false-positive rates, we can do significantly better. The relative tradeoffs between diversity of representation and the impact on discrimination ability is an important area of future investigation.

6 DISCUSSION

In the previous section we described a prototype network intrusion-detection system based on the architecture and mechanisms of natural immune systems. Although LISYS is not a complete model of the immune system nor a production quality intrusion-detection system, it illustrates many of the principles elucidated in section 2. We identified the principles of redundancy, distribution, autonomy, adaptability, diversity, use of disposable components, and modularity as key features required for successful "computation in the wild."

Of these, LISYS is weakest in the autonomy component. There are a number of ways in which LISYS could be made more autonomous. One way would be to model the so-called constant region of antibody molecules. Detectors in LISYS represent generic immune cells, combining properties of T cells, B cells, and antibodies. However, as described to date they model only the recognition capabilities of immune cells (receptors) and a small amount of state. A simple extension would be to concatenate a few bits to each detector to specify a response (analogous to different antibody *isotypes* which comprise an antibody's "constant region"). This feature would constitute a natural extension to LISYS, although it leaves open the important question of how to interpret the response bits—that is, what responses do we want LISYS to make?

A second approach to adding autonomy to LISYS involves the use of a second signal. Costimulation in LISYS is an example of a second signal, in which a second independent source must confirm an alarm before a response can be initiated. There are several examples of second signals in immunology, including helper T cells (which confirm a B cell's match before the B cell can be activated) and the complement system used in the early immune response. The complement cascade provides a general and early indication of trouble, and operates in combination with highly specific T-cell responses. One possibility for LISYS would be to incorporate the pH system described earlier to provide the second signal. pH can detect anomalous behavior on a host (e.g., during intrusion attempts) and could serve as a generic indicator of "damage" in conjunction with the specific detectors of LISYS. Thus, when a LISYS detector became activated it would respond only if pH confirmed that the system was in an anomalous state. Either of these two approaches would represent an important next step toward autonomy.

7 CONCLUSIONS

The implementation we described is based on similarities between computation and immunology. Yet, there are also major differences, which it is wise to respect. The success of all analogies between computing and living systems ultimately rests on our ability to identify the correct level of abstraction—preserving what is essential from an information-processing perspective and discarding what is not. In the case of immunology, this task is complicated by the fact that natural immune systems handle fundamentally different data streams than those handled by computers. In principle, a computer vision system or a speech-recognition system would take as input the same data as a human does (e.g., photons or sound waves). In contrast, regardless of how successful we are at constructing a computer immune system, we would never expect or want it to handle pathogens in the form of actual living cells, viruses, or parasites. Thus, the level of abstraction for computational immunology is necessarily higher than that for computer vision or speech, and there are more degrees of freedom in selecting a modeling strategy.

Our study of how the architecture of the immune system could be applied to network security problems illustrates many of the important design principles introduced in the first part of this chapter. Many of the principles discussed here are familiar, and in some cases have long been recognized as desirable for computing systems, but in our view, there has been little appreciation of the common underpinnings of these principles, little or no rigorous theory, and few working implementations that take them seriously. Hofmeyr's artificial immune system is, however, an initial step in this direction. As our computers and the software they run become more complex and interconnected, properties such as robustness, flexibility and adaptability, diversity, self reliance and autonomy, and scalability can only become more important to computer design.

ACKNOWLEDGMENTS

The authors gratefully acknowledge the support of the National Science Foundation (grants CDA-9503064, and ANIR-9986555), the Office of Naval Research (grant N00014-99-1-0417), Defense Advanced Projects Agency (grant AGR F30602-00-2-0584), the Santa Fe Institute, and the Intel Corporation.

We thank the many members of Adaptive Computation group at the University of New Mexico for their help over the past ten years in developing these ideas.

REFERENCES

[1] Ackley, D. H. "ccr: A Network of Worlds for Research." In *Artificial Life V. Proceedings of the Fifth International Workshop on the Synthesis and Simulation of Living Systems*, edited by C. G. Langton and K. Shimohara, 116–123. Cambridge, MA: The MIT Press, 1996.

[2] Balthrop, J., F. Esponda, S. Forrest, and M. Glickman. "Coverage and Generalization in an Artificial Immune System." In *GECCO-2002: Proceedings of the Genetic and Evolutionary Computation Conference*, edited by W. B. Langdon, E. Cantu-Paz, K. Mathias, R. Roy, D. Davis, R. Poli, K. Balakrishnan, V. Hanovar, G. Rudolph, J. Wegener, L. Bull, M. A. Potter, A. C. Schultz, J. F. Miller, E. Burke and N. Jonaska, 3–10. New York: Morgan Kaufmann, 2002.

[3] Balthrop, J., S. Forrest, and M. Glickman. "Revisiting Lisys: Parameters and Normal Behavior." In *CEC-2002: Proceedings of the Congress on Evolutionary Computing.* Los Alamitos, CA: IEEE Computer Society Press, 2002.

[4] International Business Machines Corporation. *AIX Version 4.3 System Management Guide: Operating System and Devices*, chapter Understanding Paging Space Allocation Policies. IBM, 1997. ⟨http://www.rs6000.ibm.com/doc_link/en_US/a_doc_lib/aixbman/baseadmn/pag_space_under.htm⟩.

[5] Curtis, Helena, and N. Sue Barnes. *Biology.* 5th ed. New York: Worth Publishers, Inc., 1989.

[6] D'haeseleer, P., S. Forrest, and P. Helman. "An Immunological Approach to Change Detection: Algorithms, Analysis and Implications." In *Proceedings of the 1996 IEEE Symposium on Computer Security and Privacy.* Los Alamitos, CA: IEEE Computer Society Press, 1996.

[7] D'haeseleer, Patrik. "An Immunological Approach to Change Detection: Theoretical Results." In *Proceedings of the 9th IEEE Computer Security Foundations Workshop.* Los Alamitos, CA: IEEE Computer Society Press, 1996.

[8] Esponda, Carlos Fernando, and Stephanie Forrest. "Defining Self: Positive and Negative Detection." In *Proceedings of the IEEE Symposium on Security and Privacy.* Los Alamitos, CA: IEEE Computer Society Press, 2001.

[9] Esponda, Carlos Fernando. "Detector Coverage with the r-Contiguous Bits Matching Rule." Technical Report TR-CS-2002-03, University of New Mexico, Albuquerque, NM, 2002.

[10] Forrest, S. *Emergent Computation.* Cambridge, MA: MIT Press, 1991.

[11] Forrest, S., S. Hofmeyr, A. Somayaji, and T. Longstaff. "A Sense of Self for Unix Processes." In *Proceedings of the 1996 IEEE Symposium on Computer Security and Privacy.* Los Alamitos, CA: IEEE Computer Society Press, 1996.

[12] Forrest, S., A. S. Perelson, L. Allen, and R. Cheru Kuri. "Self-Nonself Discrimination in a Computer." In *Proceedings of the 1994 IEEE Symposium on Research in Security and Privacy.* Los Alamitos, CA: IEEE Computer Society Press, 1994.

[13] Forrest, S., A. Somayaji, and D. H. Ackley. "Building Diverse Computer Systems." In *Sixth Workshop on Hot Topics in Operating Systems.* Los Alamitos, CA: IEEE Computer Society Press, 1998.

[14] Heberlein, L. T., G. V. Dias, K. N. Levitte, B. Mukherjee, J. Wood, and D. Wolber. "A Network Security Monitor." In *Proceedings of the IEEE Symposium on Security and Privacy.* Los Alamitos, CA: IEEE Computer Society Press, 1990.

[15] Herrold, R. P., and the RPM community. "www.rpm.org" (documentation). March 2002. ⟨http:/www.rpm.org/⟩.

[16] Hofmeyr, S. A., and S. Forrest. "Architecture for an Artificial Immune System." *Evol. Comp. J.* **8(4)** (2000): 443–473.

[17] Hofmeyr, Steven A. "An Immunological Model of Distributed Detection and Its Application to Computer Security." Ph.D. thesis, University of New Mexico, Albuquerque, NM, 1999.

[18] Mukherjee, B., L. T. Heberlein, and K. N. Levitt. "Network Intrusion Detection." *IEEE Network* (1994): 26–41.

[19] Napster. March 2002. ⟨http://www.napster.org/⟩.

[20] Osmond, D. G. "The Turn-Over of B-Cell Populations." *Immunology Today* **14(1)** (1993): 34–37.

[21] Percus, J. K., O. Percus, and A. S. Perelson. "Predicting the Size of the Antibody Combining Region from Consideration of Efficient Self/Non-self Discrimination." *PNAS* **90** (1993): 1691–1695.

[22] Pfenning, A., S. Garg, A. Puliafito, M. Telek, and K. Trivedi. "Optimal Software Rejuvenation for Tolerating Software Failures." *Performance Evaluation* **27/28(4)** (1996): 491–506.

[23] Schneider, M. "Self-Stabilization." *ACM Comput. Surveys* **25(1)** (1993): 45–67.

[24] Simon, Herbert A. *The Sciences of the Artificial.* Boston, MA: MIT Press, 1969.

[25] Somayaji, A., and S. Forrest. "Automated Response using System-Call Delays." In *Usenix Security Syposium*, 185–198. ACM Press, 2000.

[26] Somayaji, Anil. "Operating System Stability and Security through Process Homeostasis." Ph.D. thesis, University of New Mexico, Albuquerque, NM, 2002.

[27] Tonegawa, S. "Somatic Generation of Antibody Diversity." *Nature* **302** (1983): 575–581.

[28] Wierzchon, S. T. "Discriminative Power of the Receptors Activated by *k*-Contiguous Bits Rule." *J. Comp. Sci. & Tech.* **1(3)** (2000): 1–13.

[29] Wierzchon, S. T. "Generating Optimal Repertoire of Antibody Strings in an Artificial Immune System." In *Intelligent Information Systems*, edited by M. A. Klopotek, M. Michalewicz, and S. T. Wierzchon, 119–133. Heidelberg, New York: Springer-Verlag, 2000.

[30] Wierzchon, S. T. "Deriving Concise Description of Non-self Patterns in an Artificial Immune System." In *New Learning Paradigm in Soft Computing*, edited by S. T. Wierzchon, L. C. Jain, and J. Kacprzyk, 438–458. Heidelberg, New York: Springer-Verlag, 2001.

Robustness and the Internet: Design and Evolution

Walter Willinger
John Doyle

The objective of this chapter is to provide a historical account of the design and evolution of the Internet and use it as a concrete starting point for a scientific exploration of the broader issues of robustness in complex systems. To this end, we argue that anyone interested in complex systems should care about the Internet and its workings, and why anyone interested in the Internet should be concerned about complexity, robustness, fragility, and their tradeoffs.

1 INTRODUCTION

Despite the widespread use of the Internet and its impact on practically every segment of society, its workings remain poorly understood by most users. Nevertheless, more and more users take it for granted to be able to boot up their laptops pretty much anywhere (e.g., cafes, airports, hotels, conference rooms) and connect to the Internet to use services as mundane as e-mail or Web browsing or as esoteric as music- or movie-distribution and virtual reality games. The few times the users get a glimpse of the complexity of the infrastructure that supports such ubiquitous communication are when they experience various "networking" problems (e.g., the familiar "cannot connect" message, or unacceptably poor performance), because diagnosing such problems typically exposes certain aspects of the underlying network architecture (how the components of the network infrastructure interrelate) and network protocols (standards governing the exchange of data).

Consider, for example, a user sitting in a cafe and browsing the Web on her laptop. In terms of infrastructure, a typical scenario supporting such an application will include a wireless access network in the cafe; an Internet service provider that connects the cafe to the global Internet; intermediate service providers that agree to carry the user's bytes across the country or around the globe, through a myriad of separately administered, autonomous domains; and another service provider at the destination that hosts the server with the Web page requested by the user. As for protocols, successful Web browsing in this setting will require, at a minimum, standards for exchanging data in a wireless environment (cafe) and standards for the (possibly) different networking technologies encountered when transmitting the user's bit stream over the different domains' wired links within the Internet; a standard for assigning a (temporary) ID or address to the user's laptop so that it can be identified and located by any other device connected to the Internet; a set of rules for providing a single, authoritative mapping between names of hosts that provide Web pages such as `www.santafe.org` and their less mnemonic numerical equivalent (e.g., the address 208.56.37.219 maps to the more informative host name `www.santafe.org`); standards for routing the data across the Internet, through the different autonomous domains; a service that ensures reliable transport of the data between the source and destination and uses the available resources efficiently; and a message exchange protocol that allows the user, via an intuitive graphical interface, to browse through a collection of Web pages by simply clicking on links, irrespective of the format or location of the pages' content. The Internet's success and popularity is to a large degree due to its ability to hide most of this complexity and give users the illusion of a single, "seamlessly" connected network where the fragmented nature of the underlying infrastructure and the many layers of protocols remain largely transparent to the user.

However, the fact that the Internet is, in general, very successful in hiding from the user most of the underlying details and intricacies does not make them go away! In fact, even Internet experts admit having more and more troubles getting (and keeping) their arms around the essential components of this highly-engineered network that has all the features typically associated with large-scale systems:

- too complicated and hence ill-understood at the system level, but with often deep knowledge about many of its individual components (**complexity**);

- resilient to designed-for uncertainties in the environment or individual components (**robustness**); and
- yet full of surprising ("emergence") behavior, including a natural tendency for infrequently occurring, yet catastrophic events (**fragility**).

During the past 20–30 years, it has evolved from a small-scale research network into today's Internet—an economic reality that is of crucial importance for the national and international economies and for society as a whole.

It is the design and evolution of this large-scale, highly engineered Internet that we use in this article as a concrete example for discussing and exploring a range of issues related to the study of complexity and robustness in technology in general, and of large-scale, communication networks in particular. We argue that the **"robust, yet fragile"** characteristic is a hallmark of complexity in technology (as well as in biology; e.g., see Csete and Doyle [15] and Doyle et al. [17]), and that is not an accident, but the inevitable result of fundamental (some may view them as "obvious") tradeoffs in the underlying system design. As complex engineered systems such as the Internet evolve over time, their development typically follows a spiral of increasing complexity to suppress unwanted sensitivities/vulnerabilities or to take advantage of new opportunities for increased productivity, performance, or throughput. Unfortunately, in the absence of a practically useful and relevant "science of complexity," each step in this development toward increasing complexity is inevitably accompanied by new and unforeseen sensitivities, causing the complexity/robustness spiral to continue or even accelerate. The Internet's coming of age is a typical example where societal and technological changes have recently led to an acceleration of this complexity/robustness spiral. This acceleration has led to an increasing collection of short-term or point solutions to individual problems, creating a technical arteriosclerosis [4] and causing the original Internet architecture design to become increasingly complex, tangled, inflexible, and vulnerable to perturbations the system was not designed to handle in the first place.

Given the lack of a coherent and unified theory of complex networks, these point solutions have been the result of a tremendous amount of engineering intuition and heuristics, common sense, and trial and error, and have sought robustness to new uncertainties with added complexity, leading in turn to new fragilities. While the Internet design "principles" discussed in section 2 below constitute a modest theory itself that has benefited from fragmented mathematical techniques from such areas as information theory and control theory, key design issues for the Internet protocol architecture have remained unaddressed and unsolved. However, it is becoming more and more recognized, that without a coherent theoretical foundation, man-made systems such as the Internet will simply collapse under the weight of their patchwork architectures and unwieldy protocols. With this danger lurking in the background, we provide here a historical account of the design and evolution of the Internet, with the ambitious goal of using the Internet as a concrete example that reveals fundamental properties of complex systems and allows for a systematic study and basic understanding of the tradeoffs and spirals associated with complexity, robustness, and fragility. The proposed approach requires a sufficient understanding of the Internet's history; concrete and detailed examples illustrating the Internet's complexity, robustness, and fragility; and a theory that is powerful and flexible enough to help us sort out generally applicable design principles from historical accidents. The present article discusses the history of the design and evolution of the Internet from a complexity/robustness/fragility perspective and illustrates with detailed examples and well-documented anecdotes some generally applicable principles. In particular, when illustrating in section 3 the Internet's complexity/robustness spiral, we observe an implicit design principle at work that seeks the "right" balance for a horizontal and vertical separation/integration of functionalities through decentralized control ("horizontal") and careful protocol stack decomposition ("vertical"), and tries to achieve, in a well-defined sense, desirable global network behavior, robustness, and (sub)optimality. Motivated by this observation, the companion article [17] outlines an emerging theoretical foundation for the Internet that illustrates the broader role of theory in analysis and design of complex systems in technology and biology and allows for a systematic treatment of the horizontal and vertical separation/integration issue in protocol design.

1.1 WHY CARE ABOUT THE INTERNET?

As discussed in more detail in Doyle et al. [17], many of the existing "new theories of complexity" have the tendency of viewing the Internet as a generic, large-scale, complex system that shows organic-like growth dynamics, exhibits a range of interesting "emergent" phenomena, and whose "typical" behavior appears to be quite simple. These theories have been very successful in suggesting appealingly straightforward approaches to dealing with the apparently generic complex nature of the Internet and have led to simple models and explanations of many of the observed emergent phenomena generally associated with complexity, for example, *power-law statistics*, *fractal scaling*, and *chaotic dynamics*. While many of these proposed models are capable of reproducing certain trademarks of complex systems, by ignoring practically all Internet-specific details, they tend to generate great attention in the nonspecialist literature. At the same time, they are generally viewed as toy models with little, if any, practical relevance by the domain experts who have great difficulties in recognizing any Internet-specific features in these generic models.

We argue here that only extreme circumstances that are neither easily replicable in laboratory experiments or simulations nor fully comprehensible by even the most experienced group of domain experts are able to reveal the role of the enormous complexity underlying most engineered systems. In particular, we claim in this article that by explicitly ignoring essential ingredients of the architectural design of the Internet and its protocols, these new theories of complexity have missed out on the most attractive and unique features that the Internet offers for a genuinely scientific study of complex systems. For one, even though the Internet is widely used, and it is in general known how all of its parts (e.g., network components and protocols) work, in its entirety, the Internet remains poorly understood. Secondly, the network has become sufficiently large and complicated to exhibit "emergent phenomena"—empirical discoveries that come as a complete surprise, defy conventional wisdom and baffle the experts, cannot be explained nor predicted within the framework of the traditionally considered mathematical models, and rely crucially on the large-scale nature of the Internet, with little or no hope of encountering them when considering small-scale or toy versions of the actual Internet. While the Internet shares this characteristic with other large-scale systems in biology, ecology, politics, psychology, etc., what distinguishes it from these and other complex systems and constitutes its singlemost attractive feature is a unique capability for a strictly scientific treatment of these emergent phenomena. In fact, the reasons for why an unambiguous, thorough, and detailed understanding of any "surprising" networking behavior is in general possible and fairly readily available are twofold and will be illustrated in section 4 with a concrete example. On the one hand, we know in great detail how the individual components work or should work and how the components interconnect to create system-level behavior. On the other hand, the Internet provides unique capabilities for measuring and collecting massive and detailed data sets that can often be used to reconstruct the behavior of interest. In this sense, the Internet stands out as an object for the study of complex systems because it can in general be completely "reverse engineered."

Finally, because of the availability of a wide range of measurements, the relevance of any proposed approach to understanding complex systems that has also been claimed to apply to the Internet can be thoroughly investigated, and the results of any theory of complexity can be readily verified or invalidated. This feature of the Internet gives new insight into the efficacy of the underlying ideas and methods themselves, and can shed light on their relevance to similar problems in other domains. To this end, mistakes that are made when applying the various ideas and methods to the Internet might show up equally as errors in applications to other domains, such as biological networks. In short, the Internet is particularly appropriate as a starting point for studying complex systems precisely because of its capabilities for measurements, reverse engineering, and validation, all of which (individually, or in combination) have either explicitly or implicitly been missing from the majority of the existing "new sciences of complexity." Put differently, all the advantages associated with viewing the Internet as a complex system are lost when relying on only a superficial understanding of the protocols and when oversimplifying the architectural design or separating it from the protocol stack.

1.2 WHY CARE ABOUT COMPLEXITY/ROBUSTNESS?

By and large, the Internet has evolved without the benefit of a comprehensive theory. While traditional but fragmented mathematical tools of robust control, communication, computation, dynamical systems, and statistical physics have been applied to a number of network design problems, the Internet has largely been the result of a mixture of good design principles and "frozen accidents," similar to biology. However, in contrast to biology, the Internet provides access to the designer's original intentions and to an essentially complete record of the entire evolutionary process. By studying this historical account, we argue here that the Internet teaches us much about the central role that robustness plays in complex systems, and we support our arguments with a number of detailed examples. In fact, we claim that many of the features associated with the Internet's complexity, robustness, and fragility reveal intrinsic and general properties of all complex systems. We even hypothesize that there are universal properties of complex systems that can be revealed through a scientifically sound and rigorous study of the Internet. A detailed understanding of these properties could turn out to be crucial for addressing many of the challenges facing today's network and the future Internet, including some early warning signs that indicate when technical solutions that, while addressing particular needs, may severely restrict the future use of the network and may force it down an evolutionary dead-end street.

More importantly, in view of the companion paper [17] where the outline of a nascent coherent theoretical foundation of the Internet is discussed, it should be possible to use that theory to evaluate the Internet's evolutionary process itself and to help understand and compare it with similar processes in other areas such as evolutionary biology, where the processes in question can also be viewed as mixtures of "design" (i.e., natural selection is a powerful constraining force) and "frozen accidents" (i.e., mutation and selection both involve accidental elements). Thus, a "look over the fence" at, for example, biology, can be expected to lead to new ideas and novel approaches for dealing with some of the challenges that lie ahead when evolving today's Internet into a future "embedded, everywhere" world of total automation and network interconnectivity. Clearly, succeeding in this endeavor will require a theory that is powerful enough to unambiguously distinguish between generally applicable principles and historical accidents, between theoretically sound findings and simple analogies, and between empirically solid observations and superficial data analysis. In the absence of such a theory, viewing the Internet as a complex system has so far been less than impressive, resulting in little more than largely irrelevant analogies and inappropriate usage and careless analyses of available measurements.

1.3 BACKGROUND AND OUTLOOK

While none of the authors of this article had (nor have) anything to do with any design aspects of the original (nor present) Internet, much of the present discussion resulted from our interactions with various members of the DARPA-funded *NewArch Project* [4], a collaborative research project aimed at revisiting the current Internet architecture in light of present realities and future requirements and developing a next-generation architecture toward which the Internet can evolve during the next 10–20 years. Some of the members of the *NewArch*-project, especially Dave Clark, were part of and had architectural responsibilities for the early design of the Internet and have played a major role in shaping today's Internet architecture. They have written extensively about the thought process behind the early DARPA Internet architecture, about the design philosophy that has shaped the development of the Internet protocols, and about how a constantly changing Internet landscape requires a constant probing of the status-quo and advancing of new design principles for tomorrow's Internet (see for example, Clark [9], CSTB [32], Saltzer et al. [38], Blumenthal and Clark [2], and Clark et al. [11]). Another valuable source that provides a historical perspective about the evolution of the Internet architecture over time is the archive of *Requests for Comments (RFCs)* of the *Internet Engineering Task Force (IETF)* [23]. This archive offers periodic glimpses into the thought process of some of the leading Internet architects and engineers about the health of the original architectural design in view of the Internet's growing pains (for some relevant RFCs, see Carpenter and Austein [8], Kaat [25], Carpenter [7], Carpenter and Austein [6], and Clark et al. [10]). In combination, these different sources of information provide invaluable insights into the process by which the original Internet architecture has been designed and has evolved as a result of internal and external changes that have led to a constant demand for new requirements, and this article uses these resources extensively.

The common goal between us "outsiders" and the *NewArch*-members ("insiders") has been a genuine desire for understanding the nature of complexity associated with today's Internet and using this understanding to move forward. While our own efforts toward reaching this goal has been mainly measurement- and theory-driven (but with an appreciation for the need for "details"), the insiders approach has been more bottom-up, bringing an immense body of system knowledge, engineering intuition, and empirical observations to the table (as well as an appreciation for "good" theory). To this end, the present discussion is intended as a motivation for developing a theory of complexity and robustness that bridges the gap between the theory/measurement-based and engineering/intuition-based approaches in an unprecedented manner and results in a framework for dealing with the complex nature of the Internet that (i) is technically sound, (ii) is consistent with measurements of all kinds, (iii) agrees fully with engineering intuition and experience, and (iv) is useful in practice and for sketching viable architectural designs for an Internet of the future. The development of such a theoretical foundation for the Internet and other complex systems is the topic of the companion article [17].

2 COMPLEXITY AND DESIGNED-FOR ROBUSTNESS IN THE INTERNET

Taking the initial technologies deployed in the pre-Internet area as given, we discuss in the following the original requirements for the design of a "network of networks." In particular, we explore in this section how robustness, a close second to the top internetworking requirement, has influenced, if not shaped, the architectural model of the Internet and its protocol design. That is, starting with a basic packet-switching network fabric and using our working definition of robustness (i.e., **being resilient to designed-for uncertainties in the environment or individual system components**), we address here the question of how the quest for robustness—primarily in the sense of (i) flexibility to changes in technology, use of the network, etc., and (ii) survivability in the face of component failures—has impacted how the elements of the underlying networks interrelate, and how the specifics of exchanging data in the resulting network have been designed.[1] Then we illustrate how this very design is evolving over time when faced with a network that is itself undergoing constant changes due to technological advances, changing business models, new policy decisions, or modified market conditions.

2.1 THE ORIGINAL REQUIREMENTS FOR AN INTERNET ARCHITECTURE

When viewed in terms of its hardware, the Internet consists of *hosts* or endpoints (also called end systems), *routers* or internal switching stations (also referred to as gateways), and *links* that connect the various hosts and/or routers and can differ widely in speed (from slow modem connection to high-speed backbone links) as well as in technology (e.g., wired, wireless, satellite communication). When viewed from the perspective of *autonomous systems* (ASs), where an AS is a collection of routers and links under a single administrative domain, the network is an internetwork consisting of a number of separate subnetworks or ASs, interlinked to give users the illusion of a single connected network (network of networks, or "internet"). A network architecture is a framework that aims at specifying how the different components of the networks interrelate. More precisely, paraphrasing [4], a *"network architecture is a set of high-level design principles that guides the technical design of a network, especially the engineering of its protocols and algorithms. It sets a sense of direction—providing coherence*

[1]There are other important aspects to this quest for robustness (e.g., interoperability in the sense ensuring a working Internet despite a wide range of different components, devices, technologies, protocol implementations, etc.), but they are not central to our present discussion.

and consistency to the technical decisions that have to be made and ensuring that certain requirements are met." Much of what we refer to as today's Internet is the result of an architectural network design that was developed in the 1970s under the auspices of the Defense Advanced Research Project Agency (DARPA) of the U.S. Department of Defense, and the early reasoning behind the design philosophy that shaped the Internet protocol architecture (i.e., the "Internet architecture" as it became known) is vividly captured and elaborated on in detail in Clark [9].

Following the discussion in Cslark [9], the main objective for the DARPA Internet architecture some 30 years ago was *internetworking*—the development of an "effective technique for multiplexed utilization of already existing interconnected (but typically separately administered) networks." To this end,s the fundamental structure of the original architecture (how the components of the networks interrelate) resulted from a combination of known technologies, conscientious choices, and visionary thinking, and led to "a packet-switched network in which a number of separate networks are connected together using packet communications processors called gateways which implement a store and forward packet forwarding algorithm" [9]. For example, the store and forward packet-switching technique for interconnecting different networks had already been used in the ARPANET and was reasonably well understood. The selection of packet switching over circuit switching as the preferred multiplexing technique reflected networking reality (i.e., the networks to be integrated under the DARPA project already deployed the packet-switching technology) and engineering intuition (e.g., data communication was expected to differ fundamentally from voice communication and would be more naturally and efficiently supported by the packet-switching paradigm). Moreover, the intellectual challenge of coming to grips with integrating a number of separately administered and architecturally distinct networks into a common evolving utility required bold scientific approaches and innovative engineering decisions that were expected to advance the state-of-the-art in computer communication in a manner that alternative strategies such as designing a new and unified but intrinsically static "multi-media" network could have largely avoided.

A set of second-level objectives, reconstructed and originally published in Clark [9], essentially elaborates on the meaning of the word "effective" in the all-important internetworking requirement and defines a more detailed list of goals for the original Internet architecture. Quoting from Clark [9], these requirements are (in decreasing order of importance),

- *Robustness*: Internet communication must continue despite loss of networks or gateways/routers.
- *Heterogeneity (services)*: The Internet must support multiple types of communications services.
- *Heterogeneity (networks)*: The Internet architecture must accommodate a variety of networks.

- *Distributed management*: The Internet architecture must permit distributed management of its resources.
- *Cost*: The Internet architecture must be cost effective.
- *Ease of attachment*: The Internet architecture must permit host attachment with a low level of effort.
- *Accountability*: The resources used in the Internet architecture must be accountable.

While the top-level requirement of internetworking was mainly responsible for defining the basic structure of the common architecture shared by the different networks that composed the "network of networks" (or "Internet" as we now know it), this priority-ordered list of second-level requirements, first and foremost among them the robustness criterion, has to a large degree been responsible for shaping the architectural model and the design of the protocols (standards governing the exchange of data) that define today's Internet. This includes the Internet's well-known "hourglass" architecture and the enormously successful "fate-sharing" approach to its protocol design [32], both of which will be discussed in more detail below.

In the context of the discussions in Clark [9], "robustness" usually refers to the property of the Internet to be resilient to uncertainty in its components and usage patterns, and—on longer time scales (i.e., evolution)—to unanticipated changes in networking technologies and network usage. However, it should be noted that "robustness" can be (and has been) interpreted more generally to mean "to provide some underlying capability in the presence of uncertainty." It can be argued that with such a more general definition, many of the requirements listed above are really a form of robustness as well, making robustness the basic underlying requirement for the Internet. For example, in the case of the top internetworking requirement, access to and transmission of files/information can be viewed as constituting the "underlying capability," while the "uncertainty" derives from the users not knowing in advance when to access or transmit what files/information. Without this uncertainty, there would be no need for an Internet (e.g., using the postal service for shipping CDs with the requested information would be a practical solution), but when faced with this uncertainty, a robust solution is to connect everybody and provide for a basic service to exchange files/information "on demand" and without duplicating everything everywhere.

Clearly, the military context in which much of the early design discussions surrounding the DARPA Internet architecture took place played a significant role in putting the internetworking/connectivity and robustness/survivability requirements at the top and relegating the accountability and cost considerations to the bottom of the original list of objectives for an Internet architecture that was to be useful in practice. Put differently, had, for example, *cost* and *accountability* been the two over-riding design objectives (with some concern for internetworking and robustness, though)—as would clearly be the case in today's market- and business-driven environment—it is almost certain that the resulting network would exhibit a very differently designed architecture and protocol structure. Ironically though, the very commercial success and economic reality of today's Internet that have been the result of an astonishing resilience of the original DARPA Internet architecture design to revolutionary changes—especially during the last ten or so years and in practically all aspects of networking one can think of—have also been the driving forces that have started to question, compromise, erode, and even damage the existing architectural framework. The myriad of network-internal as well as network-external changes and new requirements that have accompanied "the Internet's coming of age" [32] has led to increasing signs of strains on the fundamental architectural structure. At the same time, it has also led to a patchwork of technical long-term and short-time solutions, where each solution typically addresses a particular change or satisfies a specific new requirement. Not surprisingly, this transformation of the Internet from a small-scale research network with little (if any) concern for cost, accountability, and trustworthiness into an economic power house has resulted in a network that is (i) increasingly driving the entire national and global economy, (ii) experiencing an ever-growing class of users with often conflicting interests, (iii) witnessing an erosion of trust to the point where assuming untrustworthy end-points becomes the rule rather than the exception, and (iv) facing a constant barrage of new and, in general, ill-understood application requirements and technology features. As illustrated with some specific examples in section 3 below, each of these factors creates new potential vulnerabilities for the network which in turn leads to more complexity as a result of making the network more robust.

2.2 THE DARPA INTERNET ARCHITECTURE

When defining what is meant by "the Internet architecture," the following quote from Carpenter [6] captures at the same time the general reluctance within the Internet community for "dogmas" or definitions that are "cast in stone" (after all, things change, which by itself may well be the only generally accepted principle[2]) and a tendency for very practical and to-the-point working definitions: *"Many members of the Internet community would argue that there is no [Internet] architecture, but only a tradition, which was not written down for the first 25 years (or at least not by the [Internet Architecture Board]). However, in very general terms, the community believes that [when talking about the Internet*

[2]On October 24, 1995, the Federal Networking Council (FNC) unanimously passed a resolution defining the term "Internet." The definition was developed in consultation with members of the Internet community and intellectual property rights communities and states: *The FNC agrees that the following language reflects our definition of the term "Internet." "Internet" refers to the global information system that—(i) is logically linked together by a globally unique address space based on the Internet Protocol (IP) or its subsequent extensions/follow-ons; (ii) is able to support communications using the Transmission Control Protocol/Internet Protocol (TCP/IP) suite or its subsequent extensions/follow-ons, and/or other IP-compatible protocols; and (iii) provides, uses or makes accessible, either publicly or privately, high-level services layered on the communications and related infrastructure described herein.*

architecture] the goal is connectivity, the tool is the Internet Protocol, and the intelligence is end-to-end rather than hidden in the network." To elaborate on this view, we note that connectivity was already mentioned above as the top-level requirement for the original DARPA Internet architecture, and the ability of today's Internet to give its users the illusion of a "seamlessly" connected network with fully transparent political, economic, administrative, or other sort of boundaries is testimony for the architecture's success, at least as far as the crucial internetworking/connectivity objective is concerned. In fact, it seems that connectivity is its own reward and may well be the single-most important service provided by today's Internet. Among the main reasons for this ubiquitous connectivity are the "layering principle" and the "end-to-end argument," two guidelines for system design that the early developers of the Internet used and tailored to communication networks. A third reason is the decision to use a single universal logical addressing scheme with a simple (net, host) hierarchy, originally defined in 1981.

2.2.1 Robustness as in Flexibility: The "Hourglass" Model. In the context of a packet-switched network, the motivation for using the *layering principle* is to avoid implementing a complex task such as a file transfer between two end hosts as a single module, but to instead break the task up into subtasks, each of which is relatively simple and can be implemented separately. The different modules can then be thought of being arranged in a vertical stack, where each layer in the stack is responsible for performing a well-defined set of functionalities. Each layer relies on the next lower layer to execute more primitive functions and provides services to the next higher layer. Two hosts with the same layering architecture communicate with one another by having the corresponding layers in the two systems talk to one another. The latter is achieved by means of formatted blocks of data that obey a set of rules or conventions known as a *protocol.*

The main stack of protocols used in the Internet is the five-layer *TCP/IP protocol suite* and consists (from the bottom up) of the physical, link, internetwork, transport, and application layers. The *physical layer* concerns the physical aspects of data transmission on a particular link, such as characteristics of the transmission medium, the nature of the signal, and data rates. Above the physical layer is the *link layer.* Its mechanisms and protocols (e.g., signaling rules, frame formats, media-access control) control how packets are sent over the raw media of individual links. Above it is the *internetwork layer*, responsible for getting a packet through an *internet*; that is, a series of networks with potentially very different bit-carrying infrastructures and possibly belonging to different administrative domains. The *Internet Protocol (IP)* is the internetworking protocol for TCP/IP, and its main task is to adequately implement all the mechanisms necessary to knit together divergent networking technologies and administrative domains into a single virtual network (an "internet") so as to enable data communication between sending and receiving hosts, irrespective of where in the network they are. The layer above IP is the *transport layer*, where the most commonly used *Transmission Control Protocol (TCP)* deals, among other issues, with end-to-end congestion control and assures that arbitrarily large streams of data are reliably delivered and arrive at their destination in the order sent. Finally, the top layer in the TCP/IP protocol suite is the *application layer*, which contains a range of protocols that directly serve the user; e.g., TELNET (for remote login), FTP (the *File Transfer Protocol* for transferring files), SMTP (*Simple Mail Transfer Protocol* for e-mail), HTTP (the *HyperText Transfer Protocol* for the World Wide Web).

This layered modularity gave rise to the "hourglass" metaphor for the Internet architecture [32]—the creation of a multilayer suite of *protocols*, with a generic packet (datagram) delivery mechanism as a separate layer at the hourglass' waist (the question of how "thin" or "fat" this waist should be designed will be addresses in section 2.2.2 below). This abstract bit-level network service at the hourglass' waist is provided by IP and ensures the critical separation between an ever more versatile physical network infrastructure below the waist and an ever-increasing user demand for higher-level services and applications above the waist. Conceptually, IP consists of an agreed-upon set of features that has to be implemented according to an Internet-wide standard, must be supported by all the routers in the network, and is key to enabling communication across the global Internet so that networking boundaries and infrastructures remain transparent to the users. The layers below the waist (i.e., physical, and link) deal with the wide variety of existing transmission and link technologies and provide the protocols for running IP over whatever bit-carrying network infrastructure is in place ("IP over everything"). Aiming for a somewhat narrow waist reduces for, or even removes from the typical user the need to know about the details of and differences between these technologies (e.g., Ethernet local area networks [LAN], asynchronous transfer mode [ATM], frame relay) and administrative domains. Above the waist is where enhancements to IP (e.g., TCP) are provided that simplify the process of writing applications through which users actually interact with the Internet ("everything over IP"). In this case, providing for a thin waist of the hourglass removes from the network providers the need to constantly change their infrastructures in response to a steady flow of innovations happening at the upper layers within the networking hierarchy (e.g., the emergence of "killer apps" such as e-mail, the Web, or Napster). The fact that this hourglass design predated some of the most popular communication technologies, services, and applications in use today—and that within today's Internet, both new and old technologies and services can coexist and evolve—attests to the vision of the early architects of the Internet when deciding in favor of the layering principle. "IP over everything, and everything over IP" results not only in enormous robustness to changes below and above the hourglass' waist but also provides the flexibility needed for constant innovation and entrepreneurial spirit at the physical substrate of the network as well as at the application layer.

2.2.2 Robustness as in Survivability: "Fate-Sharing." Layered network architectures are desirable because they enhance modularity; that is, to minimize duplication of functionality across layers, similar functionalities are collected into the same layer. However, the layering principle by itself lacks clear criteria for assigning specific functions to specific layers. To help guide this placement of functions among the different layers in the Internet's hourglass architecture, the original architects of the Internet relied on a class of arguments, called the *end-to-end arguments*, that expressed a clear bias against low-level function implementation. The principle was first described in Saltzer et al. [38] and later reviewed in Carpenter [6], from where the following definition is taken:

> *"The basic argument is that, as a first principle, certain required end-to-end functions can only be performed correctly by the end-systems themselves. A specific case is that any network, however carefully designed, will be subject to failures of transmission at some statistically determined rate. The best way to cope with this is to accept it, and give responsibility for the integrity of communication to the end systems. [...]*
> *To quote from [Saltzer et al. [38]], 'The function in question can completely and correctly be implemented only with the knowledge and help of the application standing at the endpoints of the communication system. Therefore, providing that questioned function as a feature of the communication system itself is not possible. (Sometimes an incomplete version of the function provided by the communication system may be useful as a performance enhancement.)' "*

In view of the crucial robustness/survivability requirement for the original Internet architecture (see section 2.1), adhering to this end-to-end principle has had a number of far-reaching implications. For one, the principle strongly argues in favor of an hourglass-shaped architecture with a "thin" waist, and the end-to-end mantra has been used as a constant argument for "watching the waist" (in the sense of not putting on more "weight," i.e., adding non-essential functionalities). The result is a lean IP, with a minimalistic set of generally agreed-upon functionalities making up the hourglass' waist: algorithms for storing and forwarding packets, for routing, to name but a few. Second, to help applications to survive under failures of individual network components or of whole subnetworks, the end-to-end argument calls for a protocol design that is in accordance with the principle of "soft state." Here "state" refers to the configuration of elements within the network (e.g., routers), and "soft state" means that *"operation of the network depends as little as possible on persistent parameter settings within the network"* [32]. As discussed in Carpenter [6], *"end-to-end protocol design should not rely on the maintenance of state (i.e., information about the state of the end-to-end communication) inside the network. Such state should be maintained only in the endpoints, in such a way that the state can only be destroyed when the endpoint itself breaks."* This feature has been coined "fate-sharing" by

D. Clark [9] and has two immediate consequences. On the one hand, because routers do not keep persistent state information about on-going connections, the original design decision of choosing packet-switching over circuit-switching is fully consistent with this fate-sharing philosophy. The control of the network (e.g., routing) is fully distributed and decentralized (with the exception of key functions such as addressing), and no single organization, company, or government owns or controls the network in its entirety. On the other hand, fate-sharing places most "intelligence" (i.e., information about end-to-end communication, control) squarely in the end points. The network's (i.e., IP's) main job is to transmit packets as efficiently and flexibly as possible; everything else should be done further up in the protocol stack and hence further out at the fringes of the network. The result is sometimes referred to as a "dumb network," with the intelligence residing at the network's edges. Note that the contrast to the voice network with its centralized control and circuit switching technology, where the intelligence (including control) resides within the network ("smart network") and the endpoints (telephones) are considered "dumb" could not be more drastic!

2.2.3 Robustness through Simplicity: "Internet Transparency." Much of the design of the original Internet has to do with two basic and simple engineering judgments. On the one hand, the design's remarkable flexibility to changes below and above the hourglass' waist is arguably due to the fact that network designers admittedly never had—nor will they ever have—a clear idea about all the ways the network can and will be used in the future. On the other hand, the network's surprising resilience to a wide range of failure modes reflects to a large degree the network designers' expectations and experiences that failures are bound to happen and that a careful design aimed at preventing failures from happening is a hopeless endeavor, creating overly complex and unmanageable systems, especially in a scenario (like the Internet) that is expected to undergo constant changes. Instead, the goal was to design a system that can "live with" and "work around" failures, shows graceful degradation under failure while still maintaining and providing basic communication services; and all this should be done in a way that is transparent to the user. The result was an Internet architecture whose design reflects the soundness and aimed-for simplicity of the underlying engineering decisions and includes the following key features:

- a connectionless packet-forwarding layered infrastructure that pursues robustness via "fate-sharing,"
- a least-common-denominator packet delivery service (i.e., IP) that provides flexibility in the face of technological advances at the physical layers and innovations within the higher layers, and
- a universal logical addressing scheme, with addresses that are fixed-sized numerical quantities and are applied to physical network interfaces.

This basic structure was already in place about 20 years ago, when the Internet connected just a handful of separate networks, consisted of about 200 hosts, and when detailed logical maps of the entire network with all its links and individual host computers were available. The idea behind this original structure is captured by what has become known as the "end-to-end transparency" of the Internet; that is, *"a single universal logical addressing scheme, and the mechanisms by which packets may flow between source and destination essentially unaltered"* [7]. In effect, by simply knowing each other's Internet address and running suitable software, any two hosts connected to the Internet can communicate with one another, with the network neither tampering with nor discriminating against the various packets in flight. In particular, new applications can be designed, experimented with, and deployed without requiring any changes to the underlying network. Internet transparency provides flexibility, which in turn guarantees innovation.

2.3 COMPLEXITY AS A RESULT OF DESIGNED-FOR ROBUSTNESS

The conceptual simplicity portrayed by the hourglass metaphor for the Internet's architectural design can be quite deceiving, as can the simple description of the network (i.e., IP) as a universal packet delivery service that gives its users the illusion of a single connected network. Indeed, when examining carefully the designs and implementations of the various functionalities that make up the different protocols at the different layers, highly engineered, complex internal structures emerge, with layers of feedback, signaling, and control. Furthermore, it also becomes evident that the main reason for and purpose of these complex structures is an over-riding desire to make the network robust to the uncertainties in its environment and components for which this complexity was deemed necessary and justified in the first place. In the following, we illustrate this robustness-driven root cause for complexity with two particular protocols, the transport protocol TCP and the routing protocol BGP.

2.3.1 Complexity-Causing Robustness Issues in Packet Transport: TCP. IP's main job is to do its best ("best effort") to deliver packets across the network. Anything beyond this basic yet unreliable service (e.g., a need to recover from lost packets; reliable packet delivery) is the application's responsibility. It was decided early on in the development of the Internet protocol architecture to bundle various functionalities into the transport layer, thereby providing different transport services on top of IP. A distinctive feature of these services is how they deal with transport-related uncertainties (e.g., lost packets, delayed packets, out-of-order packets, fluctuations in the available bandwidth) that impact and constrain, at a minimum, the reliability, delay characteristics, or bandwidth usage of the end-to-end communication. To this end, TCP provides a reliable sequenced data stream, while the User Datagram Protocol UDP—by trading reliability for delay—provides direct access to the basic but unreliable service of

IP.[3] Focusing in the following on TCP,[4] what are then the internal structures, and how complex do they have to be, so that TCP can guarantee the service it promises its applications?

For one, to assure that arbitrarily large streams of data are reliably delivered and arrive at their destination in the order sent, TCP has to be designed to be robust to (at least) lost and delayed packets as well as to packets that arrive out of order. When delivering a stream of datas to a receiver such that the entire stream arrives in the same order, with no duplicates, and reliably even in the presence of packet loss, reordering, duplication, and rerouting, TCP splits the data into *segments*, with one segment transmitted in each packet. The receiver acknowledges the receipt of segments if they are "in order" (it has already received all data earlier in the stream). Each acknowledgment packet (ACK) also implicitly acknowledges all of the earlier-in-the-stream segments, so the loss of a single ACK is rarely a problem; a later ACK will cover for it, as far as the sender is concerned. The sender runs a timer so that if it has not received an ACK from the receiver for data previously sent when the timer expires, the sender will conclude that the data (or all of the subsequent ACKs) was lost and retransmit the segment. In addition, whenever a receiver receives a segment that is out of order (does not correspond to the next position in the data stream), it generates a "duplicate ACK," that is, another copy of the same ACK that it sent for the last in-order packet it received. If the sender observes the arrival of three such duplicate ACKs, then it concludes that a segment must have been lost (leading to a number of out-of-order segments arriving at the receiver, hence the duplicate ACKs), and retransmits it *without* waiting first for the timer to expire.[5]

Next, guaranteeing reliability without making use of the available network resources (i.e., bandwidth) would not be tolerated by most applications. A key requirement for attaining good performance over a network path, despite the uncertainties arising from often highly intermittent fluctuations of the available bandwidth, is that the sender must in general maintain several segments "in flight" at the same time, rather than just sending one and waiting an entire round trip time (RTT) for the receiver to acknowledge it. However, if the sender has too many segments in flight, then it might overwhelm the receiver's ability to store them (if, say, the first is lost but the others arrive, so the receiver

[3]Originally, TCP and IP had been a single protocol (called TCP), but conflicting requirements for voice and data applications soon argued in favor of different transport services running over "best effort" IP.

[4]Specifically, we describe here a version of TCP called TCP Reno, one of the most widely used versions of TCP today. For more details about the various versions of TCP, see e.g. Peterson and Davie [35]).

[5]Imagine, for example, that segments 1 and 2 are sent, that 1 is received and 2 is lost. The receiver sends the acknowledgment ACK(1) for segment 1. As soon as ACK(1) is received, the sender sends segments 3, 4, and 5. If these are successfully received, they are retained by the receiver, even though segment 2 is missing. But because they are out of order, the receiver sends back three ACK(1)'s (rather than ACK(3), ACK(4), ACK(5)). From the arrival of these duplicates, the sender infers that segment 2 was lost (since the ACKs are all for segment 1) and retransmits it.

cannot immediately process them), or the network's available capacity. The first of these considerations is referred to as *flow control*, and in TCP is managed by the receiver sending an *advertised window* informing the sender how many data segments it can have in flight beyond the latest one acknowledged by the receiver. This mechanism is termed a "sliding window," since each ACK of new data advances a window bracketing the range of data the sender is now allowed to transmit. An important property of a sliding window protocol is that it leads to *self-clocking*. That is, no matter how fast the sender transmits, its data packets will upon arrival at the receiver be spaced out by the network to reflect the network's current carrying capacity; the ACKs returned by the receiver will preserve this spacing; and consequently the window at the sender will advance in a pattern that mirrors the spacing with which the previous flight of data packets arrived at the receiver, which in turn matches the network's current carry capacity.

TCP also maintains a *congestion window*, or CWND, that controls how the sender attempts to consume the path's capacity. At any given time, the sender confines its data in flight to the lesser of the advertised window and CWND. Each received ACK, unless it is a duplicate ACK, is used as an indication that data has been transmitted successfully, and allows TCP to increase CWND. At startup, CWND is set to 1 and the *slow start* mechanism takes place, where CWND is increased by one segment for each arriving ACK. The more segments that are sent, the more ACKs are received, leading to exponential growth. (The *slow start* procedure is "slow" compared to the old mechanism which consisted of immediately sending as many packets as the advertised window allowed.) If TCP detects a packet loss, either via duplicate ACKs or via timeout, it sets a variable called the slow start threshold, or SSTHRESH to half of the present value of CWND. If the loss was detected via duplicate ACKs, then TCP does not need to cut back its rate drastically: CWND is set to SSTHRESH and TCP enters the *congestion avoidance* state, where CWND is increased linearly, by one segment per RTT. If the loss was detected via a timeout, then the self-clocking pattern has been lost, and TCP sets CWND to one, returning to the slow start regime in order to rapidly start the clock going again. When CWND reaches the value of SSTHRESH, *congestion avoidance* starts and the exponential increase of CWND shifts to a linear increase.[6]

Clearly, these designed-for features that make TCP robust to the randomly occurring but fully expected failure modes in end-to-end communication (i.e., packet loss, packet delay, out-of-order packets, congestion episodes) are not free but come at a price, namely increased complexity. This complexity reveals itself in the type of adopted engineering solutions which in this case include explicit

[6]Congestion control as described here using packet loss as a congestion signal and an *additive-increase-multiplicative-decrease*-type congestion control mechanism at the end points was not part of the original TCP protocol but was proposed in Jacobson [24] and was subsequently added in the late 1980s in response to observed congestion collapse episodes in the Internet; see also section 3.2.1 below.

and implicit signaling (use of ACKs and packet loss), heavy reliance on timers and (hopefully) robust parameter estimation methods, extensive use of control algorithms, feedback loops, etc. The result is a protocol that has performed remarkably well as the Internet has scaled up several orders of magnitude in size, load, speed, and scope during the past 20 or so years (see section 3.1). Much of this engineering achievement is due to TCP's abilities to allocate network resources in a more or less "fair" or socially responsible manner and nevertheless achieve high network utilization through such cooperative behavior.

2.3.2 Complexity-Causing Robustness Issues in Packet Routing: BGP. Transport protocols such as TCP "don't do routing" and rely fully on IP to switch any packet anywhere in the Internet to the "correct" next hop. Addressing and routing are crucial aspects that enable IP to achieve this impressive task. As for *addressing*, each network uses a unique set of addresses drawn from a single universal logical addressing space. Each device on the Internet has a unique address that it uses to label its network interface. Each IP packet generated by any of these devices has a source and destination address, where the former references the local interface address and the latter gives the corresponding interface address of the intended recipient of the packet. When handing packets over within the network from router to router, each router is able to identify the intended receiver of each packet. Maintaining sufficient and consistent information within the network for associating the identity of the intended recipient with its location inside the network is achieved by means of *routing protocols*; that is, a set of distributed algorithms that are part of IP and that the routers run among themselves to make appropriate routing decisions. The routing protocols are required to maintain both local and global (but not persistent) state information, for each router must not only be able to identify a set of output interfaces that can be used to move a packet with a given destination address closer to its destination, but must also select an interface from this set which represents the best possible path to that destination. Robustness considerations that play a role in this context include randomly occurring router or link failures and restoration of failed network components or adding new components to the network. The routing protocols in use in today's Internet are robust to these uncertainties in the network's components, and the detection of and routing around failed components remains largely invisible to the end-to-end application—the Internet sees damage and "works" (i.e., routes) around it. The complexity in protocol design that ensures this remarkable resilience to failures in the physical infrastructure of the Internet is somewhat reduced by a division of the problem into two more manageable pieces, where the division is in accordance with separation of the Internet into ASs: each AS runs a local internal routing protocol (or *Interior Gateway Protocol (IGP))*, and between the different ASs, an inter-network routing protocol (or *Exterior Gateway Protocol (EGP))* maintains connectivity and is the glue that ties all the ASs together and ensures communication across AS boundaries. To illustrate the engineers' approaches to tackling this routing problem and discuss the resulting complex internal structures, we focus in the following on the *Border Gateway Protocol (BGP)*, the de-facto standard inter-network routing protocol deployed in today's Internet.[7]

In a nutshell, BGP is a "path-vector" routing protocols, and two BGP-speaking routers that exchange routing information dynamically with BGP use TCP as its transport protocol. As a distance-vector protocol, each BGP-speaking router determines its "best" path to a particular destination separately from other BGP-speaking routers. Each BGP-speaking router selects the "next hop" to use in forwarding a packet to a given destination based on paths to those destinations advertised by the routers at the neighboring ASs. Routers exchange paths to destinations in order to facilitate route selection based on *policy*: ASs apply individual, local policies when selecting their preferred routes, usually based on the series of ASs that a given route transits. This feature enables an administratively decentralized Internet—using these policies, ASs can direct traffic to ASs with whom they have business relationships, where traditional network routing protocols would have selected the shortest path. As the network undergoes changes (e.g., link failures, provisioning of new links, router crashes, etc.), BGP uses *advertisement* and *withdrawal* messages to communicate the ensuing route changes among the BGP routers. An advertisement informs neighboring routers that a certain path to a given destination is used and typically includes a number of attributes, such as an AS-path that lists all ASs the advertisement message has traversed, starting with the originating destination AS. A withdrawal is an update message indicating that a previously advertised destination is no longer available. Advertisements can function as implicit withdrawals if they contain a previously announced destination. A basic feature of BGP is that when a BGP-speaking router advertises to its neighboring router that it has a path for reaching a particular destination, the latter can be certain that the former is actively using that path to reach the destination.

In short, BGP uses distributed computation to tackle the inter-domain routing problem and transmits the necessary information by setting up reliable BGP sessions between any two BGP-speaking routers that have to communicate with one another. Thus, by relying on update messages from their neighbors, BGP-speaking routers find the best paths to the different destinations in an iterative manner, which in turn requires careful attention to potential problems with route instabilities (i.e., oscillations) and slow convergence. For example, to rate-limit advertisements, BGP uses timers associated with the *Minimum Route Advertisement Interval (MRAI)* parameter. The value of MRAI is configurable, although the recommended default value is 30 seconds. When a BGP-speaking router sends a route advertisement for a given destination to a neighbor, it starts an

[7] It is not our intention here to provide a historical account of the development of routing protocols in the Internet. For the purpose of our discussion, any IGP or EGP that were used in the past or are currently in use would be appropriate; we simply use (version 4 of) BGP because it is the prevalent EGP in today's Internet and is creating a rapidly growing body of literature [37, 41].

instance of this timer. The router is not allowed to send another advertisement concerning this destination to that neighbor until the corresponding timer has expired. This rate limiting attempts to dampen some of the oscillation inherent in the distance-vector approach to routing. Note that while waiting for the MRAI timer to expire, the router in question may receive many update messages from its neighbors and run its best path selection algorithm internally numerous times. As a result, the router can privately enumerate many alternative choices of its best path without burdening its neighbors with all the (uninformative) intermediate steps. Thus, using the MRAI timer reduces the number of updates needed for convergence but adds some delay to the update messages that are sent. Overall, the BGP specifications [37] explicitly mention five timers (all with recommended default values, but required to be configurable). They either ensure that requests for network resources (e.g., BGP session) will time out if not re-requested periodically, or they attempt to mitigate certain known disadvantages (e.g., route oscillation) associated with relying on distance-vector or similar kinds of routing protocols. In general, determining default values for these timers has been mostly guess work, and little (if anything) is known about their effects (individually, or in terms of "feature interactions") on the dynamics of BGP in today's Internet. In practice, however, these engineering solutions to the routing problem have produced routing protocols that have made network operations extremely resilient to hardware failures, and the ability of the Internet to "route around" failures in a way that is largely transparent to the user is one of the big success stories of the DARPA Internet architecture design.

3 THE INTERNET'S COMPLEXITY/ROBUSTNESS SPIRAL "IN ACTION"

Starting out some 30 years ago as a small-scale research network that relied on a novel, but still developing, technology, the Internet has experienced a well-documented and widely publicized transformation into a crucial force that is increasingly driving the national and international economy and is fueled by enormous IT investments on the order of billions of dollars. As for the original Internet architecture discussed in the previous section, this transformation has been accompanied by constant probing and soul-searching by parts of the Internet research community about the role and relevance of architecture or architectural design principles in such a rapidly changing environment. In the following, we briefly discuss some of the changes that have affected various aspects of the Internet and illustrate with examples their impact on the Internet architecture[8]. In particular, we point out a very typical, but in the long term potentially quite dangerous engineering approach to dealing with network-internal

[8]In this context, it is particularly illuminating to directly compare [6] with [8] and obtain a first-hand assessment of the changing nature of the architectural principles of the Internet.

and -external changes, namely responding to demands for improved performance, better throughput, or more robustness to unexpectedly emerging fragilities with increasingly complex designs or untested short-term solutions. While any increase in complexity has a natural tendency to create further and potentially more disastrous sensitivities, this observation is especially relevant in the Internet context, where the likelihood for unforeseen "feature interactions" in the ensuing highly engineered large-scale structure drastically increases as the network continues to evolve. The result is a "complexity/robustness spiral" (i.e., robust design $\rightarrow$ complexity $\rightarrow$ new fragility $\rightarrow$ make design more robust $\rightarrow$ ...) that—without reliance on a solid and visionary architecture—can easily and quickly get out of control. This section is a reminder that the Internet is only now beginning to experience an acceleration of this complexity/robustness spiral and, if left unattended, can be fully expected to experience arcane, irreconcilable, and far-reaching robustness problems in the not-too-distant future. Such a perspective may be useful when discussing a new architectural design for the Internet, where the current arguments range from completely abandoning the original architecture and starting from scratch, to gradually evolving it; i.e., sustaining and growing it in response to the new and/or changing conditions associated with the Internet's coming of age.

3.1 THE CHANGING INTERNET

For the last 25 or so years, the Internet has experienced changes in practically all aspects of networking one can think of. These changes are well-known and well-documented (see for example Floyd and Paxson [21]) and are generally associated with the Internet's exponential growth and ever-increasing degree of heterogeneity any which way on looks. For example, in terms of its size as measured by the number of hosts connected to the Internet, the number grow from a handful of hosts at the end of 1970 to about 100,000,000 in late 2000—an increase of eight orders of magnitude. In terms of distinct networks that the Internet glues together, the number was 3 in the mid-1970s and has increased to more than 100,000 by mid-2001. As far as ASs are concerned, the global Internet currently consists of more than 10,000 separate ASs, all of which are interlinked in a way that gives users the illusion of a single connected network. A closer look at the communication links within the Internet reveals the sort of heterogeneity the network is dealing with at the physical and technological levels. For one, there are wired, wireless, and satellite links; in the wired world, there are legacy copper wires (e.g., for dial-up modems) and state-of-the-art optical fiber (e.g., for high-speed access and backbone links). The bandwidth of these links can be as low as a few kilobits per second (Kbps) or can reach beyond the Mbps or even Gbps range, varying over an impressive 5–7 orders of magnitude. With satellite links being possibly part of an end-to-end communication, latencies can be 2–3 orders of magnitude larger than for typical land-based wired/wireless communication. A myriad of communication technologies (e.g., LANs, ATM, frame relay, WDM), most of which did not even exist when the original Internet architecture was conceived, can be found at the network access layer where they give rise to a patchwork of widely different types of networks, glued together by IP. IP traffic volume has experienced sustained exponential growth for the last decade or two, with the amount of data traffic in all U.S. networks recently exceeding the total amount of voice traffic. At the transport and application layers, in addition to accommodating an increasing number of new and different protocols, a direct consequence of the Internet's explosive growth has been a wide range of different implementations of one and the same version of a given protocol—some of them fully consistent with and others explicitly violating the relevant standards.

From the perspective of users, services and applications, Internet transparency (see section 2.2.3) can be viewed as the main enabler and engine of innovation at the higher layers because it has allowed users to develop, experiment with, and deploy new applications without third-party involvement. Using "best efforts," the network simply forwards packets, irrespective of ownership (*who* is sending) and content (*what* is sent), and any enhancements to this basic service must be part of the application and is hence the responsibility of the application designer, not of the network. The end-to-end application neither expects nor relies on support from the intermediate networks that goes beyond basic IP, and in return, the application is ensured that its packets will neither be restricted nor tampered with in flight across the Internet. The fact that at the IP layer, the network is "open" and "free" in the sense of a "commons" has made it possible for killer applications such as e-mail, the Web, or Napster-like services to emerge, flourish, and dominate, despite their often esoteric and ad-hoc origins. They would most likely never have materialized, had the early designers of the Internet decided to compromise this transparency or "openness" by optimizing the network for some known (e.g., telephony) or some to-be-designed application or use. In turn, this commons has created a new competitive market place where an increasing number of increasingly diverse users form fast-growing new communities and interest groups with often vastly different or even opposing objectives, transform existing organizations and businesses, start new companies and develop novel business models, and impact society and world economy as a whole.

With the explosive growth and extreme heterogeneity in its user base and in the ways it has been used, the Internet has experienced a "changing of the guards"—the emergence of powerful new players and the diminishing role of previously dominating constituents and decision makers. For example, by decommissioning the NSFNet in April of 1995,[9] NSF created a business opportunity that allowed commercial Internet Service Providers (ISP) to enter and

[9]In 1986, the National Science Foundation (NSF) established the National Science Foundation Network (NSFNet) to link six of the nations super-computer centers through 56 kbps lines. In 1990, the ARPANET was officially dissolved, and responsibility for the Internet was passed to the NSFNet. NSF managed the NSFNet from 1987 to 1995, during the first period of explosive growth in the Internet.

ultimately dominate the Internet business. It also meant a withdrawal of NSF as one of the major policy makers and pioneers for Internet-related activities and research and a transfer of power and responsibility to the private sector. This transfer was accompanied by a re-evaluation of the rank-ordered requirements on the original list of design objectives for an Internet architecture (see section 2.1). In particular, while accountability was on the original list—way at the bottom, though—it clearly never had a major impact on shaping the architectural design. However, with the power and responsibility for operating, managing, and controlling the global Internet in the hands of private companies that try to compete in the open market place and run a profitable business, accountability in the sense of accurately measuring and efficiently collecting information that is key for running a viable business takes center stage. Similarly, while not even on the list of the original objectives, security and trustworthiness have increasingly become two of the key new requirements, capable of influencing the success (or lack thereof) of new applications and services as they compete in the market place. While in the early days of the Internet, everybody knew one another or relied on the other side to behave as desired, such trustworthy behavior can clearly no longer be assumed; in fact, we have already moved beyond the point where we no longer trust most of the Internet hosts we interact with. In theory, anyone connected to the Internet is vulnerable to malicious attacks and at risk that valuable information will be lost, stolen, corrupted, or misused, or that individual computer, whole intra-networks, or the entire Internet infrastructure will be disrupted. Many of the successful and widely publicized virus or denial-of-service attacks have exploited flaws and vulnerabilities inadvertently created and distributed by the system vendors and have demonstrated the potential for large-scale economic consequences, especially for enterprises such as eBay and Amazon for which the Internet has reached "mission-critical" status.[10] Accountability, security, and trustworthiness are examples of requirements that played little or no role in the design of the original Internet architecture but have become more and more crucial during the Internet's coming of age, reaching a similar level of importance that was given to the connectivity and robustness requirements during the early design stages of the Internet architecture. However, incorporating these emerging requirements as "afterthoughts" into the existing architecture has created considerable conceptual and technical challenges and no satisfactory solution exists to date [11]. For other emerging requirements such as various types of quality of service or mobility, where the latter is largely motivated by the increasing penetration of wireless communication and requires the Internet architecture to support ubiquitous mobility, similar observations apply.

---

[10]Distributed denial-of-service is an example of the fragility created by the end-to-end and trusted endpoint principles. Here the robustness of a simple network creates the fragility to trusted endpoints. In fact, one can argue that the biggest fragility in the Internet is the designed-for easy access that the endpoints are granted.

3.2 FRAGILITIES THAT LEAD TO INCREASED COMPLEXITY: PROTOCOLS DESIGN

The story is well told how the original DARPA Internet architecture gave rise to a network that has been remarkably successful in handling the various "scaling" or growth-related challenges mentioned above, including (i) the sustained exponential growth in the number of users, hosts, and networks that make up the Internet, (ii) the exponential growth in link bandwidth and the increasing heterogeneity in networking technologies, and (iii) the extreme heterogeneity associated with the ways the network is used. This engineering achievement is testimony to the visionary thinking of the original architects of the Internet that gave rise to a network which has succeeded in providing a highly resilient packet delivery service despite experiencing sustained exponential growth over extended periods in time and with respect to a number of key networking components. In fact, a recently published document by a committee of networking experts concluded that despite all the internal and external changes that the Internet has faced during the past 30 or so years, *" [...] the Internet is fundamentally healthy and that most of the problems and issues discussed [...] can be addresses and solved by evolutionary changes within the Internet's current architectural framework and associated processes"* [32].

What are then the main reasons for why the original design of the Internet architecture has enabled the network to repeatedly adapt to and continue normal operation despite of the various scaling-related challenges that it has encountered in the process of "growing up"? Furthermore, is it possible to distill the essence of the many engineering solutions that have been proposed within or outside of the framework of the original architectural design and that have been implemented in the network to address problems related to the observed scaling phenomena? To answer these and related questions, we take in the following a closer look at the evolution in time of three of the key protocols in today's Internet: TCP, HTTP, and BGP.

3.2.1 Toward an Ever More Efficient TCP. Skipping the first important TCP-related change, namely the separation in the late 1970s of TCP into an internetwork layer protocol (IP) and transport-layer protocol (TCP), in version 4 of the respective protocols, RFC-793 [36] describes the functionalities that TCP is required to perform in order to achieve its intended goal— *"to be used as a highly reliable host-to-host protocol between hosts in packet-switched computer communication networks, and in interconnected systems of such networks."* Moreover, [36] specifies how the required functionalities (i.e., basic data transfer, reliability, flow control, multiplexing, connections, and precedence and security) are implemented and what engineering solutions are used to achieve the aimed-for robustness of TCP. These solutions consist of a range of techniques from control theory, including signaling and feedback (in the form of ACKs), window-based flow control algorithms (so as to not overwhelm the receiver with packets), timers (e.g., for retransmissions), and low-pass filters (for simple RTT estimation used in the context of determining the retransmit timer). The complexity resulting from implementing these engineering solutions makes TCP robust to the uncertainties that are inherent in basic packet transmission (i.e., damaged, lost, corrupted, duplicated, or out-of-order packets, and apparent mismatch between sending and receiving rates) and for which such complexity was selected in the first place.

The "fragility" or vulnerability of this design to rare or unexpected perturbations became apparent in October of 1986, when the network had the first of what became a series of "congestion collapses"—completely unexpected, sudden and drastic (e.g., factor-of-thousand) drops in throughput [24]; while the network continued to transmit packets, the majority of them were retransmission, thereby lowering the effective throughput over the affected links. Subsequently, Van Jacobson [24] identified in the then-prevailing TCP implementations various shortcomings that contributed to these episodes of severe congestion; for example, connections that don't get to equilibrium, i.e., running in a stable mode with a full window of data in transit, either because they are too short or because they restarted after a packet loss. To make TCP robust to this new fragility, he also proposed in Jacobson [24] a number of engineering solutions that were soon thereafter incorporated into RFC-1122 [3], which lists the requirements for a conformant implementation of TCP as of 1989.[11] The proposed solutions consisted of adding new levels or modifying the existing layer of signaling, feedback, and control; their implementations relied on using new algorithms, especially for RTT variance estimation, for performing exponential retransmit timer backoff, for slow-start, and for dynamic window sizing on congestion (also known as congestion avoidance). The result has been a more complex TCP design that appears to be very resilient to Internet congestion and hence to the fragility for which this increased complexity was deliberately designed for. As we will illustrate in more detail below, this more complex TCP design is inevitably prone to new fragilities, especially in terms of unexpected feature interactions with higher-layer protocols. In this sense, the evolution of TCP illustrates very clearly the intrinsically "robust, yet fragile" character of complex systems design and serves as a concrete example of the complexity/robustness spiral "in action".

Another aspect that not only fuels but typically accelerates this complexity/robustness spiral for TCP is the ever-present push for more efficient and improved throughput, performance, or productivity. In an environment like the Internet, where assumptions underlying the basic TCP design and usage can (and have) suddenly become invalid—even though they appeared at some point in time to be rock-solid and fundamentally sound—as a result of the changing nature of the network, striving for improved efficiency in TCP operations is like trying and hit a constantly moving target. Changes in the key assumptions

[11]Since 1989, a number of additional changes to TCP have been made, but they are not central to our discussion and will not be discussed further.

are generally associated with reduced TCP efficiency and have been a constant source for embellishments to TCP. Without going into the details of these embellishments, there are roughly three separate categories. First, there are the TCP improvements that leave the basic TCP design in tact but attempt to improve TCP's efficiency by using existing "knobs" (i.e., parameter settings) or known "rules" (e.g., selective acknowledgment, or SACK) to make TCP more aggressive when it comes to grabbing available bandwidth. The second category consists of a number of different flavors of TCP (e.g., NewReno, Tahoe, Vegas), where the different flavors are characterized in terms of the type of algorithms that are used for such mechanisms as retransmission, congestion avoidance, and slow start. In contrast to these first two categories, which only alter the actions of the end hosts (by making the proposed changes to the TCP protocol) but not that of the routers, the third type of TCP improvements requires TCP "helpers" (e.g., random early detection, or RED; explicit congestion notification, or ECN). While residing within the network (i.e., in each router) and thus expanding the hourglass' waist, these "helpers" assist TCP in optimizings end-to-end performance. The optimization typically involves some sort of active queueing management (AQM) such as altering the queueing behavior internal to each router by means of, for example, selectively marking or dropping packets.

Another fragility has come with the increasing penetration of wireless communication. TCP was designed for wired networks, and its design makes numerous assumptions that are typical for wired environments. For example, TCP assumes that packet loss is an indication of network congestion, rather than corrupted bits. To provide effective internetworking between the wired and wireless worlds, TCP has been augmented with auxiliary protocols for (link-layer) local coding and retransmission, but the interactions, for example, between link-level retransmission and transport-level retransmission remain largely ill understood.

Common to all these examples is that the improvements and modifications typically result in a more complex TCP design which, in turn, manifests itself in the implementation of additional mechanisms, more sophisticated algorithms, and more versatile signaling and feedback strategies. While some of these modifications succeeded in improving robustness without introducing significant new fragilities (e.g., SACK can be viewed as a step toward a more robust retransmission policy because it uses information about what data the receiver has received; or NewReno was a step toward more robust congestion control, avoiding the unnecessary timeouts of Reno), others turned out to be more robust in some respects, and less so in others. For example, in conventional congestion control, the congestion indication has been a dropped packet. This means that either one very conservatively infers a dropped packet from the timeout of a conservatively-tuned retransmit timer and suffers the performance costs (i.e., the pre-1988 approach), or one has less conservative indications of congestion (e.g., three duplicate acknowledgments, and less conservative retransmit timers), which in turn require sophisticated mechanisms that make these indications robust to environments with reordered packets, arbitrary delays, multiple losses from a window of data, etc. (i.e., the approach adopted in 1988). While ECN is essentially a step toward an indication of congestion that is more robust than the traditionally considered indication consisting of a dropped packet, it could also be the cause of new fragilities because ECN-related information has the potential of being used in yet unknown ways or giving rise to ill-understood interactions when deployed in large-scale networks. After all, aspects that include the size of the network, finite link capacities, and design features drive the dynamics of protocols such as TCP and couple the different simultaneous connections sharing a given link in intricate and complex ways, introducing significant and complicated interactions in time, among active connections, and across the different networking layers. In any case, the question whether achieving a given amount of additional robustness was worth the price of the resulting additional complexity always requires careful consideration, especially when justifying proposed improvements that have been specifically designed to enhance TCP performance by exploiting (potentially transient) changes in the assumptions that underlied the early design of TCP.

3.2.2 Improving Web Efficiency: HTTP. The "robust, yet fragile" character of the Internet can also be observed when moving from the transport layer to the application layer. To this end, we briefly consider the evolution of the design of the *Hypertext Transfer Protocol (HTTP)*. HTTP is an application-layer transfer protocol that relies on TCP for transport and is used by the World Wide Web (WWW) to retrieve information from distributed servers. In contrast to TCP where a considerable effort went into getting the design "right" before deploying the protocol in the network, HTTP is often referred to as a "success disaster"—an insatiable demand (fueled by the explosive popularity of the Web) for a protocol that had escaped into the real world, without any systematic evaluation of its design and without a careful assessment of its potential side effects. Fittingly, RFC-1945 [1], a document that describes most of the component specifications and implementations for HTTP (or HTTP/1.0, as it is commonly referred to), only appeared in 1996, when there were already millions of Web users and when Web-related traffic had started to dominate in the Internet.

Conceptually, HTTP/1.0 is a simple request-response protocol—the client establishes a connection to the remote server and then issues a request; in turn, the server processes the request, returns a response, and closes the connection. Much of the complexity in the design specifications of HTTP/1.0 as documented in [1] is due to ensuring ease-of-use, scalability, and flexibility. Robustness in HTTP/1.0 is largely addressed by relying on TCP for data transport, and therein lies the cause for a fragility in its design that became more and more obvious with the explosive growth of the Web. This fragility became known as the "World Wide Wait" phenomenon—a slowdown of Web performance to the point where user-perceived end-to-end performance becomes unacceptable. The root cause of this phenomenon was already known in 1994, when it was pointed out in Spero [39] that certain design aspects of HTTP/1.0 interacted badly with TCP, and that these feature interactions caused problems with user-perceived latency as well as with server scalability. However, the problem was essentially ignored until after 1996, when the Internet's performance problems suddenly became a reality due to the Web's explosive growth.

To illustrate the unanticipated design mismatch between HTTP/1.0 and TCP, note that a typical Web page includes embedded images. Since HTTP/1.0 cannot ask for multiple objects (e.g., embedded images) within the same request (and hence within the same TCP connection), fetching each object corresponds to a separate request, and hence separate TCP connection. To make things worse, while much of TCP's original design was concerned with long-lived connections (e.g., transferring large files), actual TCP connections, and in particular TCP connections commonly encountered in an HTTP/1.0 message exchange, are short-lived (i.e., consisting of just a few packets). As a result, most objects are transferred before their corresponding TCP connections complete their slow start phase; that is, much of HTTP/1.0 traffic is transported when TCP is at its least efficient. Thus, instead of relying on the strengths of TCP, HTTP/1.0 succeeded in the opposite, i.e., designing for its weaknesses. This and numerous other fragilities in the design of HTTP/1.0 have been addressed and resolved in RFC-2616 [19] that specifies the requirements for version 1.1 of HTTP, i.e., HTTP/1.1. [19] describes a considerably more complex design than that given in Berners-Lee et al. [1], and while succeeding in aligning HTTP operations more closely with TCP (e.g., through the introduction of persistent connections and pipelining), the new protocol supports features and provides for mechanisms that may well give rise to new fragilities in the form of unexpected feature interactions (i.e., with IP-related functionalities), unanticipated performance bottlenecks, or cascading failure events. For more details about HTTP, its evolution from HTTP/1.0 to HTTP/1.1, protocol descriptions for both versions, etc., we refer to [26], where further feature interactions between HTTP and TCP are discussed as well.

3.2.3 Coping with Scale: BGP. BGP was built on experience gained with EGP and EGP usage in the NSFNet backbone. Each AS uses an interior routing system to maintain a coherent view of the connectivity within the AS, and uses BGP to maintain adjacency information with it's neighboring ASs, thereby creating a global view of the connectivity of the system. As far as BGP is concerned, most of the "fragilities" encountered during its evolution from EGP to (version 4) of BGP and its numerous subsequent modifications have been due to scaling issues, e.g., exponential growth of the number of Internet hosts, ASs, routing table entries, or the size of individual ASs. For example, as far as the latter property is concerned, originally, BGP deployments were configured such that all BGP speakers within the same AS had be fully meshed so that any external routing information had to be redistributed to all other routers within that AS. This feature of BGP as specified in Rekhter and Li [37] led to well-documented scaling problems and was subsequently rectified by the introduction and specification of BGP "route reflection" (e.g., see Steward [41]).

A more serious scaling-related fragility of BGP that does not seem to have an easy solution concerns the problem caused by the explosive growth of the sizes of BGP routing tables, especially in the backbone. Recall that the routing tables used by BGP contain network-wide connectivity information. Connectivity is expressed as a preference for "shortest paths" to reach any destination address, modulated by the connectivity policies used by the different ASs. Each entry in the routing table refers to a distinct route. The attributes of the route are used to determine the "best" path from a given AS to the AS that is originating the route. Here, determining the "best" path means using an AS as the basic element of computing and finding out which routing advertisement and associated next AS hop address are the most preferred. The elements of the BGP routing domain are routing entries, expressed as a span of addresses; all addresses within each routing entry share a common origin AS and a common connectivity policy. The total size of the BGP routing table is thus a metric of the number of distinct routes within the Internet, where each route describes a contiguous set of addresses which share a common origin AS and a common reachability policy. As the size of the Internet increases and more and more devices need addresses and seek connectivity—resulting in a growing number of distinct provider networks and ASs and, in turn, an ever increasing number of distinct connectivity policies—the task of maintaining coherent reachability information becomes more formidable and results in ever larger routing tables [22]. However, the limited capabilities (e.g., memory) of even the latest routing devices mandate that this growth must be managed, and herein lies a vulnerability that the original BGP design did not anticipate.

This routing table scaling crises became acute in 1994/95 when the explosive demand for Internet address space (largely fueled by an insatiable demand for top-level domain names for Web businesses) threatened to overwhelm the capacities of the then-available routers. This fear led to the deployment of *Classless Inter-Domain Routing (CIDR)*[12] which provides some control over the number of routes a router has to remember. This was achieved by assigning addresses in a hierarchical manner that forces addresses to be assigned in blocks. The idea behind CIDR was to support hierarchical provider address aggregation; that is, a network provider is assigned an address block from an address registry, and the provider announces this entire block into the exterior routing domain as a single routing entry with a single connectivity policy. Customers of this provider use a sub-allocation from this address block, and these smaller routing entries are aggregated by the provider and are not directly passed into the exterior routing domain. By providing a new set of mechanisms and functionalities for supporting CIDR, this engineering solution was incorporated into BGP in version 4 (BGP4)

[12]Originally, the IP address space included a concept of *classes* of networks (Classes A–E, with a flat space of host addresses within each class), and a standard engineering solution is—as soon as scaling problems arise—to add hierarchy where there was none.

and has somewhat contained the observed growth of the routing tables. However, more recent measurements suggest another onset of rapid routing table growth around year 2000 [22].

While scaling has been (and is likely to continue to be) the root cause for many of the BGP-induced fragilities, there are examples of other types of BGP-related vulnerabilities that add to the picture of a "live" Internet complexity/robustness spiral. In particular, the observed slow convergence of BGP [27] remains a topic of current research, and while it is suspected to be influenced by certain BGP design features (e.g., choice of timer values, tie-breaking techniques), as of yet, no solutions have been proposed or implemented. There have been many examples of modifications to and extension of BGP4 (e.g., AS confederations, BGP communities) that increase the complexity of the protocol by providing new functionalities and, at the same time, attempt to reduce the management complexity of maintaining the Internet.

3.3 FRAGILITIES THAT LEAD TO INCREASED COMPLEXITY: FEATURE INTERACTIONS

There has been an increasing awareness for the need of a BGP that is secure against disruptions such as accidental BGP mis-configurations (which, ironically, become more likely as BGP as a protocol becomes more complex and difficult to configure) and malicious attacks that are directly or indirectly targeted at compromising local or global connectivity information stored in the routing tables. In this context, recent work describing the impact of the propagation of Microsoft worms (such as Code Red and Nimda) on the stability of the global routing system is particularly illuminating [12, 31, 42]. These studies offer compelling evidence for a strong correlation between periods of worm propagation and BGP message storm and discuss some of the undesirable network behavior observed during the worms' probing and propagation phases, including congestion-induced failures of BGP sessions due to timeouts, flow-diversity-induced failures of BGP sessions due to router CPU overloads, pro-active disconnection of certain networks (especially corporate networks and small ISPs), and failures of other equipment at the Internet edge (e.g., DSL routers and other devices). They also illustrate an argument we made earlier (see section 1.1), namely that only extreme circumstances that are neither easily replicable in a laboratory environment nor fully comprehensible by even the most diligent group of networking experts are able to shed light on the enormous complexity and hidden vulnerabilities underlying large-scale engineered structures such as the Internet-wide routing system. Compared to point failures in the core Internet infrastructure, such as power outages in telco facilities or fiber cuts (see for example [33] for an account of how the Internet performed on 9/11/01), which are fully expected to occur and represent classic instances of designed-for vulnerabilities, these worm-induced instabilities of the Internet's routing system define a completely novel class of BGP-related fragilities. These new problems will require timely and effective solutions, or next-generation worms causing cascading BGP storms that will seriously degrade not only the access side of the network but also its core loom as real possibilities [40].

3.4 FRAGILITIES THAT LEAD TO INCREASED COMPLEXITY: ARCHITECTURE DESIGN

Recall that a key feature of the original Internet architecture design was a single universal logical addressing scheme, with addresses that are fixed-sized (i.e., 32 bits) numerical quantities and are applied to physical network interfaces (for naming the node and for routing to it). With 32 bits of address space, a maximum of some 4 billion different hosts can be addressed, although in practice, due to some inefficiency in address space allocation, a few hundred million addressable devices is probably a more realistic limit. In any case, while at the time of the design of the original Internet architecture, 4 billion (or 400 million) was considered to be a safe enough upper limit that justified the "hard-wiring" of 32 bits of address space into the Internet's design, the explosive growth of the network soon identified the very design assumption of a single fixed-sized address space as the main culprit behind a very serious but completely unpredicted fragility—the potential exhaustion of available IP address space. The alarming rate of IP address depletion as a result of the tremendous demand for and popularity of top-level domain names for Web businesses increased the overall awareness for this fragility that was simply a non-issue in the original design of the Internet architecture. Starting in the early 1990s, when the problem of IP address space exhaustion was first discussed, either in terms of being imminent or as a real possibility for the not-too-distant future, a number of engineering solutions have been considered for dealing with this problem. In the following, we illustrate some of these solutions and discuss how the various proposed designs contribute to an acceleration of the Internet's complexity/robustness spiral.

3.4.1 Keep Architecture, Change IP: IPv6.

The obvious and straight-forward solution to the (real or perceived) IP address space exhaustion problem is to increase the address space. Version 6 of IP, or IPv6 as it became known, takes this approach, and its design includes 128 bits of address space. IPv6 is specified in RCF-1883 [16] and provides a number of enhancements over the existing IP (i.e., IPv4). The appealing feature of IPv6 is that conceptually, it simply represents a new generation of the IPv4 technology and fully conforms to the original Internet architecture. Unfortunately, the simplicity of this solution is deceiving, and the complications (and costs) arising from a potential transition of the Internet from IPv4 to IPv6 are substantial and have been one of the major road blocks for a widespread deployment of IPv6 to date.

While IPv6 promises to restore the simplicity as well as expand the functionality of the original Internet architecture, the increased complexity and new fragilities associated with the next-generation IP protocol are due to the need for a smooth transition period during which IPv4 and IPv6 have to be able to coexist. In the absence of viable alternatives, such a transition period will result in doubling the number of service interfaces, will require changes below and above the hourglass' waist, and will inevitably create interoperability problems. Historically, the Internet has experienced only one such comparable transition phase, namely on January 1983, when there was a transition of the ARPANET host protocol from the original Network Control Protocol (NCP) to TCP/IP [28]. As a "flag-day" style transition, all hosts were required to convert simultaneously, and those that didn't convert had to communicate via rather ad-hoc mechanisms. The transition was carefully planned over several years before it actually took place and went very smoothly. However, what worked in 1983 when the network connected some 300 hosts that required changes simply does not scale to a network with a few 100,000,000 hosts. While the Internet has responded well to incremental deployment of new technologies "out" at the edges of the network and "up" in the protocol stack, scalable solutions to the problem of deploying new technologies that require changes "inside" the network (e.g., at the IP-layer) are largely nonexistent.

3.4.2 Change Architecture, Keep IP: NAT. An alternative approach to dealing with the IP address space exhaustion problem is to work within the framework of the current IP protocol (IPv4) and allow, if necessary, for new features or capabilities, even if they explicitly compromise the original Internet architecture. For example, address reuse is an obvious method for solving the address depletion problem (at least in the short term), and motivated the introduction of Network Address Translators (NAT) [18]. A NAT is a network device that is placed at the border of a stub domain or intranet (e.g., private corporate network) and has a table consisting of pairs of local IP addresses and globally unique addresses. Since the IP addresses inside the stub domain are not globally unique and are reused in other domains, NATs were envisioned to ease the growth rate of IP address use. A significant advantage of NAT boxes is that they can be deployed without any changes to the network (i.e., IP). However, NAT boxes directly violate the architectural principle of "Internet transparency" (see section 2.2.3), whereby addresses in packets are not compromised during transport and are carried unchanged from source to destination. While the arguments in favor of or against Internet transparency often take on religious tones, the loss of transparency due to NAT is not just academic (see also Blumenthal and Clark [2] for a more thorough assessment of this problem in the context of the "end-to-end arguments"). In fact, the deployment of NATs adds up to complexity in applications design, complexity in network configuration, complexity in security mechanisms, and complexity in network management. For example, NATs break all applications that assume and rely on globally unique and stable IP addresses. If the full range of Internet applications is to be used, complex work-around scenarios are needed; e.g., NATs have to be coupled with Application Layer Gateways (ALG), inserted between application peers to simulate a direct connection when some intervening protocol or device prevents direct access. ALGs terminate the transport protocol and may modify the data stream before forwarding; e.g., identify and translate packet addresses. Thus, new applications may need NAT upgrades to achieve widespread deployment, which could have far-reaching implications on the Internet's ability in the future to support and guarantee flexibility and innovation at the application layer.

The NAT technology was introduced in 1994 with the specific concern that *"NAT has several negative characteristics that make it inappropriate as a long term solution, and may make it inappropriate even as a short term solution"* [18]. Even though the need to experiment and test what applications may be adversely affected by NAT's header manipulations before there was any substantial operational experience was articulated loud and clear in Egevang and Francis [18], networking reality is that there has been widespread deployment and use of NATs in the Internet. Reasons for this popularity of NAT as a technology[13] include (i) lower address utilization and hence a delay for IPv6 roll-out, (ii) ease of renumbering when providers change, (iii) masks global address change, and (iv) provides a level of managed isolation and security (typically as part of an already installed firewall system) between a private address realm and the rest of the Internet. In this sense, NAT is another example of a "success disaster," but in comparison to HTTP, the architectural and practical implications are possibly more severe. For one, NAT boxes move the Internet from the traditional peer-to-peer model, where any host is in theory addressable, toward a client-server model, in which only certain hosts are addressable from the global Internet. The potential that NAT—as a solution to the address depletion problem—severely curtails future usability of the network and directs the evolution of the Internet into a dead-end street looms as a real possibility. Furthermore, due to their immense popularity and widespread deployment, NAT boxes make diagnosing network faults much harder and introduce many more single points of failure, which in turn increases the fragility of entire network.

4 OUTLOOK

Despite the drastic changes that the Internet has experienced during the transition from a small-scale research network to a mainstream infrastructure, its ability to scale (i.e., adapt to explosive growth), to support innovation (i.e., foster technological progress below and above the hourglass' waist), and to ensure flexibility (i.e., follow a design that can and does evolve over time in response to changing conditions) has been remarkable. The past 20–30 years have been testimony to the ingenuity of the early designers of the Internet architecture, and the fact that by and large, their original design has been maintained and has guided

[13]Beyond its intended use, the NAT technology has recently also been used as an alternative to BGP for route-optimization in multi-homing, where businesses use two or more diverse routes to multiple ISPs to increase robustness to peering link failures.

the development of the network through a "sea of change" to what we call today's Internet is an astounding engineering achievement. However, as argued in section 3 above, the Internet is only now beginning to experience changes that pose serious threats to the original architectural design. Moreover, these threats have a tendency to create a patchwork of technical solutions that, while addressing particular needs, may severely restrict the future use of the network and may force it down an evolutionary dead-end street. To correct this development, theoretical foundations for large-scale communication networks and an analytical theory of protocol design are needed for formulating and addressing pertinent questions such as "What are fundamentally sound and practically relevant architectural design guidelines for escaping the presently observed complexity/robustness spiral and for developing effective engineering approaches that contain the inevitable fragilities of the anticipated network infrastructures, but don't automatically increase complexity?" or "What sort of protocol design principles will lead to what sort of tradeoffs between complexity and robustness, and will these designs ensure provably scalable protocols?" or "Which part of the design space do the currently used protocols occupy, and how far from optimal are they?"

To this end, note that one way to calibrate a theory of the Internet and the level of comfort that it provides for understanding today's Internet is by the number of instances with which networking researchers observe "emerging phenomena"—measurement-driven discoveries that come as a complete surprise, baffle (at first) the experts, cannot be explained nor predicted within the framework of the traditionally considered mathematical models, and rely crucially on the large-scale nature of the Internet, with little or no hope of encountering them when considering small-scale or toy versions of the actual Internet. At the same time, we are here interested in theoretical foundations of the Internet that allow for a systematic and integrated treatment of both the "horizontal" (spatially distributed) and the "vertical" (protocol stack) aspect of the Internet—a requirement that emerges very clearly from our discussions about the original design and evolution of the Internet in the previous sections. To this end, we revisit the arguments made in section 1 for why the Internet is unique as a starting point for a scientific study of complex systems and illustrate them in the context of a "classic" example of an Internet-related emergent phenomenon—self similar Internet traffic.

Intuitively, the self-similarity discovery refers to the empirical finding that measured traffic rates on links in the Internet (i.e., number of packets or bytes that traverse a given link per time unit) exhibit *self-similar* (or "fractal-like") behavior in the sense that a segment of the traffic rate process measured at some time scale looks or behaves just like an appropriately scaled version of the traffic rate process measured over a different time scale; see Leland et al. [29] and the important follow-up studies [34] and [14]. More precisely, these and numerous subsequent empirical studies described pertinent statistical characteristics of the temporal dynamics of measured traffic rate processes and provided ample evidence that measured Internet traffic is consistent with *asymptotic second-order self-similarity* or, equivalently, *long-range dependence (LRD)*; i.e., with autocorrelations $r(k)$ that decay like a power for sufficiently large lags k:

$$r(k) \sim c_1 k^{-\beta}, \quad \text{as } k \to \infty,$$

where $0 < \beta < 1$ and c_1 is a positive finite constant. These empirical findings were in stark contrast to conventional wisdom and to what traditionally used models assumed about actual Internet traffic, namely exponentially-fast decaying autocorrelations and, in turn, a classical white noise behavior when measuring traffic rates over large time scales.[14] Moreover, the ease with which it was possible to distinguish between assumed and actual traffic rate processes was surprising as well as striking.

The observed self-similar characteristic of Internet traffic constitutes a classic "emergent" phenomenon because, as an ubiquitous scaling property, it is apparently not put in by hand or design and thus invites a range of potential explanations, from being a consequence of dynamical instabilities or bifurcations (where the details of the system's design, architecture, and components are largely irrelevant), all the way to arising naturally within the confines of the Internet's hourglass protocol architecture (where it can be understood in terms of underlying architectural guidelines and design principles). Consider for example the following mathematical construction that fits in well with the layering architecture of the Internet. At the application layer, *sessions* (i.e., FTP, HTTP, TELNET) arrive at random (i.e., according to some stochastic process) on the link and have a "lifetime" or session length during which they exchange information. This information exchange manifests itself for example at the IP layer, where from the start until the end of a session, IP packets are transmitted in some bursty fashion. Thus, at the IP layer, the aggregate link traffic measured over some time period (e.g., 1 hour) is made up of the contributions of all the sessions that—during the period of interest— actively transmitted packets. Mathematically, this construction is known to give rise to LRD or, equivalently, asymptotic second-order self-similarity, provided the session arrivals follow some renewal process (e.g., Poisson) and, more importantly, the distribution $F(x)$ of the session "lifetimes" T are *heavy-tailed with infinite variance*; that is, as $x \to \infty$,

$$1 - F(x) = P[T > x] \sim c_2 x^{-\alpha},$$

where $1 < \alpha < 2$ and c_2 is a positive finite constant. Originally due to Cox [13], the main ingredient of this construction is the heavy-tailedness of the session durations. Intuitively, this property implies that there is no "typical" session length but instead, the session durations are *highly variable* (i.e., exhibit infinite

[14]Note that contrary to widely-held beliefs, LRD does not preclude smooth traffic; in fact, depending on the mean, variance, and β-value of the underlying process, modeled traffic rates can range from highly bursty to very smooth. In particular, traffic can be smooth and can exhibit LRD at the same time, thereby enriching the conventional perspective that tends to equate smooth traffic with Poisson-type traffic.

variance) and fluctuate over a wide range of time scales, from milliseconds to seconds, minutes, and hours. It is this basic characteristics at the application layer that, according to Cox's construction, causes the aggregate traffic at the IP layer to exhibit self-similar scaling behavior.[15] Moreover, the LRD parameter β of the aggregate traffic can be shown to be linearly related to the heavy-tail index α of the session lifetimes or sizes; in fact, we have $\beta = \alpha - 1$ (e.g., see Cox [13], Mandelbrot [30], and Willinger et al. [44].

What makes the Internet unique in terms of approaching the study of complex systems from a genuinely scientific perspective is its unmatched ability for validating in an often unambiguous, thorough, and detailed manner any proposed explanation of or new hypothesis about an emergent phenomenon. In fact, it is the ability of the Internet for providing enormous opportunities for measuring and collecting massive and detailed relevant data sets and for completely "reverse engineering" unexpected discoveries and surprises that gives rise to unique capabilities for developing and validating new theories for the complex Internet. Take, for example, Cox's construction. First note that it clearly identifies the type of additional measurements that should be collected and/or analyzed to test for the presence of the cause (i.e., heavy-tailed property) that is claimed to explain the phenomenon of interest (i.e., self-similar scaling). These measurements range from the application layer, where relevant variables include session durations (in seconds) and sizes (in number of bytes or packets), to the transport layer (where the items of interest are TCP connection durations and sizes), all the way to the network layer (where IP flow durations and sizes constitute the traffic variables of interest; here, an IP flow is made up of successive IP packets that satisfy certain common features). With the exception of some session-related variables (see for example Willinger and Paxson [43]), these layer-specific measurements can be readily extracted from the packet-level traces that are typically collected from individual links within the network. Indeed, while heavy-tailed distributions had been essentially unheard of in the networking context prior to the "emergence" of self-similarity, the numbers of empirical studies in support of the ubiquitous nature of heavy tails in measured Internet traffic at the different layers have steadily increased since then and have identified the heavy-tailed characteristic of individual traffic components as an *invariant* of Internet traffic for the past 10 or so years, despite the sometime drastic changes the network has undergone during that period; see for example Crovella and Bestavros [34], Paxson and Floyd [44], and Willinger et al. [14].

The implications for this networking-based explanation of the self-similarity phenomenon are far-reaching and unexpected. On the one hand, the fact that we can explain self-similar scaling in terms of the statistical properties of the individual sessions, connections, of flows that make up the aggregate link traf-

[15]A closely related earlier construction, originally due to Mandelbrot [30], relies on the notion of an *on/off process* (or, more generally, a *renewal-reward process*), but uses the same basic ingredient of heavy-tailedness to explain the self-similarity property of the aggregate link traffic.

fic shows the power of measurements that are available on the Internet to scientifically validate any given model or explanation and to "reverse engineer" unexpected discoveries and surprises. It also suggests that the LRD nature of network traffic is mainly caused by user/application characteristics (i.e., Poisson arrivals of sessions, heavy-tailed distributions with infinite variance for the session durations/sizes)—heavy-tailed files or Web objects (or other application-specific items) are streamed out onto the network in a TCP-controlled manner to create strong temporal correlations (i.e., LRD) in the aggregate traffic seen on individual links within the network.[16] On the other hand, the fact that the underlying heavy-tailed property of the various traffic-related entities appears to be a robust feature of network traffic suggests a number of new questions: Should the network be redesigned to handle self-similar aggregate traffic/heavy-tailed connections? If so, can it be redesigned? Is this self-similar/heavy-tailed aspect of network traffic a permanent feature of application demand or an historical artifact? If the former holds true, what causes application demands to be intrinsically heavy-tailed; i.e., where do the heavy tails come form? Attempts to answer these and other questions have motivated the development of a nascent theory of the Internet that is firmly based on some of the general principles reported in this chapter and derived from a study of the design of the Internet and the evolutionary path that it has taken to date. Developing and illustrating this theory will be the main focus of the companion paper [17].

ACKNOWLEDGMENTS

We wish to thank Sally Floyd and Tim Griffin for a critical reading of the original draft of this chapter and for providing valuable feedback. Their comments forced us to sharpen a number of our arguments concerning the tradeoff in Internet design between robustness and fragility, but we fully realize the limited success of some of our attempts at making this tradeoff more obvious or explicit.

REFERENCES

[1] Berners-Lee, T., R. Fielding, and H. Frystyk. "Hypertext Transfer Protocol—HTTP/1.0." Network Working Group RFC-1945, May 1996.

[2] Blumenthal, M. S., and D. D. Clark. "Rethinking the Design of the Internet: The End-to-End Argument vs. the Brave New World." *ACM Trans. Internet Tech.* **1(1)** (2001): 70–109.

[3] Braden, B., ed. "Requirements for Internet Hosts—Communication Layers." Network Working Group RFC-1122, October 1989.

[16] Note that without TCP or some TCP-type congestion control, the highly variable session workloads are capable of creating aggregate link traffic that can be very different from what we observe in today's Internet.

[4] Braden, B., D. Clark, S. Shenker, and J. Wroclawski. "Developing a Next-Generation Internet Architecture." Information Sciences Institute, University of Southern California. 2000. ⟨http://www.isi.edu/newarch/DOCUMENTS/WhitePaper.ps⟩.

[5] Carlson, J. M., and J. Doyle. "Highly Optimized Tolerance: A Mechanism for Power Laws in Designed Systems." *Phys. Rev. E* **60** (1999): 1412–1428.

[6] Carpenter, B., ed. "Architectural Principles of the Internet." Network Working Group RFC-1958, June 1996.

[7] Carpenter, B. "Internet Transparency." Network Working Group RFC-2775, February 2000.

[8] Carpenter, B., and P. Austein, eds. "Recent Changes in the Architectural Principles of the Internet." Internet Draft, IAB, February 2002.

[9] Clark, D. D. "The Design Philosophy of the DARPA Internet Protocols." *ACM Comp. Commun. Rev.* **18(4)** (1988): 106–114.

[10] Clark, D., L. Chapin, V. Cerf, R. Braden, and R. Hobby. "Towards the Future Internet Architecture." Network Working Group RFC-1287, December 1991.

[11] Clark, D. D., K. R. Sollins, J. Wroclawski, and R. Braden. "Tussle in Cyberspace: Defining Tomorrow's Internet." In *Proceedings of the ACM SIGCOMM 2002 Conference on Applications, Technologies, Architectures, and Protocols for Computer Communication*, held August 19–23, 2002, in Pittsburgh, PA, 347–356. ACM, 2002.

[12] Cowie, J., A. Ogielski, B. J. Premore, and Y. Yuan. "Global Routing Instabilities During Code Red II and Nimda Worm Propagation." Renesys Corporation, Oct. 2001. ⟨http://www.renesys.com/projects/bgp_instability⟩.

[13] Cox, D. R. "Long-Range Dependence: A Review." In *Statistics: An Appraisal*, edited by H. A. David and H. T. David, 55–74. Malden, MA: Iowa State University Press, Blackwell Publishing, 1984.

[14] Crovella, M., and A. Bestavros. "Self-Similarity in World Wide Web Traffic: Evidence and Possible Causes." *IEEE/ACM Trans. Network.* **5** (1997): 835–846.

[15] Csete, M. E., and J. Doyle. "Reverse Engineering of Biological Complexity." *Science* **295(5560)** (2002): 1664–1669.

[16] Deering, S., and R. Hinden. "Internet Protocol, Version 6 (IPv6) Specification." Network Working Group RFC-1883, December 1995.

[17] Doyle, J., S. Low, J. Carlson, F. Paganini, G. Vinnicombe, and W. Willinger. "Robustness and the Internet: Theoretical Foundations." Preprint, 2002.

[18] Egevang, K., and P. Francis. "The IP Network Address Translator (NAT)." Network Working Group, RFC-1631, May 1994.

[19] Fielding, R., J. Gettys, J. C. Mogul, H. Frystyk, L. Masinter, P. Leach, and T. Berners-Lee. "Hypertext Transfer Protocol—HTTP/1.1." Network Working Group RFC-2616, June 1999.

[20] Floyd, S., and V. Jacobson. "The Synchronization of Periodic Routing Messages." *Proceedings of the ACM SIGCOMM 1993 Conference on Applica-*

tions, Technologies, Architectures, and Protocols for Computer Communication, held September 13–17, 1993, in San Francisco, CA, 33–44. ACM, 1993.

[21] Floyd, S., and V. Paxson. "Difficulties in Simulating the Internet." *IEEE/ACM Trans. Network.* **9(4)** (2001): 392–403.

[22] Huston, G. "Architectural Requirements for Inter-Domain Routing in the Internet." Internet-Draft, Internet Architecture Board Working Group of the IETF, February 2001. ⟨http://www1.ietf.org/mail-archive/ietf-announce/Current/msg11417.html⟩.

[23] Internet Engineering Task Force. ⟨http://www.ietf.org/rfc.html⟩.

[24] Jacobson, V. "Congestion Avoidance and Control." *ACM Comp. Comm. Rev.* **18(4)** (1988): 314–329.

[25] Kaat, M. "Overview of 1999 IAB Network Layer Workshop." Network Working Group RFC-2956, October 2000.

[26] Krishnamurthy, B., and J. Rexford. *Web Protocols and Practice.* Boston, MA: Addison-Wesley, 2001.

[27] Labovitz, C., A. Ahuja, A. Bose, and F. Jahanian. "Delayed Internet Routing Convergence." In *Proceedings of the ACM SIGCOMM 2000 Conference on Applications, Technologies, Architectures, and Protocols for Computer Communication*, held August 28 to September 1, 2000, in Stockholm, Sweden, 175–187. ACM, 2000.

[28] Leiner, B. M., V. G. Cerf, D. D. Clark, R. E. Kahn, L. Kleinrock, D. C. Lynch, J. Postel, L. G. Roberts, and S. Wolff. "A Brief History of the Internet." Internet Society. 2002. ⟨http://www.isoc.org/internet/history/brief.shtml⟩.

[29] Leland, W. E., M. S. Taqqu, W. Willinger, and D. V. Wilson. "On the Self-Similar Nature of Ethernet Traffic (Extended Version). *IEEE/ACM Trans. Network.* **2** (1994): 1–15.

[30] Mandelbrot, B. B. "Long-Run Linearity, Locally Gaussian Processes, H-Spectra and Infinite Variances." *Intl. Econ. Rev.* **10** (1969): 82–113.

[31] Moore, D., C. Shannon, and K. Claffy. "Code-Red: A Case Study on the Spread and Victims of an Internet Worm." *Proceedings of Internet Measurement Workshop 2002*, held November 6–8, 2002, in Marseille, France. ACM and USENIX, 2002. ⟨http://:www/icir.org/vern/imw-2002/proceedings.html⟩.

[32] Computer Science and Telecommunications Board (CSTB), National Research Council. *The Internet's Coming of Age.* Washington, DC: National Academy Press, 2001.

[33] Computer Science and Telecommunications Board (CSTB), National Research Council. *The Internet under Crisis Conditions: Learning from September 11.* Washington, DC: National Academy Press, 2003.

[34] Paxson, V., and S. Floyd. "Wide-Area Traffic: The Failure of Poisson Modeling." *IEEE/ACM Trans. Network.* **3** (1995): 226–244.

[35] Peterson, L. L., and B. L. Davie. *Computer Networks, a Systems Approach.* San Francisco, CA: Morgan Kaufmann, 1996.

[36] Postel, J., ed. "Transmission Control Protocol." Network Working Group RFC-793, September 1981.

[37] Rekhter, Y., and T. Li. "A Border Gateway Protocol 4 (BGP-4)." Network Working Group RFC-1771, March 1995.

[38] Saltzer, J. H., D. P. Reed, and D. D. Clark. "End-to-End Arguments in System Design." *ACM Trans. Comp. Sys.* **2(4)** (1984): 277–288.

[39] Spero, S. "Analysis of HTTP Performance." July 1994. ⟨http://www.w3.org/Protocols/HTTP/1.0/HTTPPerformance.html⟩.

[40] Staniford, S., and V. Paxson. "How to Own the Internet in Your Spare Time." In *Proceedings of the 11th USENIX Security Symposium*, held August 5–9, 2002, 149–167. Berkeley, CA: USENIX, 2002. ⟨http://www.icir.org/vern/papers/cdc-usenix-sec02/⟩.

[41] Stewart, J. W. *BGP4, Inter-Domain Routing in the Internet.* Boston, MA: Addison-Wesley, 1999.

[42] Wang, L., X. Zhao, D. Pei, R. Bush, D. Massey, A. Mankin, S. F. Wu, and L. Zhang. "Observation and Analysis of BGP Behavior under Stress." In *Proceedings of Internet Measurement Workshop 2002*, held November 6–8, 2002, in Marseille, France. ACM and USENIX, 2002. ⟨http://www.icir.org/vern/imw-2002/proceedings.html⟩.

[43] Willinger, W., and V. Paxson. "Where Mathematics Meets the Internet." *Notices of the AMS* **45** (1998): 961–970.

[44] Willinger, W., M. S. Taqqu, R. Sherman, and D. V. Wilson. "Self-Similarity through High-Variability: Statistical Analysis of Ethernet LAN Traffic at the Source Level." *IEEE/ACM Trans. Network.* **5** (1997): 71–86.

Robustness and the Internet: Theoretical Foundations

John C. Doyle
Steven H. Low
Fernando Paganini
Glenn Vinnicombe
Walter Willinger
Pablo Parrilo

While control and communications theory have played a crucial role throughout in designing aspects of the Internet, a unified and integrated theory of the Internet as a whole has only recently become a practical and achievable research objective. Dramatic progress has been made recently in analytical results that provide for the first time a nascent but promising foundation for a rigorous and coherent mathematical theory underpinning Internet technology. This new theory directly addresses the performance and robustness of both the "horizontal" decentralized and asynchronous nature of control in TCP/IP as well as the "vertical" separation into the layers of the TCP/IP protocol stack from application down to the link layer. These results generalize notions of source and channel coding from information theory as well as decentralized versions of robust control. The new theoretical insights gained about the Internet also combine with our understanding of its origins and evolution to provide a rich source of ideas about complex systems in general. Most surprisingly, our deepening understanding from genomics and molecular biology has revealed that at the network and protocol level, cells and organisms are strikingly similar to technological networks, despite having completely different material substrates, evolution, and development/construction.

1 INTRODUCTION

Many popular technological visions emphasize ubiquitous control, communications, and computing, with systems requiring high levels of not only autonomy and adaptation, but also evolvability, scalability, and verifiability. With current technology these are profoundly incompatible objectives, and both biology and nanotechnology create additional novel multiscale challenges. The rigorous, practical, and unified theoretical framework essential for this vision has until recently, proven stubbornly elusive. Two of the great abstractions of science and technology have been the separation, in both theory and applications, of (1) controls, communications, and computing from each other, and (2) the systems level from its underlying physical substrate. These separations have facilitated massively parallel, wildly successful, explosive growth in both mathematical theory and technology, but left many fundamental problems unresolved and resulted in a poor foundation for future systems of systems in which these elements must be integrated.

As discussed in Willinger and Doyle [35], much of the success of the Internet has been a result of adhering more or less faithfully over time to a set of fundamental network design principles adopted by the the early builders of the Internet (e.g., layering, fate-sharing, end-to-end). In this sense, these "principles" constitute a modest "theory" of the Internet. This theory is brilliant in the choices that were made, but it is also shallow in addressing only limited and high-level aspects of the full Internet protocol design problem. Seeking decomposition of hard problems into simpler subproblems will always be an important design strategy and is at the heart of the role of protocols, but an integrated and unified theory is required to do this in a rigorous and robust manner. Nevertheless, it is pedagogically useful in introducing new theoretical concepts to decompose the key ideas in a way that both reflects the Internet protocol stack itself and builds on and makes analogies with familiar ideas. In particular, Internet researchers have drawn on all aspects of engineering systems theory, including information theory (source and channel separation, data compression, error correction codes, rate distortion), control theory (feedback, stability, robustness), optimization (duality), dynamical systems (bifurcation, attractors), and computational complexity (P-NP-coNP). Each offers some perspective on the Internet, but no one view has so far provided a foundation for the protocol suite as a whole, and in particular the horizontal, distributed, asynchronous, and dynamic nature of control at each layer, or the vertical interaction of the various layers of the protocol stack. In the following, we give a brief review of these myriad contributions and will sketch the new directions by making analogies using more familiar ideas.

2 THE INFORMATION THEORY ANALOGY

A familiar analogy with information theory is that the overall network resource allocation problem, from server-client interactions through congestion control and routing to network provisioning, can be decomposed or separated into the TCP/IP protocol stack. Conventional information theory does not address this problem because of the intrinsic role that dynamics and feedback play in TCP/IP. Fortunately, a source/channel mice/elephant picture of Web and Internet traffic is emerging that begins to address these issues. New results not only complete the explanation of existing "self-similar" Internet traffic and its connection with application traffic, but are a promising starting point for a more complete source/channel coding theory analogous to that from Shannon information theory for conventional communication problems, though necessarily differing greatly in detail. Here the issue is the "vertical" separation problem: the optimality of separating the application level source vs. the TCP/IP level "channel" of the protocol stack, and then the optimal "coding" of the source and channel. The most familiar idea is that the protocol separation is, at best, optimal only in an asymptotic sense, where the asymptotes here involve large network load and capacity, as opposed to Shannon's large block size, and even in these asymptotic regimes there may be unavoidable suboptimalities. The source coding aspect is the most well-developed, with a variety of results, and models and theories of various levels of detail.

Many central aspects of internetworking that might seem related to source and channel coding problems in networks have received almost no theoretical treatment. For example, if the Web sites and clients browsing them are viewed collectively as a single aggregate "source," then this source involves both feedback and geometry as users interactively navigate hyperlinked content. While coding theory is relevant to file compression, the geometric and dynamic feedback aspects are less familiar. Furthermore, the "channel" losses experienced by this source are primarily due to congestion caused by traffic generated by the source itself. This traffic has long-range correlations and is self-similar on many time scales [17, 31, 36], which, in turn, can be traced to heavy-tailed file distributions in source traffic being streamed out on the net [7, 37]. These discoveries have inspired recent but extensive research in the modeling of network traffic statistics, their relationship to network protocols and the (often huge) impact on network performance. Despite these efforts, the full implications of this research have yet to be understood or exploited, and only recently has there emerged a coherent coding and control theoretic treatment of the geometric and dynamic aspects of Web and other application traffic.

3 THE HEAVY-TAILED SOURCE VS. SELF-SIMILAR CHANNEL TRAFFIC CHARACTERISTICS

Heavy-tailed source behavior and the self-similar nature of aggregate network traffic initially frustrated mainstream theorists, because both characteristics violate standard assumptions in information and queueing theory. The strongly heavy-tailed nature of traffic at the source level and nearly self-similar characteristic of aggregate traffic at the link level are quite unlike the traditionally assumed Poisson traffic models. Real network traffic exhibits long-range dependence and high burstiness over a wide range of time scales. While most files ("mice") have few packets, most packets are in large files ("elephants"). It has further been widely argued that the dominant source of this behavior is due to the heavy-tailed nature of Web and other application traffic being streamed out onto the network by TCP to create long-range correlations in packet rates. The applications naturally create bandwidth-hogging elephants and delay-sensitive mice, which coexist rather awkwardly in the current Internet. Our new treatment of this problem builds on theories from robust control [26, 32] and duality in optimization [12, 20], all with a generalized coding perspective from the HOT framework [4, 5, 8, 39] and provides a radically different view. For one, we claim that heavy-tailed source traffic must be embraced, because it is not an artifact of current applications and protocols, but is likely to be a *permanent and essential feature* of network traffic, including a majority of advanced network scenarios (the bad news?). Furthermore, we show that not only can a new theory be developed to handle heavy-tailed source traffic, but if properly exploited, heavy-tailed source behavior is, in fact, ideal for efficient and reliable transport over packet-switched networks (the good news!). We aim to show that Web and Internet traffic can be viewed as a (perhaps very unfamiliar) joint source and channel coding problem which can be treated systematically as a global optimization problem that is implemented in a decentralized manner.

That the heavy-tailed distributions associated with Web traffic are likely to be an invariant of much of the future network traffic, regardless of the application, is one important insight to be gained from this research direction, even in its currently nascent state. We expect that the current split of most traffic into mice and elephants will persist. Most files will be mice; these files make little aggregate bandwidth demand (due to their small size), but need low latency. Most of the packets come from elephants; these files demand high average bandwidth, but tolerate varying per packet latency. For example, sensor and real-time control applications naturally code into time-critical mice with measurement updates and actuator commands, against a background of elephants which update models of the dynamical environment and network characteristics. Of course, a coherent theory to make rigorous these informal observations is far from available, and the current toy model-based results are merely suggestive and encouraging.

While the empirical evidence for this mice/elephant mix in Web and other Internet traffic has received substantial attention recently, not only has no other theoretical work been done to explain it, but, in fact, the implications for congestion control have been largely ignored, except for the repeated assertions that these distributions break all the standard theories. Fortunately, this type of traffic creates an excellent blend when the network is properly controlled, and we have already made significant progress in exploring the profound implications of heavy-tailed traffic for network quality of service (QoS) issues. Thus two critical properties of networks converge in a most serendipitous manner: heavy tails are both a ubiquitous and permanent feature of network traffic, and an ideal mix of traffic, if properly controlled.

4 THE CONTROL THEORY ANALOGY

The control theory analogy derives from a duality theory of decentralized network resource allocation. This theory integrates and unifies existing approaches to TCP, and also proposes a utility and duality framework for designing new TCP/AQM strategies. This duality theory currently addresses most completely the "horizontal" decomposition of the decentralized and asynchronous nature of TCP/AQM, but it provides the framework for the mathematics for the above "vertical" theory of the protocol stack as well. The latter addresses the source/channel separation between applications and TCP/IP, as well as the further decomposition into TCP and IP routing [34], and router and link layer network provisioning [1]. A complete separation theory for the entire protocol stack with realistic modeling assumptions will likely require years to develop, but the initial results are very encouraging.

Specifically, we have shown that one can regard TCP/AQM as a distributed primal-dual algorithm carried out over the Internet by TCP sources and routers in the form of congestion control to maximize aggregate utility subject to capacity constraints at the resources [12, 14, 19, 20, 21, 24, 25, 38]. Different TCP or AQM protocols solve the same utility maximization problem with different utility functions, using different iterative algorithms. The TCP algorithm defines the objective function of the underlying optimization problem. The AQM algorithm ensures the complementary slackness condition of the constrained optimization and generates congestion measures (Lagrange multipliers) that solve the dual problem in equilibrium. The model implies that the equilibrium properties of networks under TCP/AQM control, such as throughput, delay, queue lengths, loss probabilities, and fairness, can be readily understood by studying the underlying optimization problem. Moreover, since the problem is a concave program, these properties can be efficiently computed numerically.

It is possible to go between utility maximization and TCP/AQM algorithms in both directions [19]. We can start with general utility functions, e.g., tailored to our applications, and then derive TCP/AQM algorithms to maximize aggregate utility [12, 15, 20, 23, 25]. Conversely, and historically, we can design TCP/AQM algorithms and then reverse-engineer the algorithms to determine the underlying utility functions they implicitly optimize and the associated dual problem. This is the consequence of end-to-end control: as long as the end-to-end congestion measure to which the TCP algorithm reacts is the *sum* of the constituent link congestion measures, such an interpretation is valid.

A critical recent development concerns the robustness and stability of the dynamics of TCP/AQM. These results clarify the dynamic (including instability) properties of existing protocols [9, 22], and have led to new designs which are provably, robustly scalable for large network size, capacity, and delay [6, 13, 16, 26, 27, 28, 32, 33]. The central technical issue here is the stability of a fully decentralized and asynchronous feedback control system. Here it is insufficient to take asymptotic limits, and one needs to prove robustness to arbitrary topologies and delays. The main insight from this series of work is to scale down source responses with their own round trip times and scale down link responses with their own capacities, in order to keep the gain over the feedback loop under control. It turns out that the required scaling with respect to capacity [26] is already built-in at the links if sources react to queueing delay [2, 6, 11, 21], as opposed to packet loss. Delay as a congestion signal is also observable at the source, making it possible to stabilize a TCP/IP network by only modifying the hosts but not the routers [10, 11].

The duality model clarifies that any TCP/AQM algorithms can be interpreted as maximizing aggregate utility in equilibrium and the stability theory explains how to design these algorithms so that the equilibrium is stable, under the assumption that the routing is given and fixed. Can TCP–AQM/IP, with minimum cost routing, jointly solve the utility maximization over both source rates and their routes? The dual problem of utility maximization over both source rates and routing has an appealing structure that makes it solvable by minimum cost routing using congestion prices generated by TCP–AQM as link costs. This raises the tantalizing possibility that TCP–AQM/IP may indeed turn out to maximize utility with proper choice of link costs. We show in Wang et al. [34], however, that the primal problem is NP-hard, and hence cannot be solved by minimum cost routing unless P=NP. This prompts the questions: How well can IP solve the utility maximization approximately? In particular, what is the effect of the choice of link cost on maximum utility and on routing stability? For the special case of a ring network with a common destination, we show that there is an inevitable tradeoff between utility maximization and routing stability. Specifically, link costs and minimum cost routing form a feedback system. This system is unstable when link costs are pure prices. It can be stabilized by adding a static component to the link cost. The loss in utility, however, increases with the weight on the static component. Hence, while stability requires a small weight on prices, utility maximization favors a large weight.

While duality theory again plays a key role in formulating and studying TCP–AQM/IP-related problems, the implicit assumption is that the underlying physical network infrastructure/network design is given and fixed. Can the above-described nascent mathematical theory of the Internet be extended to layers below IP by incorporating the decentralized mechanisms and forces at work for link layer network provisioning? Note that the ability to make detailed measurements at the relevant layers and an in-depth knowledge of how the individual parts work and are interconnected make it possible to unambiguously diagnose and "reverse engineer" any such proposed theory and allow for a clean separation between sound and specious approaches. To this end, we argue in Alderson et al. [1] that by focusing on the causal forces at work in network design and aiming at identifying the main economic and technological constraints faced by network designers and operators, observed large-scale network behavior (e.g., hierarchical structures, statistical graph properties) can be expected to be the natural by-product of an optimization-based approach to link-layer network provisioning. In particular, we show in Li et al. [18] that when imposing certain technological constraints that reflect physical hardware limitations (e.g., router capacity), the space of possible network designs with identical large-scale behavior (expressed in terms of the node degree distributions of the resulting graphs) divides into two very different groups. One group consists of the set of generic or "random" configurations, all of which have poor performance and high cost and low robustness. In contrast, the other group contains all non-generic or "highly-designed" configurations, and they are characterized by high performance, low cost, and high robustness.

5 COMPLEXITY ISSUES

The ultimate challenge in this overall research program is creating a theoretical infrastructure for formal and algorithmic verification of the correctness and robustness of scalable network protocols and embedded software for control of distributed systems. We must ultimately be able to *prove* that systems are robust, and at multiple levels of abstraction, including that of embedded code. The two approaches traditionally most concerned with these issues are robust control and formal software verification, which until recently have been both limited in their own domains, and largely independent. New results both radically expand and extend these methods, but also offer hope for their ultimate integration. In both cases, the main conceptual objective is to guarantee that a clearly defined set of "bad behaviors" is empty. For example, in the case of robustness analysis of linear systems, that set can correspond to a particular combination of uncertain parameters producing a large performance index. In protocol verification the bad behavior can be associated, for instance, with a deadlock condition.

Under minimal assumptions, most verification problems can be shown to belong to the computational complexity class known as co-NP, and the asymmetry between the classes NP and co-NP are as important and profound as those between P and NP (unless they are the same, which is unlikely). Problems in NP always have short proofs, while those in co-NP have short refutations. In contrast, there is no reason, in principle, to expect co-NP proofs to be short, that is, polynomial time verifiable. It is a remarkable fact that in many cases, concise arguments about robustness (or correctness) can be provided. The consequences and implications of this finding are not fully understood. A promising theoretical framework for the unification of robustness and verification methods is the systematic theory of convex relaxations for co-NP problems recently developed in Parrilo [29] and Parrilo and Strumfels [30]. The breakthrough in Parrilo's work is that the search for *short proof certificates* can be carried out in an *algorithmic way*. This is achieved by coupling efficient optimization methods and powerful theorems in semialgebraic geometry.

Short proofs and relaxations may produce solutions that are actually better (less brittle, more scalable, evolvable, etc.) than the original formulation. For instance, there are many results describing an inverse relationship between the "hardness" of a problem (or its proof length) and the distance to the set of ill-posed instances. Similarly, the IP layer was designed primarily for simplicity and robustness while performance was sacrificed. Scalability was a less explicit consideration, but one that turned out to be a very strong property of the design. Our work on the robustness analysis of TCP/AQM/IP suggests that the protocol may be *provably* robust as well, again not a consideration originally. Indeed, the TCP needed to make this proof work is not the one originally designed nor is it one currently deployed. Thus, designing for extreme robustness has produced a protocol that was remarkably scalable, and may evolve into one that is verifiable as well.

Questions in complex systems are typically nonlinear, nonequilibrium, uncertain, and hybrid, and their analysis has relied mainly on simulation. By contrast, our program focuses on rigorous analysis. One computer simulation produces one example of one time history for one set of parameters and initial conditions. Thus simulations can only ever provide counterexamples to hypotheses about the behavior of a complex system, and can never provide proofs. In technical terms, they can, in principle, provide satisfactory solutions to questions in NP, but not to questions in co-NP. Simulations can never prove that a given behavior or regularity is necessary and universal; they can, at best, show that a behavior is generic or typical. What is needed at this juncture is an effective (and scalable) method for, in essence, systematically proving robustness properties of nonlinear dynamical systems. That such a thing could be possible (especially without P=NP=co-NP in computational complexity theory) is profound and remarkable, and it is the main focus of our approach.

6 ROBUST YET FRAGILE

A key aspect of complex systems that has been a main theme of HOT [3, 4, 5, 8, 39] is the relationship between robustness and fragility. It has the computational counterpart of "dual complexity implies primal fragility." Practically speaking, the "robust yet fragile" viewpoint completely changes what is possible. Organisms, ecosystems, and successful advanced technologies are highly constrained in that they are not evolved/designed arbitrarily, but necessarily in ways that are robust to uncertainties in their environments and their component parts. The ensuing complexity makes these systems robust to the uncertainties for which such complexity was selected, but also makes the resulting systems potentially vulnerable to rare or unanticipated perturbations. This complexity is also largely cryptic and hidden in idealized laboratory settings and in normal operation, becoming acutely conspicuous when contributing to rare cascading failures or chronically, through fragility/complexity evolutionary spirals [35]. These puzzling and paradoxical features are an ongoing source of confusion to experimentalists, clinicians, and theorists alike, and have led to a rash of specious theories both in biology and more recently about the Internet. However, these "robust yet fragile" features are neither accidental nor artificial and derive from a deep and necessary interplay between complexity and robustness, modularity, feedback, and fragility. Failure to explicitly exploit the highly structured, organized, and "robust yet fragile" nature of such systems hopelessly dooms other methods to be overwhelmed by their sheer complexity.

The HOT concept was introduced to focus attention on the robust yet fragile character of complexity. This is the most essential and common property of complex systems in biology, ecology, technology, and socio-economic systems. HOT offers a new and promising framework to study not only network problems, but also put networks in a bigger context. This will be important both with the convergence of existing communication and computing networks and their widely proposed role as a central component of vast enterprise and global networks of networks including transportation, energy, and logistics. Research within the HOT framework addresses many complementary aspects of the multifaceted area of networked complex systems. Issues such as robustness, scalability, verifiability and computability can now (and must) be investigated and understood within a common framework. We believe that these different requirements are not only compatible, but can be combined in a very natural fashion. Our technologies have a priori emphasized these as separate issues, and the promise of a unified approach to simultaneously handle these critical aspects is of paramount importance. While seeking the decomposition of hard problems into simpler subproblems will continue to be an important design strategy and is at the heart of the role of protocols, an integrated and unified theory is required to do this in a rigorous and robust manner.

7 CONCLUSIONS AND OUTLOOK

The preliminary findings discussed in this article are intended to illustrate the potential and sketch some of the ingredients of the type of mathematical theory of the Internet that we envision and that would allow for a holistic treatment of the entire TCP/IP protocol stack. In particular, our recently derived TCP modifications provide a perfect example of the practical advantages of including a strong theoretical machinery as an integral part of the design rationale. They also suggest the style of proofs that we think are possible, and are consistent with the existing methods in communications, controls, and computing. We expect to be able to systematically verify the robustness and correctness of protocols, systems, and software with realistic assumptions, and to further prove approximate optimality in extreme, asymptotic scenarios, which may be severe abstraction from reality. Note we are essentially abandoning the hope of optimally robust systems as intrinsically unobtainable for problems of practical interest. Even so, we expect the expansion of the classes of problems for which we can rigorously and algorithmically verify robustness to be slow and difficult, particularly when physical layers impose resource limitations. Nevertheless, we fully expect that our approach will provide the necessary theoretical framework for designing provably robust and evolvable future network technologies. Furthermore, our approach has already demonstrated how the ensuing new insights about the Internet combine with our understanding of its origins and evolution to provide a rich source of ideas about emergence, complexity, and robustness in general.

Emergence and complexity are often used in ways that seem more intent on obfuscation than clarification. Nevertheless, it is certainly fair to say that the Internet is teeming with complexity and emergence, including power law distributions, self-similarity, and fractals. It could also be described as adaptive, self-organizing, far-from-equilibrium, nonlinear, and heterogeneous. These are all popular notions within the community of researchers loosely organized around such rubrics as Complex Adaptive Systems (CAS), New Science of Complexity (NSOC), Chaoplexity, and more specifically Self-Organized Criticality (SOC), Edge-of-Chaos (EOC), and New Science of Networks.[1] Recently, a rash of papers have offered explanations for emergent Internet phenomena from these perspectives. Comparisons with our approach have appeared elsewhere [3, 4, 8, 35] and demonstrate that these CAS/NSOC/SOC/EOC-based "explanations"—serving at best as simple null hypotheses—can be convincingly debunked and are easy to reject. In turn, they have been largely irrelevant for gaining a deeper understanding of the Internet or other large-scale communication networks.

REFERENCES

[1] Alderson, D., J. C. Doyle, R. Govindan, and W. Willinger. "Toward an Optimization-Driven Framework for Designing and Generating Realistic In-

[1]While these rubrics differ from one another, from the perspective of this paper, these differences are all quite minor.

ternet Topologies." *ACM SIGCOMM Comp. Comm. Rev.* **33(1)** (2003): 41–46.

[2] Brakmo, L. S., and L. L. Peterson. "TCP Vegas: End to End Congestion Avoidance on a Global Internet." *IEEE J. Sel. Areas in Comm.* **13(8)** (1995): 1465–1480.

[3] Carlson, J. M., and J. C. Doyle. "Complexity and Robustness." *Proc. Natl. Acad. Sci. USA* **99** (2002): 2538–2545.

[4] Carlson, J. M., and J. C. Doyle. "Highly Optimized Tolerance: A Mechanism for Power Laws in Designed Systems." *Phys. Rev. E* **60** (1999): 1412–1428.

[5] Carlson, J. M., and J. C. Doyle. "Highly Optimized Tolerance: Robustness and Design in Complex Systems." *Phys. Rev. Lett.* **84(11)** (2000): 2529–2532.

[6] Choe, H., and S. H. Low. "Stabilized Vegas." In the electronic proceedings of IEEE Infocom, held April 2003, in San Francisco, CA. ⟨http://netlab.caltech.edu/FAST/publications.html⟩.

[7] Crovella, M. E., and A. Bestavros. "Self-Similarity in World Wide Web Traffic: Evidence and Possible Causes." *IEEE/ACM Trans. Network.* **5(6)** (1997): 835–846.

[8] Doyle, J. C., and J. M. Carlson. "Power Laws, Highly Optimized Tolerance and Generalized Source Coding." *Phys. Rev. Lett.* **84(24)** (2000): 5656–5659.

[9] Hollot, C. V., V. Misra, D. Towsley, and W. B. Gong. "Analysis and Design of Controllers for AQM Routers Supporting TCP Flows." *IEEE Trans. Automatic Control* **47(6)** (2002): 945–959.

[10] Jin, C., D. Wei, S. H. Low, G. Buhrmaster, J. Bunn, D. H. Choe, R. L. A. Cottrell, J. C. Doyle, H. Newman, F. Paganini, S. Ravot, and S. Singh. "Fast Kernel: Background Theory and Experimental Results." In *First International Workshop on Protocols for Fast Long-Distance Networks*, February 2003.

[11] Jin, C., D. X. Wei, and S. H. Low. "FAST TCP: Motivation, Architecture, Algorithms, Performance." In the electronic proceedings of IEEE Infocom, held April 2003, in San Francisco, CA. ⟨http://netlab.caltech.edu/FAST/publications.html⟩.

[12] Kelly, F. P., A. Maulloo, and D. Tan. "Rate Control for Communication Networks: Shadow Prices, Proportional Fairness and Stability." *J. Oper. Resh. Soc.* **49(3)** (1998): 237–252.

[13] Kunniyur, S., and R. Srikant. "Designing AVQ Parameters for a General Topology Network." In the Proceedings of the Asian Control Conference, held September 2002, in Singapore. Papers will be consdiered for publication in the Asian Journal of Control and ISA Transactions.

[14] Kunniyur, S., and R. Srikant. "End-to-End Congestion Control: Utility Functions, Random Losses and ECN Marks." *IEEE/ACM Trans. Network.* **11(5)** (2003): 689–702.

[15] Kunniyur, S., and R. Srikant. "End–to–End Congestion Control Schemes: Utility Functions, Random Losses and ECN Marks." In *Proceedings of IEEE Infocom*, March 2000. ⟨http://www.ieee-infocom.org/2000/papers/401.ps⟩.

[16] Kunniyur, S., and R. Srikant. "A Time-Scale Decomposition Approach to Adaptive ECN Marking." *IEEE Trans. Automatic Control*, June 2002.

[17] Leland, W. E., M. S. Taqqu, W. Willinger, and D. V. Wilson. "On the Self-Similar Nature of Ethernet Traffic." *IEEE/ACM Trans. Network.* **2(1)** (1994): 1–15.

[18] Li, L., R. Tanaka, D. Alderson, S. L. Low, W. Willinger, and J. C. Doyle. "The Importance of Annotated Graphs in Internet Topology Generation." Manuscript, unpublished, 2003. Submitted to ⟨http://netlab.caltech.edu⟩.

[19] Low, S. H. "A Duality Model of TCP and Queue Management Algorithms." *IEEE/ACM Trans. Network.* **11(4)** (2003): 525–536. ⟨http://netlab.caltech.edu⟩.

[20] Low, S. H., and D. E. Lapsley. "Optimization Flow Control, I: Basic Algorithm and Convergence." *IEEE/ACM Trans. Network.* **7(6)** (1999): 861–874.

[21] Low, S. H., L. L. Peterson, and L. Wang. "Understanding Vegas: A Duality Model." *J. of ACM* **49(2)** (2002): 207–235.

[22] Low, S. H., F. Paganini, J. Wang, and J. C. Doyle. "Linear Stability of TCP/RED and a Scalable Control." *Comp. Networks J.* (2003): to appear.

[23] Massoulie, L., and J. Roberts. "Bandwidth Sharing: Objectives and Algorithms." In *Infocom'99*, March 1999.

[24] Massoulie, L., and J. Roberts. "Bandwidth Sharing: Objectives and Algorithms." *IEEE/ACM Trans. Network.* **10(3)** (2002): 320–328.

[25] Mo, J., and J. Walrand. "Fair End-to-End Window-based Congestion Control." *IEEE/ACM Trans. Network.* **8(5)** (2000): 556–567.

[26] Paganini, F., J. C. Doyle, and S. H. Low. "Scalable Laws for Stable Network Congestion Control." In *Proceedings of Conference on Decision and Control*, December 2001.

[27] Paganini, F., Z. Wang, J. C. Doyle, and S. H. Low. "Congestion Control for High Performance, Stability and Fairness in General Networks." Submitted for publication, March 2003.

[28] Paganini, F., Z. Wang, S. H. Low, and J. C. Doyle. "A New TCP/AQM for Stability and Performance in Fast Networks." In *Proc. of IEEE Infocom*, April 2003.

[29] Parrilo, P. A. "Structured Semidefinite Programs and Semialgebraic Geometry Methods in Robustness and Optimization." Ph.D. thesis, California Institute of Technology, Pasadena, CA, 2000.

[30] Parrilo, P. A., and B. Sturmfels. "Minimizing Polynomial Functions." "DIMACS volume of the Workshop on Algorithmic and Quantitative Aspects of Real Algebraic Geometry in Mathematics and Computer Science." Submitted.

[31] Paxson, V., and S. Floyd. "Wide-Area Traffic: The Failure of Poisson Modeling." *IEEE/ACM Trans. Network.* **3(3)** (1995): 226–244.
[32] Vinnicombe, G. "On the Stability of Networks Operating TCP-like Congestion Control." In *Proc. of IFAC*, 2002.
[33] Vinnicombe, G. "Robust Congestion Control for the Internet." Submitted for publication, 2002.
[34] Wang, J., L. Li, S. H. Low, and J. C. Doyle. "Can TCP and Shortest-Path Routing Maximize Utility?" In *Proc. of IEEE Infocom*, April 2003.
[35] Willinger, W., and J. C. Doyle. "Robustness and the Internet: Design and Evolution." This volume.
[36] Willinger, W., and V. Paxson. "When Mathematics Meets the Internet." *Notices of the AMS* **45(8)** (1998): 961–970.
[37] Willinger, W., M. S. Taqqu, R. Sherman, and D. V. Wilson. "Self-Similarity through High Variability: Statistical Analysis of Ethernet LAN Traffic at the Source Level." *IEEE/ACM Trans. Network.* **5(1)** (1997): 71–86.
[38] Yaiche, H., R. R. Mazumdar, and C. Rosenberg. "A Game Theoretic Framework for Bandwidth Allocation and Pricing in Broadband Networks." *IEEE/ACM Trans. Network.* **8(5)** (2000).
[39] Zhu, X., J. Yu, and J. C. Doyle. "Heavy Tails, Generalized Coding, and Optimal Web Layout." In *Proceedings of IEEE Infocom*, April 2001.

Index

R

S